FLUID POWER

FLUID POWER

Theory and Applications

Third Edition

James A. Sullivan
Southern Illinois University
Carbondale, Illinois

A RESTON BOOK
PRENTICE HALL, Englewood Cliffs, New Jersey 07632

Library of Congress Cataloging-in-Publication Data

Sullivan, James A.
    Fluid power : theory and applications / James A. Sullivan.
        p.   cm.
    "A Reston book."
    Includes bibliographical references and index.
    ISBN 0-13-323080-5
    1. Fluid power technology.   I. Title.
TJ840.S882 1989
620.1'06--dc19   88-16883
                                                        CIP

Editorial/production supervision
and interior design: *Carol L. Atkins*
Cover design: *Wanda Lubelska*
Manufacturing buyer: *Robert Anderson*

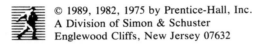 © 1989, 1982, 1975 by Prentice-Hall, Inc.
A Division of Simon & Schuster
Englewood Cliffs, New Jersey 07632

Printed in the United States of America
10 9 8 7 6 5 4 3 2 1

ISBN   0-13-323080-5

Prentice-Hall International (UK) Limited, *London*
Prentice-Hall of Australia Pty. Limited, *Syndey*
Prentice-Hall Canada Inc., *Toronto*
Prentice-Hall Hispanoamericana, S.A., *Mexico*
Prentice-Hall of India Private Limited, *New Delhi*
Prentice-Hall of Japan, Inc., *Tokyo*
Simon & Schuster Asia Pte. Ltd., *Singapore*
Editora Prentice-Hall do Brasil, Ltda., *Rio de Janeiro*

CONTENTS

FOREWORD

Fluid power (hydraulics and pneumatics), mechanical, and electrical power are the three basic means for transmitting power. Although fluid power is a very old technology, it was not until the 20th century that machinery and equipment designers discovered the many instances where the ease of control, compactness, and versatility of fluid power made it the ideal replacement for mechanical and electrical transmissions, and enabled machinery and equipment manufacturers to accomplish functions previously impossible within the limits of mechanical and electrical technology. Some common examples are articulated man-lifts, hydraulic presses, 747 flight controls, and even the actuators that blink the eyes and move the fingers on the almost-human mannequins at Disney World.

Once the advantages of fluid power became known, machinery and equipment manufacturers were quick to incorporate them into their products. The industry had an almost explosive growth to its present $6 billion level. The rapid growth of the new technology generated an unprecedented demand for educational support.

Professor James A. Sullivan, author of this text, is highly qualified in fluid power education and instruction. He is one of the dedicated cadre of teachers who were introduced to fluid power during the 1965–66 Institutes on Fluid Power Education. Since then he has enriched his knowledge, both theoretical and practical. He is currently working on validation of the occupations in fluid power, identification of job competencies for the Fluid Power Educational Foundation, and certification of Fluid Power Specialists and Mechanics. I heartily recommend this text's use by the instructor and the student.

James I. Morgan, CAE
President
National Fluid Power Association

PREFACE

This third edition contains seventeen chapters, five more than in the previous edition. Much of the material has been reorganized and improved with new illustrations of components and circuits. Three new chapters have been added to reflect the increased interest in troubleshooting and pneumatics. Additionally, the problems sections at the end of each chapter have been expanded.

As with previous editions, the present text relates theory to approved practice. When theory is explained, a concerted attempt is made to identify and solve related problems—and then to identify appropriate laboratory work, suggest how it is performed, and cite sources of related information from which other learning activities can be organized. ASTM standards and conventional tests, for example, are explained and suggested as appropriate exercises for the fluid power technician. And several maintenance procedures are laid out for use by the fluid power mechanic.

The author wishes to gratefully acknowledge the contributions of many companies and individuals who made this work possible. Included in this list are professional and standards organizations, publishers, schools and educational equipment companies, petroleum companies, original equipment manufacturers, and personal associates. Credit lines are given for these contributions whenever possible throughout the text.

James A. Sullivan
Southern Illinois University
Carbondale, Illinois

CHAPTER *1*

INTRODUCTION

1-1 INTRODUCTION

Fluid power means using pressurized fluids in a confined system to accomplish work. Both liquids and gases are fluids. Most hydraulic systems use petroleum oils, but often synthetic oils and water base fluids are used for safety reasons. Pneumatic systems use air which is exhausted to the atmosphere after doing the work.

The oil used in hydraulic systems exhibits the characteristics of a solid and provides a rigid medium to transfer power through the system. Conversely, the air used in pneumatic systems is spongy and provision must be made in its control to effect smooth operation of actuators. Air and oil are frequently combined in one system to provide the advantages of both in accomplishing the work specified.

A fluid power system accomplishes two main objectives. First, it provides substantial fluid force to move actuators in locations away from the power source where the two are connected by pipes, tubes, or hoses. A power source, for example a gasoline or diesel engine coupled to a hydraulic pump, can be housed in one area to power a cylinder or hydraulic motor one hundred feet or more away in another location. This is a decided advantage over systems using a mechanical drive train as the location of the output becomes less accessible. Second, fluid power systems accomplish highly accurate and precise movement of the actuator with relative ease. This is particularly important in such applications as the machine tool industry where tolerances are often specified to one ten thousandth of an inch and must be repeatable during several million cycles.

1-2 HISTORICAL PERSPECTIVE

The modern era in fluid power began around the turn of the century. Hydraulic applications were made to such installations as the main armament system of the USS Virginia in 1906. In this application, a variable speed hydrostatic transmission was installed to drive the main guns (see Fig. 1-1). Since that time, the marine industry has applied fluid power to cargo handling and winch systems, controllable-pitch propellors, submarine control systems, shipboard aircraft elevators, aircraft and missile launch systems, and radar-sonar drives.

Figure 1-1 USS Virginia (*courtesy of Sperry Vickers*).

1-3 CAPABILITIES

Hydraulics and pneumatics have almost unlimited applications in the production of goods and services in nearly all sectors of the country. Several industries are dependent on the capabilities that fluid power affords. Among these are agriculture, aerospace and aviation, construction, defense, manufacturing and machine tool, marine, material handling, mining, transportation, undersea technology, and public utilities, including communications transmission systems.

The world's need for food and fiber production has caused unprecedented leadership in agricultural equipment development, and particularly in applying hydraulics to solve a variety of problems. Figure 1-2 illustrates a modern four wheel drive tractor that features extensive use of hydraulic power. A 26.5 gal/min eight cylinder variable volume radial piston pump supplies fluid to a closed-center load-sensitive circuit for instant response. The pump unloads to a minimum standby pressure to reduce power consumption when demand is low. Hydraulics power the rear and front main drive and power take-off (pto) clutches, wet disc brakes, remote valves, implement hitch, draft sensing, power shift transmission, differential lock, and hydrostatic steering. A closed accumulator-style reservoir is used to supplement the flow from the charge pump during maximum demand from large bore cylinders.

Other applications to agriculture include combines, forage harvesters, backhoes, chemical sprayers, and organic fertilizer spreaders. Extensive design and

Figure 1-2 Modern wheel tractor (*courtesy of John Deere Company*).

application are also being made to wheel motors driven in remote locations to assist in marginal tractive conditions. Wheel motors consist of a hydraulic motor mounted integrally with a wheel and tire assembly. Braking is usually designed into the system. All that is required to make the wheel motor functional is mounting and hydraulic line connections back to the main control valve and power source. Auxiliary power drive wheel motors have several advantages to the agricultural industry. For example, they can be used to maintain and improve the turning ability of regular row crop tractors. They allow for high design which maintains under-axle crop clearances. They can be used to retain front axle adjustability and to provide on-the-go engagement and disengagement in forward as well as reverse.

Figure 1-3 illustrates the application of hydraulic power to the aviation and aerospace industries.

Another sector of our economy that has benefited from the brute power of hydraulics and pneumatics is the construction industry. Crawler tractors, road graders, bucket loaders, trenchers, backhoes, hydraulic shovels and pan scrapers are just a few of the many applications. Figure 1-4 shows a large (235 ton payload) off-the-road truck with many hydraulic and pneumatic components. Notice the large telescope hydraulic dump cylinders.

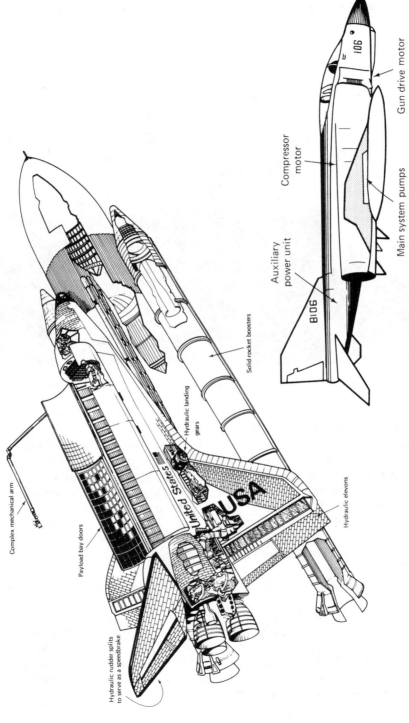

Complex mechanical arm

Payload bay doors

Hydraulic rudder splits
to serve as a speedbrake

Hydraulic landing
gears

Solid-rocket boosters

Hydraulic elevons

Auxiliary
power unit

Compressor
motor

Main system pumps

Gun drive motor

Figure 1-3 Aerospace and aviation applications of fluid power (*courtesy of Fluid Power Education Foundation*).

Figure 1-4 Off-the-road truck (*courtesy of WABCO Construction and Mining Equipment Group, An American Standard Company*).

The manufacturing and machine tool industry is very dependent on hydraulic power to provide the power and close tolerance necessary in controlled production. Figure 1-5 illustrates a 2,000 ton forging press, one of the largest self-contained presses in existence. Hydraulic fluid is supplied by 20 pumps, each of which has a capacity of 35 gal/min. At an operating pressure of 2,000 lbf/in^2 the press consumes over 800 hp. The press is driven by fluid supplied from a central hydraulic pumping system consisting of ten double-end electric motors driving the 20 hydraulic pumps.

Undersea applications of hydraulics are receiving widespread attention in research, development, and utilization. The Pisces series of submersibles, developed by a Vancouver, B.C. company, can work at depths to 6,500 ft and are engaged in a variety of work ranging from rescue to salvage operations. Figure 1-6 illustrates a two-man operation inside the sub in the control of the steering and stabilizing systems, bow thrusters, grappling hooks, net winches, mooring winches, and anchor windlasses. Also shown is the DSRV-1 developed by Lockheed for the U.S. Navy undersea research effort.

Material handling conveys products from the point of production to the consumer. Several systems are in use and nearly all of them depend on hydraulics

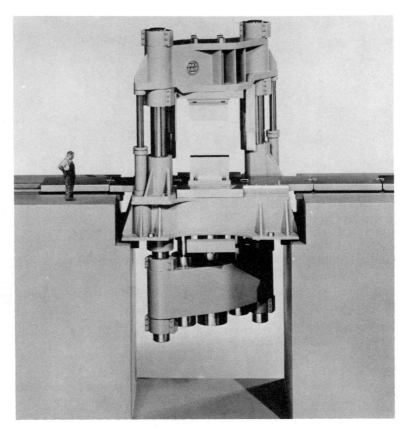

Figure 1-5 2,000-ton self-contained forging press (*courtesy of Abex Corporation, Denison Division*).

and pneumatics in the transport of goods. These include garbage trucks, conveyors, hydraulic leveling dock systems, hoists, truck and trailer tailgate systems, and even industrial robots.

Fluid power has wide application in continuous coal mining. Machines equipped with several hydraulic cylinders, jacks, and motors dig and load coal at the rate of two tons per minute. Conventional cutting and blasting are not required. Other mining applications utilizing fluid power pneumatics include rock drills, hydraulic track laying machines, shuttle cars, roof bolting machines, and conventional hydraulic jacks.

Transportation systems provide examples of the most varied uses of fluid power. Road and off-the-road vehicles are typically suspended by pneumatic tires. Air assisted brakes (both high pressure and vacuum assisted) are standard equipment. Power steering systems, either of the hydraulic assist or full power steering type, are also standard and relieve the operator of the fatigue commonly associated with physical exertion over long periods of time. Hydrostatic transmissions are very common on all types of vehicles. Suspension systems are

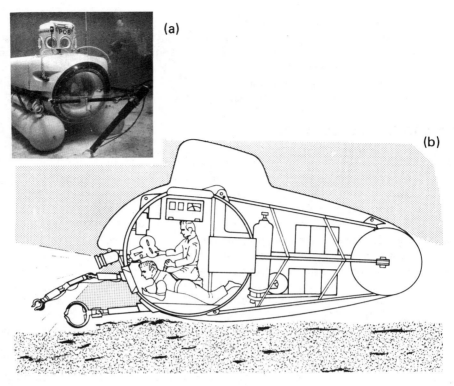

Figure 1-6 (a) U.S. Navy Undersea Research Vehicle (*courtesy of Parker-Hannifin*); (b) International operation of Undersea Research Vehicle (*courtesy of Fluid Power Education Foundation*).

dampened with hydraulic shock absorbers, and some combine pneumatics by using the compressible nature of gases as the basis for air-oil suspension systems. Hydraulic wheel drive motors are a recent addition to the transportation line. They give almost unlimited flexibility to the design which can mount the power plant in a convenient location to power individually driven wheels as they are required to support the load and provide traction sufficient to propel it in almost any direction with a variety of stepless speeds.

Utilities and communications are two industries where the use of fluid power is vital. Line utility vehicles are now a must to support personnel working above the ground. Similar vehicles are used in applications that vary throughout industry from fruit tree maintenance and picking to jumbo jet service. Almost unlimited movement is available to the operator with these systems. They can be given direction from the bucket, from the side of the truck vehicle, from the end of an umbilical cord, and even by remote signals from an electronic control box. Other applications to the communication and utilities industries include hydraulic trenchers, cable boring machines, earth augers, pipe laying machines, tampers, and many others.

Three applications given recent attention in the fluid power industry are miniature pneumatics, moving part logic, and fluidics. Miniature pneumatics makes use of small air-powered components such as cylinders and valves to carry out small assigned tasks as well as to control other large components. Moving part logic components and fluidics make use of logic elements with functions similar to several electronics counterparts, such as capacitors, resistors, and amplifiers to control other hydraulic and pneumatic systems.

1-4 CAREERS IN FLUID POWER

Career opportunities in the fluid power industry look promising. Workers function primarily at three levels: (1) mechanics and maintenance men, (2) technicians, and (3) engineers. Work at each of the three levels is diverse. Through the efforts of the National Fluid Power Association,[1] the Fluid Power Education Foundation[2], and Fluid Power Society[3], however, substantial progress has been made developing new job descriptions for fluid power mechanic, fluid power technician, and fluid power engineer for inclusion in the *Dictionary of Occupational Titles,* the nation's data base for occupational information. Two of the job titles and definitions formulated and which have been accepted by the U.S. Department of Labor are: Fluid Power Mechanic (DOT 600.281) and Fluid Power Technician (DOT 007.161).

In general, mechanics and maintenance men work with fluid power components in all industries and make up one of the fastest growing occupational groups in a nation's labor force.

In the U.S., with a labor force of 110 million workers, employment of mechanics and repairmen totals nearly three million. More than 800 thousand of these are automobile mechanics. Recent trends in licensure of this group require training in fluid power related accessories and components, including braking systems, hydromatic transmissions, power steering, air emission control systems, air conditioning, and stabilizing systems. Industry employs about 300 thousand industrial machinery repairmen, another 100 thousand millrights, 75 thousand farm equipment mechanics, and other miscellaneous mechanics. All told, nearly 30% of the mechanics and repairmen employed are in the manufacturing industries, and the majority of these are employed in plants that produce durable goods such as transportation equipment machinery, primary metals, and fabricated metal products. Another 20% are employed in shops that specialize in servicing such equipment. Most of the remaining mechanics and repairmen are employed in the transportation, construction, and public utilities industries, and by the government at all levels. Most employment opportunities for mechanics and repairmen occur in the more populous and industrialized areas.

Many mechanics and repairmen learn their skills on the job or through

[1,2] National Fluid Power Association and Fluid Power Education Foundation are located at 3333 N. Mayfair Road, Milwaukee, Wisconsin, 53222 USA.

[3] At the time of this writing, the Fluid Power Society was located at 2900 N. 117th Street, Milwaukee, Wisconsin 53226.

apprenticeship training. They are union affiliated and earn high wages. Some apprenticeship training programs require and give credit for related courses from high schools, trade and technical schools, and private schools. Training and experience in the armed services also help young men and women prepare for occupations such as aircraft hydraulic mechanic and others.

Some employers consider a formal apprentice training program to be the best way to learn skilled maintenance and repair work. An apprenticeship consists of three to five years of paid on-the-job training, supplemented each year by at least 144 hours of related classroom instruction. Formal apprenticeship agreements are registered with the state apprenticeship agency and the U.S. Department of Labor's Bureau of Apprenticeship and Training. Employers look for applicants who have mechanical aptitude and manual dexterity. Many employers prefer people whose hobbies or interests include automobile repair and model building. A high school education or the equivalent is a must. Also necessary prior to training, or as a part of training, are courses in mathematics, physics, blueprint reading, and machine shop. Many manufacturers of fluid power systems offer specialized training courses for mechanics and maintenance personnel through private schools.

Workers in most maintenance and repair occupations have several avenues of advancement. Some move into supervisory positions, such as foreman, maintenance manager, or service manager. Specialized training prepares others to advance to sales, technical writing, and technician jobs. A substantial number of servicemen have also opened their own businesses.

Fluid power technicians assist engineering teams with such work as sales, applied research, planning production, maintenance, testing, and installation. Education includes two years of post high school training containing several courses in physics, mathematics, fluid power, mechanisms, and communications skills, and culminates in an associate degree. The technician can describe ideas about machinery to superiors, clients, and peer workers, and write technical reports. Work is usually performed at a drafting desk station, in specialized laboratories, and in the shop or plant. Figure 1-7 shows an engineering technician at work in a modern fluid power design facility.

Joint efforts between the Fluid Power Society (FPS) and industry over the last several years have resulted in certification of fluid power specialists[4]. These efforts grew out of the realization that certification is closely tied to specialized technologies. Membership in certification is open to men and women who function in sales, technical aspects, and other specialties in the fluid power industry.

Engineers comprise the second largest professional occupation in this country. Of the more than 1.1 million engineers in the labor force, nearly 600 thousand are in the manufacturing industries. Engineers who work with fluid power compo-

---

[4] Certification of technicians is also done by two other groups: Institute for The Certification of Engineering Technicians, sponsored by the National Society of Professional Engineers, 2029 K Street, N.W., Washington, D.C. 20006., and American Society of Certified Engineering Technicians, National Office, 2029 K Street, N.W., Washington, D.C. 20006.

Figure 1-7 Engineering technician in a modern fluid power design facility (*courtesy of John Deere Company*).

nents and systems are found primarily in this manufacturing group. The background of most engineers who work primarily with fluid power is rooted in a mechanical engineering degree from one of the many four-year institutions located throughout the states. Another area producing several engineers for fluid power is industrial engineering. Whereas mechanical engineers are concerned primarily with the production, transmission, and use of power, industrial engineers determine the most effective methods of using the basic factors of production, manpower, and materials. They are concerned with people and things in contrast to engineers in other specialties who generally are concerned more with developmental work in subject fields such as power and mechanics. Industrial engineering is an excellent background for sales related work and customer relations in the fluid power industry.

1-5 THE USE OF UNITS

English and SI metric units are used throughout this book. The *basic* units used from the English system are length, force, and time. *Derived* units include area, pressure, velocity, volume, and flow rate.

The units of length are the foot and inch (ft, in.). The units of force are the pound and ounce (lbf, oz). These same units are used for weight. The units for time are the minute and second (min, sec). Derived units are made up of basic units. Area is derived in square feet and square inches (ft^2, $in.^2$), pressure is in pounds per square inch ($lbf/in.^2$), velocity is in feet per minute or feet per second (ft/min, ft/sec), and volume is in cubic feet or cubic inches (ft^3, $in.^3$). One gallon equals 231 in^3.

The units for distance in the SI metric system most often used are the meter and centimeter (m, cm). The unit of force is the Newton (N), and the unit of time is the second (sec). Area is derived in square meters and square centimeters (m^2, cm^2). The unit of pressure is derived in Pascals in honor of the French physicist. A Pascal equals one Newton per square meter (Pa, N/m^2). This is somewhat confusing because the Newton also is an honorary name given after the English mathematician, Isaac Newton. Because the Pascal is small, the kilo Pascal (kPa), which is 1000 Pascals, and mega Pascal (MPa), which is one million Pascals, are used often in hydraulics. Velocity in the SI metric system is derived in meters per second (m/sec). Volume is derived in cubic meters or cubic centimeters (m^3, cm^3).

Flow rate is a derived unit, but it is not the same in hydraulics as in pneumatics. In hydraulics, flow rate is a liquid measure. In the English system the units are gallons per minute (gal/min), and in SI metrics liters per minute (l/min). In pneumatics, flow is a volumetric measure: cubic feet per minute or cubic feet per second in the English system (ft^3/min, ft^3/sec), and cubic meters per minute or cubic meters per second (m^3/min, m^3/sec) in the SI metric system.

In most cases the units should be included with the numbers when calculations are made. This is because without the units, the answer would be incomplete. And unless the units are included with each number in the calculation, there would be no easy way to check through the problem to see if errors have been made. The use of units also is important when conversions are made from one system to another, from liquid to volumetric measures, or from larger to smaller units. Conversion factors for derived units are difficult to remember so if one can work through the conversion with the units themselves, it usually is easy to find the conversion between base units like inches, centimeters, pounds, and Newtons.

1-6 HOW FLUID POWER WORKS

Fluid power works in accordance with the laws governing the behavior of the fluid itself. The two most common fluids used are air in pneumatic systems and oil in hydraulic systems. The beginning of an understanding of how fluid power works is attributed to Blaise Pascal (1650), who discovered that the pressure exerted by a confined fluid acts undiminished the same in all directions at right angles to the inside wall of the container. This is known as Pascal's Law and is often thought of as the foundation of the discipline. The law can be extended to include transmission and multiplication of force, as shown in Fig. 1-8.

Fluid power works by applying a force against a movable area. In a typical fluid power system, a fluid power pump or compressor delivers fluid through the control valve to a system actuator, such as a cylinder, through lines at high pressure. Unused fluid in hydraulic systems is returned to a reservoir at low pressure for cooling, storage, and later use. Figure 1-9 illustrates a simulated

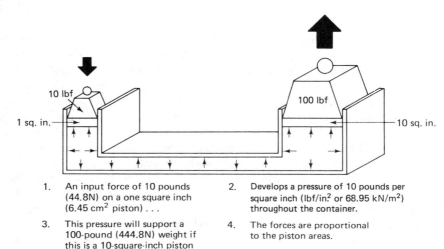

1. An input force of 10 pounds (44.8N) on a one square inch (6.45 cm² piston) . . .

2. Develops a pressure of 10 pounds per square inch (lbf/in² or 68.95 kN/m²) throughout the container.

3. This pressure will support a 100-pound (444.8N) weight if this is a 10-square-inch piston

4. The forces are proportional to the piston areas.

Figure 1-8 Transmission and multiplication of fluid power force.

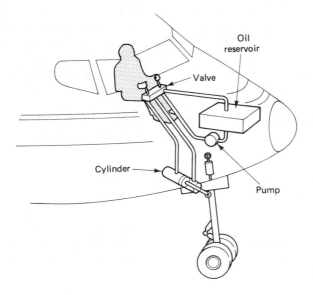

Figure 1-9 How fluid power works.

hydraulic system used for lowering and raising the nose wheel of an aircraft landing gear. Movement of the control valve to lower the wheel causes fluid to be delivered at high pressure from the pump to the blank end of the cylinder. To raise the wheel, the control valve directs fluid to the rod end of the cylinder. In both cases fluid power works by applying a force against the area of each side of the movable piston in the cylinder.

Examining the definition and the example closer

$$\text{Force} = \text{Pressure} \times \text{Area} \qquad (1\text{-}1)$$

All we need to remember is that the units of force, pressure, and area must be consistent. This means that in the English system of measurement the weight or force is in pounds (lbf), the area is in square inches (in.^{2}), and the pressure is in lbf/in.^{2}. In SI metric units, the force is in Newtons (N), the area is in square meters (m^2), and the pressure is in N/m^2, which is given the name Pascals (Pa) after that famous physicist. Conversions among the various units used for pressure around the world[5] are given in Table 1-1. Only the lbf/in.^{2} and Pa will be used here. The Pa is the recognized international unit of pressure, but the kg/cm^2 and bar are still used in some countries in the world.

TABLE 1-1 UNITS OF PRESSURE AND CONVERSION FACTORS

Convert to / Convert from	lbf/in.^{2}	kPa Multiply by	kgf/cm^2	bars
lbf/in.^{2}	1	6.895	0.07	0.069
kPa	0.145	1	0.0101	0.010
kgf/cm^2	14.22	98.07	1	0.98
bars	14.5	100	1.02	1

pascal (Pa) = 1 N/m^2
Conversions are made by multiplying the units in the left margin by the conversion factors in the boxes to arrive at the units across the top. Example: 1000 lbf/in.^{2} × 6.895 = 6895 kPa

When designing or applying fluid power to a system, the force required at the output is used to determine the pressure of the system and the cross section area of the cylinder. System pressures are also determined from the strength of components, cost factors, and safety precautions. When the system pressure is

[5] Conversions are made this way
1-lbf/in.^{2} = (1-lbf)(4.448 222-N/lbf)(1/in.^{2})(1550-in.^{2}/m^2) = 6895 N/m^2 or

lbf/in.^{2} × 6895 = N/m^2 = 6895 Pa = 6.895 kPa

known and the force that the system must apply is specified, the area of the cylinder then can be computed by solving the formula for this value.

$$\text{Area} = \frac{\text{Force}}{\text{Pressure}}$$

Example 1

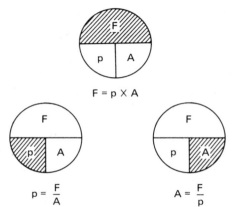

$$F = p \times A$$

$$p = \frac{F}{A}$$

$$A = \frac{F}{p}$$

Figure 1-10 Force, pressure, and area relationships.

Pressure = 1500 lbf/in.²

Area = (0.7854)(4 in.²) = 12.57 in²

Force = p × A = (1500 lbf/in.²)(12.57 in.²) = 18,855 lbf

To simplify calculations using this formula, the desired value can be determined using Fig. 1-10. By covering the desired value, the relationship between the other two is given in the proper order. That is, force equals pressure times area, pressure equals force divided by area, and area equals force divided by pressure. Look at the example that follows the figure, which asks for the solution of the force that would be applied to the cylinder rod if the cylinder has a bore of 4 in. and the pressure in the cylinder is 1500 lbf/in². Many helpful devices such as this have been developed to assist the fluid power mechanic and technician to make calculations.

1-7 SUMMARY

Fluid power is used throughout industry to accomplish work and deliver power from the source to some other location. Hydraulic and pneumatic systems have been put to a variety of uses, from the more common plant air systems and hydraulic production systems to more sophisticated systems used in airplanes, aerospace vehicles, and industrial robots. Miniature pneumatics and fluidic logic systems provide tremendous possibilities for fluid power system control.

The modern era in fluid power was heralded by the use of applications made to such installations as the main armament system on the USS Virginia in 1906. Fluid power is now considered to be a support system in all industries, including agriculture, aerospace, aviation, construction, manufacturing, materials handling, and many others.

Fluid power is a multi-billion dollar industry annually. While several factors account for the growth of the industry, the primary one is the need for more flexible power for a variety of uses in the durable goods production, agricultural, construction, and services industries.

Career opportunities are promising for fluid power mechanics, maintenance men, technicians, and engineers. Certification and extensive training are now available. This is in response to the demand for qualified personnel and the efforts of such groups as the National Fluid Power Association, Fluid Power Education Foundation, and Fluid Power Society over the last several years to give recognition to this important and growing segment of highly skilled professionals.

STUDY QUESTIONS AND PROBLEMS

1. List ten hydraulic applications and ten pneumatic applications.
2. List the components on one *hydraulic* application and give their specifications.
3. List the components of one *pneumatic* application and give their specifications.
4. Consult the Fluid Power Society to determine the current requirements to become a certified fluid power specialist.
5. List five schools in your state which offer fluid power courses or a program in fluid power leading to a degree.
6. Compute the area and volume for a cylinder with a 2-in. bore and a 12-in. stroke.
7. Compute the bore of a cylinder which must exert 10,000 lbf with a pressure limitation of 1200 lbf/in^2.
8. Compute the force available from a hydraulic cylinder with a bore of 75 mm under a pressure of 10 MPa.
9. What bore would be necessary on a cylinder operating at 15 MPa to exert a force of 65 kN?
10. A log splitter is to exert a 10,000 lbf force using a 3-in. bore cylinder. At full capacity, what will be the system pressure?
11. How much fluid is needed to stroke a cylinder with a 100 mm bore and 0.50 m stroke?
12. Construct a table of values which lists the force in lbf available from cylinders with bore diameters of 2, 4, 6, 8 and 10 in., and pressures of 400, 800, 1200, 1600, and 2000 lbf/in^2. Use a table like Table 1-2 to record the values.

TABLE 1-2 CAPACITY OF HYDRAULIC CYLINDERS IN POUNDS FORCE

Pressure (lbf/in.^{2})	400	800	1,200	1,600	2,000
Diameter (in.)					
2 in.	——	——	——	——	——
4 in.	——	——	——	——	——
6 in.	——	——	——	——	——
8 in.	——	——	——	——	——
10 in.	——	——	——	——	——

13. Construct a table using the same values as in Problem 12, but this time convert the diameter of the cylinder to mm, the pressure to MPa and the force to MN.

14. If a hydraulic cylinder has a rod diameter half the size of the bore, what will be the difference in force available if the pressure applied alternately to each end remains constant? (Clue: Try two convenient values such as 4 in. for the piston and 2 in. for the rod diameter.)

15. In Problem 14, what would happen if the pressure were applied to both sides of the piston at the same time? (Make a drawing to prove your answer.)

CHAPTER 2

BASICS
OF HYDRAULICS

2-1 INTRODUCTION

There are a number of basics that explain how a fluid power system works. One of these has to do with raising its energy level by increasing the pressure within the fluid. This is done by a pump moving fluid against the load resistance. Another explains how a small force by a person operating the hydraulic system, for example on a hydraulic jack handle, can produce a very large force on the object being lifted. Still other principles have to do with the flow of fluid in the system as it transfers fluid power from one place to another in a system. The concepts of work, power, and horsepower show what can be accomplished with a fluid power system. The output of fluid power systems is measured in fluid horsepower and this puts it all together. Finally the measurements of torque and speed show how fluid horsepower can be broken down to size components for the system.

2-2 HOW OIL TRANSMITS POWER

The two cylinders in Fig. 2-1 are connected by a piece of tubing so that when one of them retracts, the other extends. This is how fluid power is transmitted by the pressurized fluid moving from one place to another. If both cylinders have the same bore, the force at the output cylinder as well as the speed are the same as the force and speed applied to the input cylinder. There are some friction losses, of course, but in simplified terms, what enters at the input will exit at the output. This is really an application of the law of conservation of energy which simply means that energy can neither be created nor destroyed. And while you don't get something out of the system for nothing, you do get out everything that is put in, minus a small amount which is accounted for by friction and slippage in the parts. This loss usually is no more than 5–10%. Keep in mind that during the time that

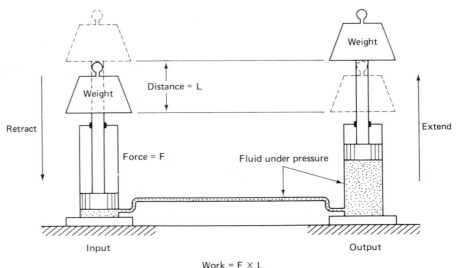

Figure 2-1 Transmission of force.

the system is transmitting fluid power from one cylinder to the other the pressure remains the same everywhere. It is the same in both cylinders and in the line between the cylinders. This is really just another demonstration of Pascal's Law that says the pressure in the system is the same everywhere. There may be a slight drop in pressure when the fluid flows from one cylinder to the other, if the line is very long, but again, this can be accounted for by flow friction losses.

2-3 WORK PERFORMED BY THE SYSTEM

Notice that work was accomplished by the output cylinder in Fig. 2-1. By this we mean that the cylinder rod moved a force F through a distance L. This is the definition of work, and the units are always in ft-lbf or m-N. The formula used is

$$\text{Work } (W) = \text{Force } (F) \times \text{Distance } (L)$$

or simply

$$W = F \times L \tag{2-1}$$

Example 1

Let's say that the weight (N) represented 1000 lbf (4448 N) and the distance was 15 in. (38 cm). Then the work from the system would be

$$W = (1000 \text{ lbf})(1.25 \text{ ft}) = 1250 \text{ ft-lbf}$$

In SI units the solution is the same but with different units:

$$W = (4448 \text{ } N)(0.381 \text{ m}) = 1695 \text{ N-m}$$

2-4 MULTIPLICATION OF FORCE

Now what happens is the input piston, called the pump, and the output piston, called the linear actuator, have different areas? Well, as might be expected, if the pump piston is smaller than the actuator piston, then the force is greater at the output. This multiplication of force isn't free, of course, and what happens is that when the force increases, the distance that the output piston rod travels is less. Since the pressure in the system and the work accomplished remain the same, there must also be a balance between the input and output forces and distances traveled. This says simply that the work input equals the work output. Figure 2-2 shows how the force is multiplied as fluid moves from the pump cylinder to the ram. We can call this a ram because the cylinder rod is at least half as large as the cylinder bore. The balance is given by the formula

$$F_1 \times L_1 = F_2 \times L_2 \qquad (2\text{-}2)$$

When three of the variables in the formula are known, the fourth can be determined simply by substitution. In practical examples, usually one of the ratios is known or three of the four quantities are given by the design requirements of the system. Let's look at an example.

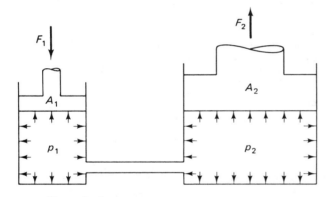

Figure 2-2 Multiplication of force.

Example 2

An operator can exert 100-lbf (444.8-N) on the pump piston. If the ram must lift 2000 lbf (8896-N) a distance of 12 in. (0.3048 m), how far would the pump piston have to travel? Look at the hydraulic jack system in Fig. 2-3. This allows the operator to move the handle up and down to lift the weight. When the handle is lifted, fluid from the reservoir is drawn into the pumping chamber at atmospheric pressure. When the handle is shoved down, the check valve to the right opens and the fluid enters the ram at high pressure. You can see that the left check valve opens on the up stroke, and the right one opens on the down stroke when the left one closes. Now to solve the problem, what must be done is to solve the formula for L_1.

$$L_1 = L_2 \times \frac{F_2}{F_1} = (12 \text{ in.}) \frac{(2000 \text{ lbf})}{(100 \text{ lbf})} = 240 \text{ in.}$$

In SI units

$$L_1 = (0.3048 \text{ m}) \frac{(8896 \text{ N})}{(4448 \text{ N})} = 6.096 \text{ m}$$

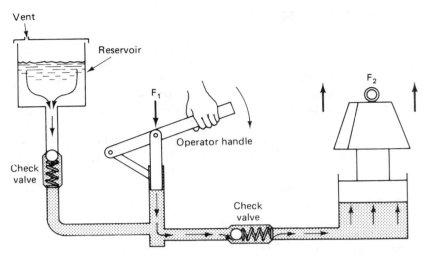

Figure 2-3 Single-acting hydraulic system.

Now all that would have to be done would be to set the stroke of the pump to some convenient unit, for example 1 in., and you would know that it would take 240 strokes to lift the weight. That's really a small price to pay for the tremendous force that can be exerted by the ram.

There is another interesting way to look at the balance between the pump and the ram, and that is to begin with the pressure balance. That is, the pressure in the system is the same everywhere. This is taken from the Bernoulli principle and applied to both the pump and the ram cylinder chambers which hold the fluid. This can be shown with a simple formula:

$$\text{Pressure} = \frac{F_1}{A_1} = \frac{F_2}{A_2} \tag{2-3}$$

Example 3

Look again at the system shown in Fig. 2-3. If the pump piston had a diameter of 1/4 in. (6.35 mm) and the ram a diameter of 1 in. (25.4 mm), how much weight could be lifted by a downward force on the pump piston of 150 lbf (667.2 N)? Here the problem is almost the same as the previous example, except that we're solving for F_2.

$$F_2 = F_1 \frac{A_2}{A_1} = (150 \text{ lbf}) \frac{(0.785 \text{ in.}^2)}{(0.049 \text{ in.}^2)} = 2403 \text{ lbf, just over a ton}$$

In SI units

$$F_2 = (667.2 \text{ N}) \frac{(506 \text{ mm}^2)}{(31.7 \text{ mm}^2)} = 10659 \text{ N, written as 10.7 KN}$$

This makes it relatively easy to see why hydraulics is the natural choice when it comes to multiplying a small force from a person to provide a large force to lift or move a heavy object. The largest hydraulic jack in Fig. 2-4, for example, has a capacity of 100 tons, yet its lightweight alloy construction permits an aver-

Basics of Hydraulics   Chap. 2

Figure 2-4 Hydraulic jacks (*courtesy of Charles S. Madan Co., Ltd.*).

age workman to carry it from place to place. It is also common with these large capacity jacks to have a progressive pump. This means that when the load is light, the larger pump piston can be selected, and as the load gets progressively heavier, the smaller pump piston is used. This saves the operator a great deal of time because when the smallest pump piston is used, the movement of the ram piston is very slow.

2-5 POWER AND HORSEPOWER

Work and power are different. Where work means to move an object from one place to another, power introduces the element of time. It is not enough to just move an object with a force through some distance. It must be done within a specific time, for example, one minute, two minutes, or an hour. This is what power means, the rate at which work is done, and the formula is

$$P = \frac{F \times L}{t} = \frac{W}{t} \qquad (2\text{-}4)$$

When the power is known it gives a convenient way to describe the output of many machines, but it doesn't give us a convenient way to compare one machine with another, for example, two hydraulic motors of the same size. James Watt had this same problem when he was trying to compare his steam engines to horses used in the mines back in the late 1800s. To solve the problem, he tested the capacity of draft horses to lift weights that were hung over pulleys. The actual layout was similar to Fig. 2-5. After several tests he estimated that a strong horse could raise 150 lbf about 3⅔ ft/sec which works out to be about 550 ft-lbf/sec. While the value of a horsepower turned out to be high for most horses, this has been the standard for rotating machinery since that time. The formula is

$$HP = \frac{F \times L}{t \times 550} = \frac{W}{t \times 550} = \frac{P}{550} \qquad (2\text{-}5)$$

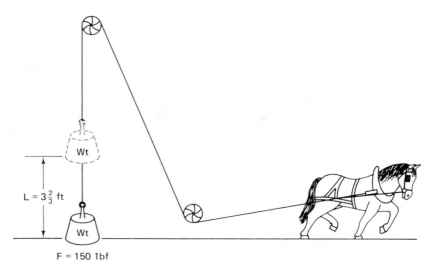

Figure 2-5 Horsepower.

and since the top and bottom have the same units, horsepower is dimensionless. Its simply called horsepower, without any units. The SI metric equivalent of power is measured in watts, and the formula for horsepower is

$$HP = \frac{P}{746} \qquad (2\text{-}6)$$

It is also dimensionless. All that really must be remembered is that there are 550 ft-lbf/sec in an English horsepower, and 746 watts in an SI metric horsepower. Let's look at a couple of examples to see how the formula works.

Example 4

What horsepower output would result if the horse in Fig. 2-5 lifts 800 lbs 20 ft in 20 seconds when pulling at full capacity. That sounds like a strong horse, but let's put the numbers in the formula to see

$$HP = \frac{(800 \text{ lbf})(20 \text{ ft})}{(20 \text{ sec})(550 \text{ ft lbf/sec})} = 1.45 \text{ hp}$$

That is a very strong horse. Notice in the formula that all the units cancel. This is important because if they don't, it indicates there's an error somewhere in the solution.

Example 5

An electric motor is to be coupled to a 3 hp hydraulic pump (Fig. 2-6). If the motor will draw 2000 watts at its rated output, will it do the job? First of all it has to be assumed the motor will operate at 100% efficiency, which it won't, but that gives us a place to start. Using the SI metric horsepower formula for the motor

$$HP = \frac{(2000 \text{ watts})}{(746 \text{ watts})} = 2.68 \text{ hp}$$

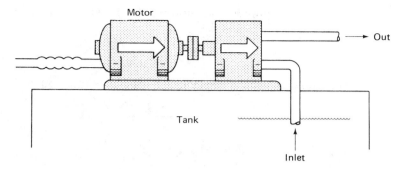

Figure 2-6 Motor-pump assembly.

No, the motor won't do the job, even at 100% efficiency. What would likely happen is that it would overload and overheat when the pump was at full capacity.

2-6 FLOW RATE IN A SYSTEM

In mechanical systems, horsepower is defined as the product of force and distance, divided by the product of time and a constant, 550 ft-lbf, because these are common units. In hydraulics, however, the common units are pressure and flow rate, and these are used to compute fluid horsepower.

In hydraulics, the flow rate Q is measured in gal/min and liter/min. It is a measure of the displacement volume of fluid V, divided by the time t. The general formula for flow rate is

$$Q = \frac{\text{Displacement}}{\text{time}} = \frac{V}{t} \qquad (2\text{-}7)$$

For hydraulic cylinders the displacement equals the product of the bore area and stroke.

$$V = A \times L$$

where V = displacement in in.^{3}/stroke

A = cross section of the bore in in.^{2}

L = cylinder stroke in in.

Pumps and motors with vanes or gears instead of pistons have a different internal geometry, but the displacement is still measured in in.^{3}/rev. In the metric system, displacement is measured in m^3/rev or mm^3/rev, depending on the size of the unit.

When the displacement of cylinders, pumps, or motors is used to compute the flow rate, the displacement in in.^{3}/stroke or in.^{3}/rev must be converted to gal/min. Thus the general formula for flow rate becomes

$$Q = \frac{V(\text{in.}^3/\text{rev}) \times 1(\text{gal})}{t(\text{min}) \times 231(\text{in.}^3)} = \frac{V}{t \times 231} \qquad (2\text{-}8)$$

For hydraulic pumps and motors, the general formula for computing the displacement is

$$Q = \frac{V \times N}{231} \qquad (2\text{-}9)$$

And if the pump or motor has cylinders, the flow rate equals

$$Q = \frac{(A \times L \times n) \times N}{231} \qquad (2\text{-}10)$$

where Q = flow rate in gal/min

$\quad L$ = stroke in in.

$\quad n$ = no. of cylinders

$\quad N$ = speed in rev/min

Example 2

Compute the displacement of a 5-cylinder piston pump having a 1⅛ in. bore and a 1 in. stroke. What would be the flow rate if the pump turns at 1200 rev/min?

$$V = \frac{3.14 \times (1\tfrac{1}{8} \text{ in.})^2 \times 1 \text{ in.} \times 5 \text{ cyls/rev}}{4} = 4.97 \text{ in.}^3/\text{rev}$$

Finally, computing the flow rate

$$Q = \frac{V \times N}{231} = \frac{(4.97 \text{ in.}^3/\text{rev}) \times (1200 \text{ rev/min})}{231 \text{ in.}^3/\text{gal}} = 25.8 \text{ gal/min}$$

2-7 FLUID HORSEPOWER

When a cylinder must transmit power in two directions, a system like that shown in Fig. 2-7 can be used. The components of the system are the reservoir, gear pump, pressure relief valve to prevent overloading, the four-way control valve, double-acting cylinder, and the connecting tubing. When the control valve is in the neutral position, the oil circulates through the pump, center spool, and back to the reservoir and the system is idling. When the hand-operated control valve is shifted, the system begins to perform useful work. Shifting the operator handle to the left directs oil from the pump at high pressure to the cap end of the cylinder and moves the load to the right. At the same time, oil leaving the rod end of the cylinder returns to the reservoir through the control valve at low pressure. Shifting the operator handle to the right directs oil to the rod end of the cylinder which returns the load to its original position and again the oil from the blank end of the cylinder returns to the reservoir at low pressure through the control valve. What happens when the cylinder reaches the end of its stroke, or encounters an immovable object? When these situations occur, the cylinder stalls and the relief valve opens to allow the oil from the gear pump to return to the reservoir without damaging any of the components in the system. This is another great advantage of fluid power systems.

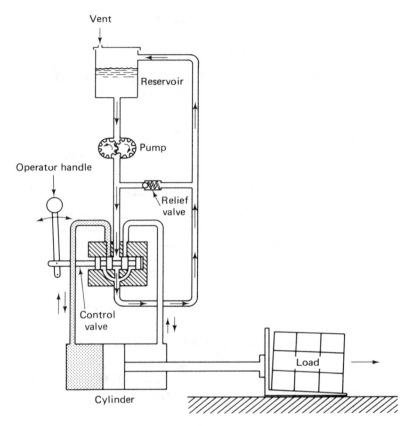

Figure 2-7 Double-acting hydraulic system.

The fluid power that a system delivers equals the pressure multiplied by the flow rate. In English units the formula is

$$FHP = \frac{p \times Q}{1714} = \frac{(\text{lbf/in.}^2)(\text{gal/min})}{1714} \qquad (2\text{-}11)$$

In SI metric units, the formula is

$$FHP = \frac{p \times Q}{44\ 760} = \frac{(\text{kPa})(\text{liters/min})}{44\ 760} \qquad (2\text{-}12)$$

Notice that the formulas are followed by the units that must be used to make the answers correct.

Example 6

A simple example would ask a question like what is the fluid horsepower of a system operating at 2000 lbf/in.^{2} delivering 5 gal/min; and the answer would be

$$FHP = \frac{(2000\ \text{lbf/in.}^2)(5\ \text{gal/min})}{1714} = 5.83\ \text{hp}$$

Completing the same example again in SI metric units, notice how they are converted from the English units

$$FHP =$$

$$\frac{(2000 \text{ lbf/in.}^2)(4.448 \text{ N/lbf})(1550 \text{ in.}^2/\text{m}^2)(5 \text{ gal/min})(3.785 \text{ l/gal})(0.001 \text{ m}^3/\text{l})}{(44\ 760)(\text{m} \cdot \text{N/min})}$$

and

$$FHP = 5.83 \text{ hp}$$

Or, stated in SI metric units, what is the horsepower transmitted by a system delivering 23 liters/min at a pressure of 13 793 kPa. Substituting in the SI metric formula

$$FHP = \frac{(13\ 793)(18.9 \text{ liters/min})}{(44\ 760)} = 5.83 \text{ hp}$$

which is another way of saying the same thing. For convenience, Tables 2-1 and 2-2 are used to simplify computing the FHP when the pressure and flow rate are known.

Perhaps the easiest way to convert $p \times Q$ in English units to SI metric units is to multiply by 26.1.

Example 7

Converting a flow rate of 5 gal/min at 2000 lbf/in.^{2} to a flow rate in liters/min at a certain pressure in kPa

$$(5 \text{ gal/min})(2000 \text{ lbf/in.}^2)(26.1)(1/\text{gal})(\text{N/lbf})(\text{in.}^2/\text{m}^2)(\text{kN/N}) = 261\ 000$$

The conversion of the units seems overpowering so just to keep it simple remember to use the conversion factor of 26.1 and refer to this example only when a refresher about the units is necessary.

2-8 TORQUE AND TORQUE HORSEPOWER

Torque

Pumps receive a twisting effort called torque to make them turn. Motors deliver a torque at their output shaft. Torque simply means to make an effort to twist or turn and is measured as a force multiplied by the distance along the radius of the shaft.

Figure 2-8 illustrates a physical example of torque where a person exerts a 50 lbf pulling force on a pipe wrench 12 in. from the center of the pipe. Thus the pipe receives a torque of 50 lbf-ft.

You may have noticed that torque is measured in the same units as work, but conceptually they are different. Work requires that the force move an object some distance. Thus, to perform work, movement is required. In Fig. 2-8 when the pipe becomes tight the wrench will stop even though it will continue to exert a turning force on the pipe. The same is true with fluid power applications. A gear motor may stall, for example, and continue to deliver a torque at its output shaft. In fact, the torque will typically increase at stall because the pressure will increase to the relief valve setting. This is another advantage of fluid power. Stalling does

TABLE 2-1 FLUID HORSEPOWER (ENGLISH UNITS)

$$\text{From the formula FHP} = \frac{p \times Q}{1714}$$

Gallons/minute	100 lbf/in.²	200 lbf/in.²	300 lbf/in.²	400 lbf/in.²	500 lbf/in.²	750 lbf/in.²	1000 lbf/in.²	1500 lbf/in.²	2000 lbf/in.²	2500 lbf/in.²	3000 lbf/in.²
0.5	0.029	0.058	0.088	0.117	0.146	0.219	0.292	0.438	0.583	0.729	0.875
1.0	0.058	0.117	0.175	0.233	0.292	0.438	0.583	0.875	1.167	1.459	1.750
1.5	0.088	0.175	0.263	0.350	0.438	0.656	0.875	1.313	1.750	2.188	2.625
2.0	0.117	0.233	0.350	0.467	0.583	0.875	1.167	1.750	2.334	2.917	3.501
2.5	0.146	0.292	0.438	0.583	0.729	1.094	1.459	2.188	2.917	3.646	4.376
3.0	0.175	0.350	0.525	0.700	0.875	1.313	1.750	2.625	3.500	4.376	5.251
3.5	0.204	0.408	0.613	0.817	1.021	1.532	2.042	3.063	4.084	5.105	6.126
4.0	0.233	0.467	0.700	0.933	1.167	1.750	2.333	3.501	4.667	5.834	7.001
5	0.292	0.583	0.875	1.167	1.459	2.188	2.917	4.376	5.834	7.293	8.751
6	0.350	0.700	1.050	1.400	1.750	2.625	3.501	5.251	7.001	8.751	10.502
7	0.408	0.817	1.225	1.634	2.042	3.063	4.084	6.126	8.168	10.210	12.252
8	0.467	0.933	1.400	1.867	2.333	3.501	4.667	7.001	9.335	11.669	14.002
9	0.525	1.050	1.575	2.100	2.625	3.938	5.251	7.876	10.502	13.127	15.753
10	0.583	1.167	1.750	2.333	2.917	4.376	5.834	8.751	11.669	14.586	17.503

TABLE 2-2 FLUID HORSEPOWER (SI METRIC UNITS)

From the formula $FHP = \dfrac{p \times Q}{44\,760}$

Liters/minute	500 kPa	1000 kPa	1500 kPa	2500 kPa	3500 kPa	5000 kPa	7500 kPa	10 000 kPa	12 500 kPa	15 000 kPa	17 500 kPa
1	0.011	0.022	0.034	0.056	0.078	0.112	0.168	0.224	0.280	0.336	0.392
2	0.022	0.045	0.067	0.112	0.156	0.223	0.335	0.447	0.558	0.670	0.782
3	0.034	0.067	0.101	0.168	0.235	0.335	0.501	0.667	0.835	1.002	1.169
4	0.045	0.089	0.134	0.223	0.313	0.447	0.670	0.893	1.117	1.340	1.563
5	0.056	0.112	0.168	0.279	0.391	0.559	0.838	1.117	1.397	1.676	1.955
10	0.112	0.223	0.335	0.558	0.782	1.117	1.676	2.235	2.793	3.352	3.911
15	0.168	0.335	0.503	0.838	1.173	1.676	2.513	3.351	4.188	5.026	5.864
20	0.223	0.447	0.670	1.117	1.564	2.234	3.351	4.468	5.585	6.702	7.819
25	0.279	0.559	0.838	1.396	1.955	2.793	4.189	5.585	6.982	8.378	9.774
30	0.335	0.670	1.005	1.676	2.346	3.351	5.027	6.702	8.378	10.054	11.730
35	0.391	0.782	1.172	1.955	2.737	3.910	5.865	7.819	9.774	11.729	13.684
40	0.447	0.894	1.340	2.234	3.128	4.462	6.702	8.937	11.171	13.404	15.639
45	0.503	1.005	1.508	2.513	3.519	5.027	7.540	10.054	12.567	15.080	17.593
50	0.559	1.117	1.676	2.793	3.910	5.585	8.378	11.171	13.963	16.756	19.549

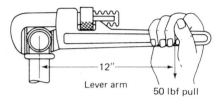

Torque = 50 1bf × 1 ft = 50 1bf − ft   **Figure 2-8**   Torque.

not damage the system. Look at Fig. 2-9. The sprocket receives a torque from a hydraulic motor to give the chain a linear motion. There is a torque at the sprocket whether the chain is moving or not. What is important to understanding the concept of torque is that the force is delivered at right angles to the radius at some distance from the center of the sprocket. The formula for torque is

$$T = F \times D$$

Notice that the torque is measured in lbf-ft or *N-m*, where work is measured in ft-lbf or *m − N*. This is done by general agreement to keep the two separate when calculations are made.

Torque Horsepower

Knowing the torque gives us a valuable tool to compute the horsepower of turning machinery such as pumps and motors. There are two formulas that are used the

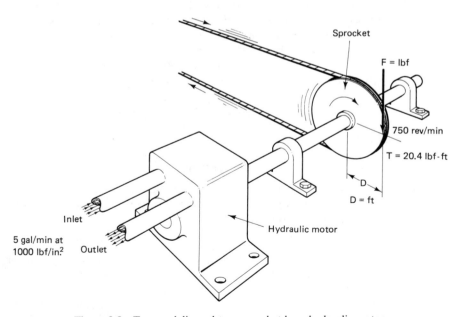

Figure 2-9   Torque delivered to a sprocket by a hydraulic motor.

most. In English units,

$$THP = \frac{\frac{2\pi N \, T}{60}}{550} = \frac{2\pi N \, T}{33{,}000} = \frac{N \times T}{5252} \tag{2-13}$$

Where N is rev/min and T is in lbf-ft.

In SI metric units

$$THP = \frac{\frac{2\pi N \, T}{60}}{746} = \frac{2\pi N \, T}{44\ 760} = \frac{N \times T}{7121} \tag{2-14}$$

Where N is rev/min and T is measured in $N \cdot m$.

Example 8

Now let us look again at a motor receiving the delivery of 5 gal/min at a pressure of 2000 lbf/in.[2] from the last example, and make a computation for the torque that might be delivered at the output shaft turning at 500 rev/min. This would assume 100% efficiency, of course, but that matter will be dealt with later. All that has to be done is to solve the torque horsepower formula for torque T. Here is the computation.

$$T = \frac{(THP)(33{,}000)}{2\pi N} = \frac{(5.83)(33{,}000 \text{ ft-lbs/min})}{(2)(3.14)(500 \text{ rev/min})} = 61.27 \text{ lbf-ft.}$$

Notice that torque is measured in lbf-ft, where work is measured in ft-lbf.

2-9 TORQUE HORSEPOWER AND FLUID HORSEPOWER RELATIONSHIPS

The question about what the difference is between fluid horsepower and torque horsepower may have occurred to you during the previous discussion. Well, there really isn't any, a horsepower is a horsepower; just the variables in the formula are different. For that reason, the two can be set equal to each other and used to solve for any one of the variables if the other three are known, and frequently they are. Setting the two equal in English units

$$FHP = \frac{p \times Q}{1714} = \frac{N \times T}{5252} = THP \tag{2-15}$$

The SI metric equivalent is

$$FHP = \frac{p \times Q}{44\ 760} = \frac{N \times T}{7122} = THP \tag{2-16}$$

The formulas can be used to size a fluid power component when the pressure and flow rate are known, and the desired speed or torque are specified by the application. Let's try two applications.

Example 9

The pressure of the components in a fluid power system have a rating of 2000 lbf/in.[2] and we wish to use a hydraulic motor which could turn a drum hoist similar to that in Fig. 2-10 at 20 rpm with a torque of 2000 lbf-ft. What would be the flow rate and horsepower of the motor? First, we should set the two horsepower formulas equal and solve for the flow rate, and then follow up by solving for the horsepower.

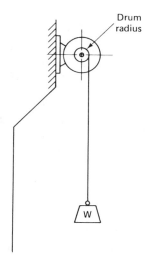

Drum
radius

W

Figure 2-10 Hoisting drum.

In English units

$$Q = \frac{(1714)(20 \text{ rev/min})(2000 \text{ lbf-ft})}{(5252)(2000 \text{ lbf/in.}^2)} = 6.5 \text{ gal/min}$$

and the horsepower would be

$$FHP = \frac{(2000 \text{ lbf/in.}^2)(6.5 \text{ gal/min})}{(1714)} = 7.6 \text{ hp}$$

The metric equivalent would be solved either by knowing the values for the pressure and torque in SI units, or by converting the pressure to kPa and torque to N · m. Solving the problem that way

$$Q = \frac{(44\ 760)(20 \text{ rev/min})(2712 \text{ N} \cdot \text{m})}{(7122)(13\ 793 \text{ kPa})} = 24.71 \text{ liters/min}$$

and the horsepower would be

$$FHP = \frac{(13\ 793 \text{ kPa})(24.71 \text{ liters/min})}{(44\ 760)} = 7.6 \text{ hp}$$

which is the same answer.

Example 10

Another problem is to consider how much torque might be expected from a hydraulic motor turning at 750 rpm, operating at 1000 lbf/in.², and receiving fluid at 5 gal/min (see Fig. 2-9). Here we are asking about expected torque. The problem could just as easily describe a pump as well as a motor, except for losses which will be described later. Now look at the solution. In English units, using Eq. 2-15

$$T = \frac{5252 \times p \times Q}{1714 \times N} = \frac{5252 \times 1000 \times 5}{1714 \times 750} = 20.4 \text{ lbf-ft}$$

Losses will be covered in the discussion of pumps, motors, and cylinders, but a few words here will explain the general concept. Simply put, it takes more than the required fluidpower output to turn a hydraulic pump input shaft. Take another look at the motor pump unit in Fig. 2-6. There are losses in the driving motor, in the coupling, and in the pump itself where slippage and friction eat away at the input. The same is true of hydraulic motors which receive more fluid power than they deliver as mechanical power at the output shaft. It comes down to this. If a hydraulic pump is 85% efficient it will take about 118% at the mechanical input to give 100% of the fluid horsepower specified at the output. The same is true of a hydraulic motor which receives 100% of the requirement in fluid horsepower at the input, but will deliver only 85% of that to the output shaft as mechanical horsepower. It's not a great problem as long as it's understood that in actual applications the losses must be figured in.

2-10 SUMMARY

What makes a fluid power system is the pressurized fluid. When the fluid moves a cylinder, work is accomplished. If the movement is timed, the horsepower can be calculated using James Watt's horsepower formula, the fluid horsepower formula, or the torque horsepower formula. Torque and twisting effort mean the same thing. Torque and work have the same units but conceptually they're different. Work requires movement, torque does not. If the shaft moves, torque horsepower can be calculated. Three advantages of fluid power systems are: (1) tremendous force multiplication occurs, (2) no harm is done to the system should it stall, and (3) torque output continues even when the hydraulic motor is stalled.

STUDY QUESTIONS AND PROBLEMS

1. The cylinder in Fig. 2-11 is used to raise the weight 9 in. If the bore of the cylinder is 2½ in. and the pressure is 1800 lbf/in.², how much work is accomplished?

2. An automobile lift raises a car weighing 3000 lbf, 6 ft off the shop floor (Fig. 2-12). Assuming the lift itself weighs 500 lbf and that friction is negligible, what is the work necessary to raise the car in ft-lbf and m-N?

3. If the bore diameter of the lift cylinder in Problem 2 is 6 in., what will be the pressure in the system when it comes to rest in the raised position?

4. In Problem 2, what would be the power expended if the lift raises the car in 15 secs? Compute the answer in ft-lbf/min and hp.

5. If the auto lift cylinder in Problem 2 has a diameter of 8 in., how fast would it have to descend for the flow rate to become 10 gal/min? (*Clue:* set $t = 1$.)

6. What would the return flow rate in Problem 5 have to be set at for the auto to descend in 10 secs?

7. A double-acting cylinder with a 2.5-in. bore, a 1-in. diameter rod, and 14-in. stroke cycles 25 times per minute. What is the flow rate?

8. What is the flow rate to a cylinder with a 3-in. bore and 22-in. stroke that fully extends in 8 secs?

9. How many seconds should it take a hydraulic cylinder with a 4-in. bore and 24-in. stroke to fully extend if it receives fluid at 9 gal/min?

Basics of Hydraulics   Chap. 2

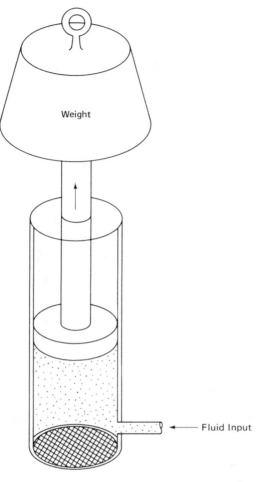

Weight

Fluid Input

Figure 2-11 Problem 1.

10. What is the bore of a cylinder that extends 18 in. in 6 seconds while receiving fluid at 4 gal/min?

11. How many in./sec should a cylinder with 1.5-in. bore extend if it receives fluid at 1 gal/min?

12. What is the minimum displacement of a gear motor turning at 800 rev/min if the flow rate to the motor is 6 gal/min?

13. What flow rate would be required to turn a gear motor with a displacement of 2 in.^{3}/rev at 1000 rev/min?

14. What is the displacement of a motor that turns 300 rev/min when the flow rate is 10 gal/min?

15. If a hydraulic motor with a displacement of 1.65 in.^{3}/rev receives fluid at 20 gal/min, what would be its maximum output speed?

16. A low-speed, high-torque piston motor with 5 cylinders having a bore of 1.5 in. and stroke of 1.75 in. turns at 200 rev/min. What is the flow rate to the motor?

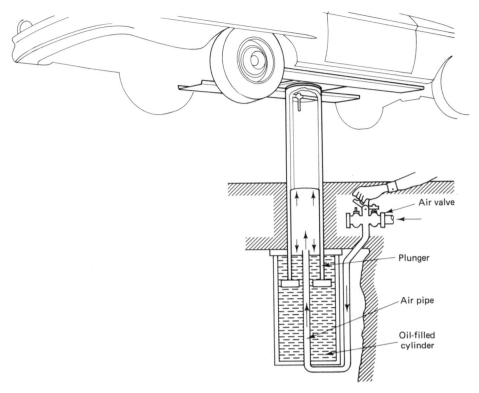

Figure 2-12 Problem 2.

17. What size cylinder connected to a 5 gal/min (22.7 l/min) pump would be required to limit the extension velocity to 2 ft/sec?

18. Compute the flow rate to a cylinder with a 4-in. bore extending at 3 ft/min.

19. Compute the fluid horsepower of a system with a 5 gal/min pump transferring fluid to a cylinder at a pressure of 1800 lbf/in².

20. A hydraulic cylinder is used to lift a load of 750 lbf a distance of 4 ft in 3 secs. Compute the *power* and *fluid horsepower*.

21. A wheel motor similar to that in Fig. 2-13 must exert a traction force of 1500 lb. If the outside diameter of the wheel is 30 in., what would be the torque from the motor?

22. Ignoring the friction, if the wheel motor in Problem 21 receives fluid from a 7 gal/min pump, and the pressure is limited to 2500 lbf/in.², how fast would the wheel motor turn and what would be the ground speed of the vehicle?

23. A 5-in.³ displacement motor is to be tested on a bench at 3000 rpm at 2500 lbf/in². Assuming 100% efficiency, estimate the FHP output of the motor?

24. A cylinder with an area of 4 in.² raises a load 2000 lbf, 2 ft using fluid supplied by a single-acting hand pump with a cylinder area of 0.25 in.² and a 3-in. stroke. Assuming no losses, compute the following:
    (a) pressure in the system
    (b) number of strokes of the hand pump
    (c) operator force exerted on the pump piston

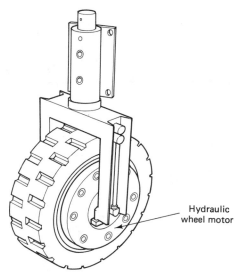

Figure 2-13 Problem 21.

25. How long would it take a pump delivering 25 gal/min to extend a 5-in. cylinder with a 30-in. stroke?

26. In Problem 25, if the rod in the cylinder were 1.5 in., how long would the cylinder take to retract?

27. An operator pumps a foot-operated single-acting floor jack through a 10-in. stroke at a rate of 40 strokes per minute. The 2000 lbf car being raised is observed to move at the rate of 12 in. per min. What is the multiplication of force?

28. A hydraulic system powered by a 10 hp motor operates at 2500 lbf/in². At maximum capacity, and assuming no losses, what is the flow through the system in gal/min and liters/min?

29. A pump with a 10 hp rating operates at 1500 rev/min. What is the torque in lbf-ft and *N-m*?

30. Assuming 100% efficiency, what displacement motor would be required to deliver 200 lbf-ft of torque at 1000 rev/min if the pressure is limited to 2500 lbf/in²?

CHAPTER 3

ENERGY IN HYDRAULIC SYSTEMS

3-1 INTRODUCTION

Hydraulic fluid flowing in a high pressure system contains *potential energy* due to its initial elevation and pressure energy developed by the pump, and *kinetic* or *flow* energy which results from the movement of the fluid through the plumbing and components.

Energy is added to the system through the pump and subtracted from the system through cylinders and hydraulic motors moving the load resistance. And since energy can neither be created nor destroyed, the total energy of the system remains constant. This is the basis for the Bernoulli equation.

The potential energy due to elevation (*EPE*) is measured in units of *head* or *pressure*. It results from the distance the reservoir is mounted above or below the pump inlet, and the specific gravity (*Sg*) of the fluid. For water and high water content fluids, the *Sg* is approximately 1.0. Petroleum base hydraulic oils have an *Sg* in the range of 0.75 to 0.90; whereas the *Sg* of synthetic fluids such as phosphate esters has a range of 1.02 to 1.20. In the English system of measurement, this establishes the specific weight of the fluid between 35 and 75 lb/ft^3, or 7 and 10 lb/gal.

The pressure energy (*PE*) of the fluid added by the pump working against the load resistance has the same units as the energy due to elevation, but it results from energy consumed by the prime mover driving the pump. Most of the energy in a hydraulic system is attributed to the *PE* supplied by the pump, and this is where the energy costs are incurred.

The kinetic energy (*KE*), which is usually a small portion of the total, accounts for flow losses. It is measured as a head loss or pressure drop across restrictions, valves, fittings, and hydraulic tubing. Thus, the *KE* term is a measure of the efficiency of the system.

3-2 PRESSURE AND HEAD

Hydraulic head is the height in feet or meters that a liquid can be elevated for a given pressure which is measured in lbf/in.² or Pascals (N/m²). Hydraulic head is sometimes referred to as pressure head. For every pressure given for a specific liquid, there is a height that can be assigned that indicates the elevated length that a column of the same fluid can attain. The term head receives widespread use in hydroelectric applications where the turbine is positioned many feet below the surface of the reservoir that provides the motive force. It also provides a means of expressing the potential energy of the liquid fluid directly if the specific weight of the fluid is known.

Frequently it is desirable to express head as pressure, and conversely, pressure as head. From the definition of pressure, we have

$$\text{Pressure } (p) = \frac{\text{Force } (F)}{\text{Area } (A)}$$

The pressure exerted by a 1 ft³ cube of water with a specific weight of 62.4 lbf/ft³ (Fig. 3-1) is

$$p = \frac{62.4 \text{ lbf/ft}^3}{144 \text{ in.}^2/\text{ft}^2} = 0.433 \text{ lbf/in.}^2 \text{ for each ft of head } (h)$$

This indicates that the force on 1 in.² of the lower surface area of 1 ft³ of water in the example is 0.433 lbf. Since the water occupies 1 ft³, 1 in.² of the lower surface area has a dimension of the base of a column of water 1 ft high. The force exerted by each of these columns with a 1-in.² base, multiplied by a 1-ft length, against the

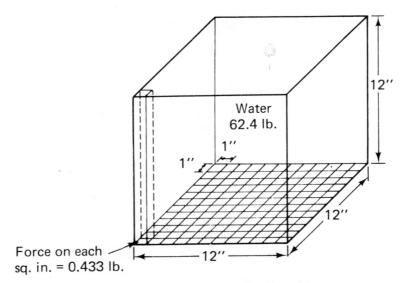

Figure 3-1   Pressure, head and specific weight.

base of the container is 0.433 lbf, and is the equivalent of 1 ft of head. So that

$$p = h \times 0.433 \qquad (3\text{-}1)$$

or solving the expression for head

$$\text{Head } (h) = \frac{\text{Pressure } (p)}{0.433}$$

$$h = \frac{p}{0.433}$$

and

$$h = p \times 2.31 \qquad (3\text{-}2)$$

In metric units since 1 m³ cube of water weighs 9802 N

$$p = 9802 \, h \, \text{Pa(N/m}^2) \qquad (3\text{-}3)$$

and

$$h = \frac{p}{9802} = 1.02 \times 10^{-4}p \text{ meters} \qquad (3\text{-}4)$$

Notice in developing the relationship of pressure, defined as a force per unit area, that the specific weight of water was used as the force per unit cubic foot and cubic meter. In fluid power systems, the medium is oil which usually weighs less than water and because of this difference in specific weight, expressions to convert pressure to head must be altered. Water has a specific gravity of 1. Oil has a specific gravity in the range of 0.75 − 0.90, and 1 ft³ of this fluid would weigh about (0.80 × 62.4) 50 lbf rather than 62.4 lbf as would water. Expressions for converting pressure to head, and head to pressure, thus become

$$p = \frac{F}{A} = \frac{Sg \times 62.4}{144}$$

and

$$p = Sg \times h \times 0.433 \qquad (3\text{-}5)$$

Rearranging

$$h = \frac{p}{Sg \times 0.433}$$

and

$$h = \frac{p \times 2.31}{Sg} \qquad (3\text{-}6)$$

In metric units

$$p = 9802 \times Sg \times h \qquad (3\text{-}7)$$

and

$$h = \frac{1.02 \times 10^{-4}p}{Sg} \qquad (3\text{-}8)$$

Several problems requiring the solution for pressure in lbf/in.^{2} (Pa), or head in ft (m), can be solved using these formulas.

Example 1

Compute the reading on a pressure gauge inserted at the base of a column of fluid 50 ft high with a specific gravity of 0.85 (Fig. 3-2).

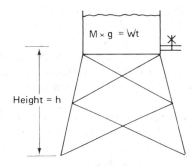

Figure 3-2 Example 1.

Solution From Eq. 3-5

$$p = Sg \times h \times 0.433$$
$$p = 0.85 \times 50 \times 0.433 = 18.4 \text{ lbf/in.}^2$$

Example 2

Convert to head in feet and meters a pressure reading of 1500 lbf/in.^{2} taken from a hydraulic pressure gauge monitoring a system using a hydraulic oil with a density of 0.78.

Solution Substituting in Eq. 3-6,

$$h = \frac{p \times 2.31}{Sg}$$

and

$$h = \frac{1500 \times 2.31}{0.78} = 4442 \text{ ft}$$

In metric units, a pressure reading of 1500 lbf/in.^{2} first must be converted to Pa using the conversion factor[1]

$$\text{lbf/in.}^2 \times 6895 = Pa$$

and then solve Eq. 3-8

$$h = \frac{(1500 \text{ lbf/in.}^2)(6895)(1.02 \times 10^{-4})}{0.78} = 1352 \text{ m}$$

[1] The exact conversion factor is lbf/in.$^2 \times (6.8947572 \times 10^3)$.

Example 3

Convert a hydraulic pressure reading of 13.79 megapascals (MPa) to meters if the oil has a Sg of 0.83.

Solution Substituting in Eq. 3-8

$$h = \frac{(1.02 \times 10^{-4})(13\ 790\ 000\ N/m^2)}{0.83} = 1695\ m$$

3-3 CONTINUITY EQUATION

For purposes of making most calculations, the flow in fluid power systems is considered to be **steady.** The flow is steady when the velocity at any period in time is constant. The velocity at successive places in the conductor, however, may change; for example, where the cross section area of a tube or hose is larger or smaller. To approximately fulfill the conditions of steady flow, changes with respect to time in pressure, temperature, fluid density, and flow rate must be almost zero.

The continuity equation governing flow in a fluid conductor is derived from the principle that during conditions of steady flow, the mass of fluid flowing past any point in the conductor is the same. For practical purposes and in the case of hydraulic fluids that are nearly incompressible

$$Q = A_1 v_1 = A_2 v_2 = \text{Constant} \qquad (3\text{-}9)$$

Where the flow rate (Q) is in gal/min, the area (A) is in.^{2}, and the velocity (v) is in feet per second (ft/sec),

$$Q(\text{gal/min}) = A(\text{in.}^2) \times v(\text{ft/sec}) \times 12(\text{in./ft}) \times 60(\text{sec/min}) \times 1/231(\text{gal/in.}^3)$$

$$Q(\text{gal/min}) = A(\text{in.}^2) \times v(\text{ft/sec}) \times 3.12$$

and

$$Q = 3.12Av \qquad (3\text{-}10)$$

The continuity equation is used to solve for the velocity, area, or flow rate at any point in a system if two of the three quantities are given.

Example 4

Referring to Fig. 3-3, fluid is flowing in a 2-in. diameter pipe at the rate of 40 gal/min. What is the velocity of the fluid in that portion of the conductor? What would be the velocity of the fluid in the 4-in. diameter manifold?

Solution (a) The 2-in. diameter pipe has a cross section area of 3.14 in^2. The fluid velocity is computed by rearranging Eq. 3-10.

$$Q = 3.12Av$$

$$v = \frac{Q}{A \times 3.12}$$

$$v = \frac{40}{3.14 \times 3.12} = 4.08\ \text{ft/sec}$$

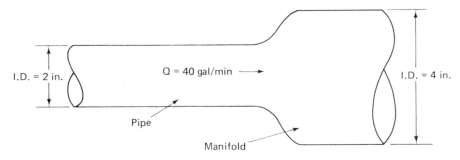

I.D. = 2 in.

Q = 40 gal/min →

I.D. = 4 in.

Pipe

Manifold

Figure 3-3 Example 4.

(b) The 4-in. diameter pipe has a cross section of 12.56 in^2. Solving the continuity Eq. 3-9 for v_2:

$$A_1 v_1 = A_2 v_2$$

$$v_2 = \frac{A_1 v_1}{A_2}$$

and

$$v_2 = \frac{3.14 \times 4.08}{12.56} = 1.02 \text{ ft/sec}$$

Notice that doubling the diameter reduces the flow rate to one-fourth of its original value.

Example 5

Fluid is flowing in a 2-in. diameter pipe at the rate of 18 gal/min (Fig. 3-4). What is the minimum diameter to which the pipe might be reduced without exceeding the maximum recommended velocity of 10 ft/sec?

Solution First, the velocity of the fluid in the 2-in. diameter pipe is determined using Eq. (3-10):

$$Q = 3.12 A v_1$$

$$v_1 = \frac{Q}{A_1 \times 3.12}$$

$$v_1 = \frac{18}{3.14 \times 3.12} = 1.84 \text{ ft/sec}$$

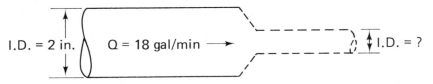

I.D. = 2 in.   Q = 18 gal/min →   I.D. = ?

Figure 3-4 Example 5.

Knowing that the area of a circle is computed from

$$A = \frac{\pi D^2}{4}$$

the value for the reduced diameter of the pipe is computed from

$$Q = A_2 \times v_2 \times 3.12$$

$$A_2 = \frac{\pi D_2^2}{4} = \frac{Q}{v_2 \times 3.12}$$

$$D_2^2 = \frac{4 \times Q}{\pi \times v_2 \times 3.12}$$

and

$$D_2 = \sqrt{\frac{4 \times 18}{3.14 \times 10 \times 3.12}} = 0.857 \text{ in.}$$

Similar values are obtained using flow nomograms, such as that shown in Table 3-1, that have been derived using the continuity equation. Knowing two values allows for the solution of the third value by placing a straightedge across the two known values and reading the third quantity directly.

Example 6

Compute the velocity in a 1-in. I.D. suction line (actual size) of a hydraulic system in which 10 gal/min is being pumped.

Solution Using Table 3-1 and placing a straightedge across the values for the 10 gal/min flow rate in the left column and the 1-in. actual I.D. pipe size in the center column, the velocity at the suction line is read as 4.1 ft/sec in the right column. This is approximately equal to the computed value.

3-4 POTENTIAL AND KINETIC ENERGY

The energy available from the prime mover is converted to a more useful form through the action of the fluid hydraulic pump. The prime mover is itself a conversion device that usually consists of a drive motor powered by electricity or chemical energy stored in hydrocarbon fuel to power a combustion engine. Although energy is also available from other sources, electric motors and internal combustion engines provide nearly all the motive force to raise the energy level of fluid power systems.

The total energy available from fluid in a system consists of the potential energy due to its initial elevation (*EPE*), the pressure potential energy (*PE*), and the kinetic energy (*KE*) due to the movement of the fluid. That is

Total Energy = Elevation Energy (*EPE*) + Pressure Energy (*PE*)

+ Velocity Energy (*KE*)

and

Total Energy = *EPE* + *PE* + *KE*   (3-11)

TABLE 3-1 CONTINUITY EQUATION NOMOGRAPH (*ENGLISH UNITS*).

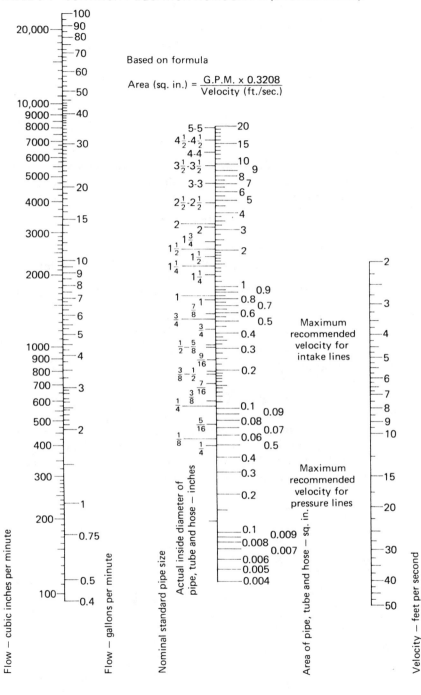

Based on formula

$$\text{Area (sq. in.)} = \frac{\text{G.P.M.} \times 0.3208}{\text{Velocity (ft./sec.)}}$$

Maximum recommended velocity for intake lines

Maximum recommended velocity for pressure lines

Flow — cubic inches per minute

Flow — gallons per minute

Nominal standard pipe size

Actual inside diameter of pipe, tube and hose — inches

Area of pipe, tube and hose — sq. in.

Velocity — feet per second

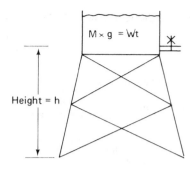

Figure 3-5 Potential energy as a force available to act through a distance (h).

The potential energy of a fluid can be equated to the force it exerts toward the center of gravity (Mg), and the distance (h) through which it is available to move on command (Fig. 3-5). This is expressed by the formula

$$\text{Potential Energy} = M \times g \times h \tag{3-12}$$

or, if the weight of the fluid rather than its mass is given,

$$\text{Potential Energy} = w \times h \quad \text{in ft-lbf} \tag{3-13}$$

Where it is more desirable to use pressure (lbf/in.²) than head (h), Eq. 3-6 can be used

$$h = \frac{p \times 2.31}{Sg}$$

and substituting

$$\text{Potential Energy} = \frac{w \times p \times 2.31}{Sg} \quad \text{in ft-lbf} \tag{3-14}$$

When fluid flows in a hydraulic system, potential energy is converted to the energy of motion, kinetic energy (Fig. 3-6). Potential energy and kinetic energy are mutually convertible. As the potential energy of the fluid is dissipated by falling from a higher point to a lower point, its kinetic energy is increased by an

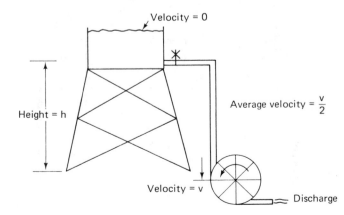

Figure 3-6 Kinetic energy as energy of motion.

Energy in Hydraulic Systems   Chap. 3

equal amount. This is a restatement of the law of conservation of energy which says, in effect, that energy can neither be created nor destroyed.

If a body of fluid at rest at some height is dropped through a distance (h), its velocity is (v) feet per second (ft/sec). The acceleration due to gravity (g) is approximately constant. The average velocity of the falling body of fluid is $(v/2)$ ft/sec since it has a starting velocity of 0 ft/sec, and a terminal velocity of (v) ft/sec. The time taken for the body to cover the distance equals the distance (h) divided by the average velocity $(v/2)$. That is,

$$t = \frac{h}{\frac{v}{2}} = \frac{2 \times h}{v}$$

where the time (t) is in sec. The acceleration constant (g) due to gravity equals the change in velocity divided by the time. That is,

$$g = \frac{v - o}{t}$$

$$g = \frac{v}{\frac{2h}{v}} = \frac{v^2}{2h}$$

and

$$\frac{v^2}{2} = g \times h$$

From the potential energy formula, Eq. 3-12,

$$\text{Potential Energy} = M \times g \times h$$

The value $(v^2/2)$ can be substituted for $(g \times h)$, and the potential energy formula converted to the kinetic energy formula. That is,

$$\text{Kinetic energy} = \frac{M \times v^2}{2}$$

Converting mass (M) to weight (w), the formula for kinetic energy becomes

$$\text{KE} = \frac{w \times v^2}{2 \times g} \quad \text{in ft-lbf} \tag{3-15}$$

and the expression for the total energy in the system now becomes

$$\text{Total Energy} = \text{Elevation} + \text{Pressure} + \text{Kinetic}$$

$$\text{Total Energy} = (w \times h) + (w \times p \times 2.31) + \frac{w \times v^2}{2 \times g} \tag{3-16}$$

where each of these terms is in ft-lbf.

Let's look at three examples to see how each of the energy terms fits into the general formula.

Example 7

A hydraulic system has a 200 gallon reservoir mounted above the pump to produce a positive head at the inlet. This will prevent the pump from cavitating, particularly at start-up. If the static pressure at the inlet is to be 10 lbf/in², and the fluid has a Sg of 0.87, how high must the average fluid level be above the pump inlet, and what is the energy of elevation of the fluid in the reservoir?

Solution   The problem, which is illustrated in Fig. 3-7, asks first for the elevation pressure and then the total elevation energy (the first term in Eq. 3-11). The elevation pressure can be computed using Eq. 3-6 as follows:

$$h = \frac{p \times 2.31}{Sg}$$

$$h = \frac{10 \text{ lbf/in.}^2 \times 2.31 \frac{\text{ft}}{\text{lbf/in.}^2}}{0.87}$$

and

$$h = 26.6 \text{ ft}$$

The units of the second term may at first look confusing until one notices that they are derived from 1/0.433 in Eq. 3-6. That is,

$$\frac{2.31}{1} = \frac{1}{\frac{0.433 \text{ lbf/in.}^2}{\text{ft}}} = \frac{\text{ft}}{\text{lbf/in.}^2}$$

Finally, solving for the elevation energy using Eq. 3-13

$$EPE = w \times h = 200 \text{ gal} \times 0.87 \times 8.34 \text{ lbf/gal} \times 26.6 \text{ ft}$$

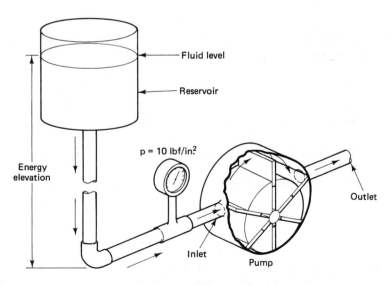

Figure 3-7   Example 7.

and

$$EPE = 38,601 \text{ ft-lbf}$$

Example 8

If friction and kinetic or flow energy in Example 7 are ignored, how much pressure would the pump have to account for to lift a 1000 lbf load using a 2-in. bore cylinder.

Solution   Here the problem illustrated in Fig. 3-8 asks that the total pressure energy required to lift the load be equated to the combined pressure developed by the energy of elevation (*EPE*) and the pump. From Eq. 3-11

$$\text{Total Energy} = EPE + PE + KE$$

and since friction and kinetic energy are ignored, the KE term drops out and the energy developed by the pump equals

$$PE = \text{Total Energy} - EPE$$

Notice that each of the energy terms in Eq. 3-16 contains the weight of the fluid which for this example does not change and can be cancelled out allowing us to solve only for the pressure to lift the load.

$$\text{Total Pressure} = \frac{F}{A} = \frac{1000 \text{ lbf}}{3.14 \text{ in.}^2} = 318.5 \text{ lbf/in.}^2$$

Finally, subtracting the pressure of elevation (Example 8) from the total pressure,

$$p = 318.5 \text{ lbf/in.}^2 - 10.0 \text{ lbf/in.}^2 = 308.5 \text{ lbf/in.}^2$$

This assumes, of course, that the oil is not returned to the reservoir.

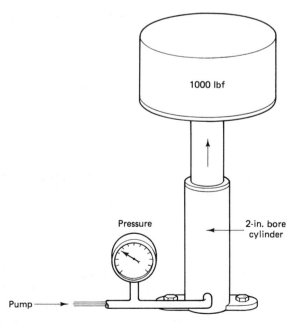

Figure 3-8   Example 8.

Example 9

In Fig. 3-9, if friction is ignored, how much kinetic energy would be dissipated if 25 gal/min were flowing through a pump outlet with an area of 1 in²?

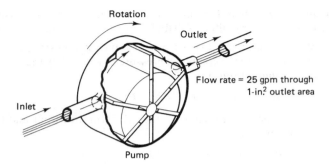

Figure 3-9 Example 9.

Solution Here the problem asks for the value of the kinetic energy term in Eq. 3-11, and it must be remembered that energy is a sum total, given in units of force and distance—for example, ft-lbf, or $m \cdot N$. Thus, the term w in Eq. 3-15 must be in force units—that is, the total weight of the fluid for some specified time—for example, a minute or an hour. Using 1 min as the total time and 0.87 as the Sg of the fluid,

$$w = 25 \text{ gal/min} \times Sg \times 231 \text{ in.}^3/\text{gal} \times 1/1738 \text{ ft}^3/\text{in.}^3 \times 62.4 \text{ lbf/ft}^3$$

and

$$w = 181.4 \text{ lbf}$$

Solving the continuity equation for the velocity

$$v = \frac{Q}{a} = \frac{25 \text{ gal/min} \times 231 \text{ in.}^3/\text{gal} \times 1/12 \text{ ft/in.} \times 1/60 \text{ min/sec}}{0.785 \text{ in.}^2}$$

and

$$v = 10.2 \text{ ft/sec}$$

Finally, solving Eq. 3-15 for the kinetic energy

$$KE = \frac{w + v^2}{2 \times g}$$

$$KE = \frac{(184.4 \text{ lbf}) \times (10.2 \text{ ft/sec})^2}{2 \times 32.2 \text{ ft/sec}^2} = 298 \text{ ft-lbf}$$

Notice that the value of the kinetic energy term is small, particularly when the velocity of the fluid is kept within limits. When the velocity increases much above 10 ft/sec, friction loss generates unnecessary heat in the system.

3-5 BERNOULLI'S EQUATION

Conservation of energy in a fluid power system is described by Bernoulli's theorem which defines the total energy of the system as constant (Fig. 3-10). For example, assuming no work is done or energy dissipated to the surroundings, in a pipe where fluid flows in a streamline through a restriction, the energy level of the

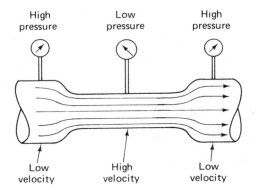

High pressure   Low pressure   High pressure

Low velocity   High velocity   Low velocity

Figure 3-10   Bernoulli's theorem.

fluid through the restriction will remain constant even though both the pressure and the velocity of the fluid change. Bernoulli's theorem explains this by equating the energy at any two points in the system. That is

$$\text{Total Energy}_{\text{(point 1)}} = \text{Total Energy}_{\text{(point 2)}}$$

and

$$(w_1 \times h) + \frac{w_1 \times p_1 \times 2.31}{Sg} + \frac{w_1 \times v_1^2}{2 \times g}$$

$$= (w_2 \times h) + \frac{w_2 \times p_2 \times 2.31}{Sg} + \frac{w_2 \times v_2^2}{2 \times g} \qquad (3\text{-}17)$$

For most calculations, the weight of the fluid doesn't change appreciably and will cancel out. And if the potential energy due to elevation is equated to Z, the equation becomes

$$Z_1 + \frac{p_1 \times 2.31}{Sg} + \frac{v_1^2}{2 \times g} = Z_2 + \frac{p_2 \times 2.31}{Sg} + \frac{v_2^2}{2 \times g} \qquad (3\text{-}18)$$

Solving problems using Bernoulli's equation requires a systematic approach. In general, the following steps are followed:

1. Diagram the system indicating at which points solutions are to be made and the direction of flow.
2. Determine the datum plane Z_2 and the elevation of Z_1. If these are approximately equal—for example, when fluid flows horizontally in a hydraulic system—they can be cancelled out.
3. Set up Bernoulli's equation with each energy component in the same units, including energy added by pumps and energy subtracted by motors.
4. Flow losses should be subtracted.
5. If the velocity of the fluid is not known, solve the continuity equation.
6. Finally, equate the total energy added and subtracted at point 1 in the system to the total energy at point 2 in the system. Typically, the equation is solved for pressure or head, including losses, from point 1 to point 2

Example 10

In Fig. 3-11, fluid with a specific gravity of 0.91 is flowing horizontally at the rate of 500 gal/min from a pipe with an inside diameter of 3 in. to one with a diameter of 1-in. The pressure at point 1 is 1000 lbf/in². Assuming no work is done or energy dissipated from the system, compute the pressure at point 2.

Solution   The direction of flow is from p_1 to p_2, and since the flow is horizontal, thus introducing no change in elevation, the quantities Z_1 and Z_2 can be cancelled from Eq. 3-18.

The energy equation for the system then is

$$\frac{p_1 \times 2.31}{Sg} + \frac{v_1^2}{2 \times g} = \frac{p_2 \times 2.31}{Sg} + \frac{v_2^2}{2 \times g}$$

The velocities v_1 and v_2 are not known and must be determined by solving the continuity equation. The velocity v_1 is determined from

$$Q = A_1 \times v_1 \times 3.12$$

$$v_1 = \frac{Q}{(A_1)(3.12)}$$

and

$$v_1 = \frac{500}{(7.065)(3.12)} = 22.68 \text{ ft/sec}$$

Solving the continuity equation for v_2

$$v_2 = \frac{Q}{(A_2)(3.12)}$$

and

$$v_2 = \frac{500}{(0.785)(3.12)} = 204.15 \text{ ft/sec}$$

Finally, solving Bernoulli's equation for p_2 where g has the value 32.2 ft/s²

$$\frac{(1000)(2.31)}{0.91} + \frac{(22.68)^2}{64.4} = \frac{(p_2)(2.31)}{0.91} + \frac{(204.15)^2}{64.4}$$

$$(2538.46 + 7.99) = \frac{(p_2)(2.31)}{0.91} + 647.16$$

and

$$p_2 = \frac{0.91(2546.45 - 647.16)}{2.31} = 748 \text{ lbf/in.}^2$$

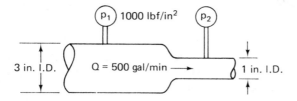

Figure 3-11   Example 10.

Here you will notice that a decrease in the size of the pipe is accompanied by a corresponding increase in velocity since the flow rate in the system is constant. The pressure also drops because the energy transfer through the system remains constant. If the size of the conductor were increased rather than decreased, the velocity would have decreased with a corresponding increase in pressure. Finally, the kinetic energy term or velocity head in Bernoulli's equation is seen to increase or decrease as the square of the velocity.

Potential energy that is added to the system in terms of head by the pump (H_a), energy extracted (H_e), and lost (H_l) also must be accounted for when computing actual rather than ideal flow conditions. These take the form of potential energy and are written in terms corresponding to head or pressure. The complete equation for Bernoulli's theorem then becomes

$$Z_1 + \frac{p_1 \times 2.31}{Sg} + \frac{v_1^2}{2 \times g} + H_a = Z_2 + \frac{p_2 \times 2.31}{Sg} + \frac{v_2^2}{2 \times g} + H_e + H_l \qquad (3\text{-}19)$$

Example 11

Fluid flows from a pump at 1000 lbf/in.^{2} at the rate of 200 gal/min horizontally to a motor operating at 750 lbf/in.^{2} (Fig. 3-12). Back pressure on the motor discharge port is 350 lbf/in^2. The fluid has a specific gravity of 0.85. Compute the energy extracted from the fluid in terms corresponding to head.

Solution Flow is horizontal, introducing no change in elevation so the quantities Z_1 and Z_2 cancel out of Eq. 3-19. Since the fluid flow rate and pipe size are constant, the velocity head terms are equal and also cancel out of the equation.

If the pressure at the inlet to the pump is considered to be 0 lbf/in.^{2} gauge, the potential energy term on the left side of the equation also becomes 0.

The energy equation for the system then becomes

$$H_a = \frac{p_2 \times 2.31}{Sg} + H_e + H_l$$

Computing the head associated with the potential energy added to the system

$$H_a = \frac{p \times 2.31}{0.85}$$

$$H_a = \frac{1000 \times 2.31}{0.85} = 2717.65 \text{ ft}$$

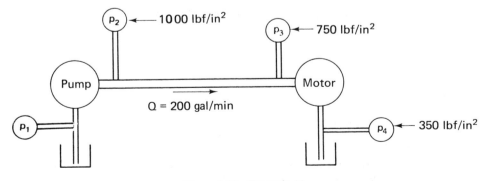

Figure 3-12 Example 11.

Finally, solving Bernoulli's equation for head associated with the energy that can be extracted from the system

$$2717.65 = 951.18 + H_e + 679.41$$

$$H_e = 2717.65 - 951.18 - 679.41 = 1087.06 \text{ ft}$$

3-6 TORRICELLI'S THEOREM

Torricelli's theorem is a special case of Bernoulli's equation derived for an ideal system.

In Fig. 3-13 a large tank holds water that follows out a small pipe at the base. Bernoulli's equation for the system is

$$Z_1 + \frac{p_1 \times 2.31}{Sg} + \frac{v_1^2}{2 \times g} = Z_2 + \frac{p_2 \times 2.31}{Sg} + \frac{v_2^2}{2 \times g}$$

Since the system is considered ideal, losses are negligible and H_l equals 0. The elevation (h) equals $(Z_1 - Z_2)$, and if Z_2 is zero, h equals Z_1.

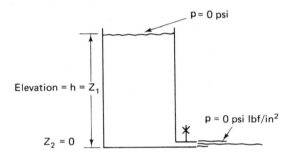

Figure 3-13 Toricelli's theorem.

If the top of the tank is open to the atmosphere, the gauge pressure is 0 at the surface and p_1 equals 0. The pressure at the surface of the outlet near the tank is also 0 since the free jet is discharged into the atmosphere.

Where the area of the surface of water in the tank is sufficiently large, the downward velocity of the surface is negligible and the velocity head (v_1) also becomes 0.

Bernoulli's equation for this case then becomes

$$Z_1 = h = \frac{v_2^2}{2 \times g}$$

and

$$v_2 = \sqrt{2gh}$$

which is a statement of Torricelli's theorem.

Example 12

Compute the velocity of a fluid flowing from the base of a water tank located 40 ft below the water level (Fig. 3-14).

Energy in Hydraulic Systems   Chap. 3

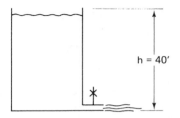

Figure 3-14   Example 12.

Solution   Using Torricelli's theorem

$$v = \sqrt{2gh}$$
$$v = \sqrt{2 \times 32.2 \times 40} = 50.7 \text{ ft/sec}$$

3-7 SUMMARY

The potential and kinetic energy in hydraulic systems can be accounted for in terms of the specific gravity of the fluid, head or pressure, and the velocity. Potential energy is influenced by the pressure resulting from elevation position and from input energy added through the hydraulic pump. Kinetic energy is the energy of motion and is influenced primarily by the velocity of the fluid through the system. In hydraulic systems, the primary factor affecting the energy level of the system is the pressure resulting from input to the system from outside sources through the pump.

Bernoulli's equation sets the energy level of the system at any two cross sections through the conductor equal to each other. It is derived from laws governing the conservation of energy and is used, typically, to determine the amount and source of energy extracted from the system as a result of friction losses, energy dissipated moving the load resistance, and energy discharged or wasted as heat.

Torricelli's theorem is a special application of Bernoulli's equation. Stated as $v = \sqrt{2gh}$, Torricelli's theorem is useful to determine flow through orifices and restrictions. The basic formula assumes frictionless flow. In practice, flow coefficient between 0.5 and 1.0 are introduced to account for friction losses associated with a particular orifice configuration.

STUDY QUESTIONS AND PROBLEMS

1. A 250 gal reservoir must be mounted in a 4 ft × 6 ft space. What is the height of the reservoir in inches?
2. How tall would a building have to be to deliver oil with a Sg of 0.82 from the roof to ground level at a pressure of 500 lbf/in²? How might this be achieved in a building half this tall?
3. What pressure would a pump have to overcome to deliver oil with a Sg of 0.90 to a tank with fluid surface at a height of 120 ft above the pump?

4. Compute the pressure readings on a dual gauge in $lbf/in.^2$ and kPa inserted in the base of a storage tank 40 ft high, full of oil that has a Sg of 0.87.

5. A pressure reading of 2250 $lbf/in.^2$ is observed on a hydraulic gauge. If the fluid has a Sg of 0.85, what would be the pressure in MPa?

6. A 200 gal hydraulic reservoir is placed on the second floor with the fluid level 12 ft above the pump inlet. If the petroleum base fluid has a Sg of 0.89, what would be the pressure at the pump inlet in $lbf/in.^2$ and ft of head? How much would the pressure increase for a synthetic fluid with a Sg of 1.15?

7. In Problem 6, assuming no friction or pumping losses, what would be the absolute and gauge pressure readings at the pump inlet if the reservoir were mounted so that the fluid level were 2 ft below the inlet? See Fig. 3-15. (*Clue:* atmospheric pressure = 0 $lbf/in.^2$ gauge = 14.7 $lbf/in.^2$ absolute.)

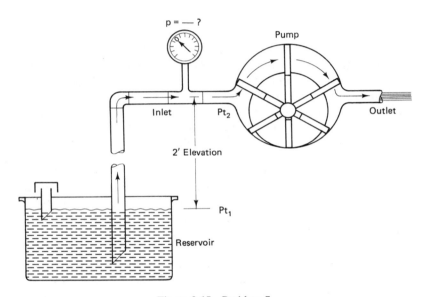

Figure 3-15   Problem 7.

8. At a maximum velocity of 15 ft/sec, how many gal/min of fluid will flow through a pipe with an I.D. of 4 in?

9. A 1-in. I.D. hydraulic tube delivers 35 gal/min. What is the fluid velocity in ft/sec?

10. What hose I.D. would be required to deliver 50 gal/min, assuming no friction losses at a maximum velocity of 8.5 ft/sec?

11. How large would the I.D. of a tube have to be to deliver 15 gal/min if the fluid velocity were held to 10 ft/sec? Assume no friction losses.

12. A flow meter in the return line to a hydraulic reservoir reads 18 gal/min. How large would the suction line have to be to limit the velocity to 3 ft/sec?

13. Fluid is flowing from a larger pipe with an I.D. of 8 in. to a smaller pipe at the rate of 500 gal/min (Fig. 3-16). What would be the minimum I.D. of the smaller pipe to limit the velocity to 10 ft/sec?

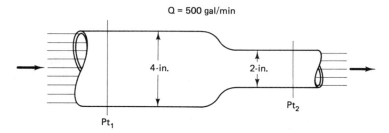

Q = 500 gal/min

4-in. 2-in.

Pt₁ Pt₂

Figure 3-16 Problem 13.

14. It is noticed that doubling the I.D. of a pipe reduces the velocity by a factor of 4. If instead, the objective were to reduce the velocity to half its original value, by what factor would the smaller diameter pipe be increased?

15. Hydraulic oil with a Sg of 0.91 is pumped horizontally in a 2-in. I.D. distribution line at the rate of 200 gal/min. A pressure gauge mounted in the pump outlet reads 2000 lbf/in.^{2}, but some distance downstream another gauge mounted in the line reads 1750 lbf/in^2. Assuming the gauge readings are accurate, how does the Bernoulli formula given by Eq. 3-19 account for this difference?

16. Fluid is flowing horizontally at the rate of 500 gal/min from a pipe with a 4-in. I.D. to a smaller one with a 2-in. I.D. If the fluid has a Sg of 0.88, and the pressure is 2000 lbf/in.^{2} at Point 1 in the 4-in pipe, compute the pressure at Point 2 in the 2-in. pipe.

17. A hydraulic pump delivers 55 gal/min of oil with a Sg of 0.95 at 2000 lbf/in.^{2} through a line with a cross section of 2 in^2. If the line is reduced to a cross section area of 1/2 in.^{2}, what change would there be in velocity and pressure?

18. Compute the velocity of a fluid which would flow from the base of a water storage tank where the level of the surface is 35 ft above the ground. Consult Fig. 3-17.

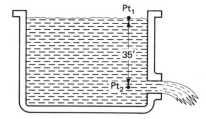

Pt₁

35′

Pt₂

Figure 3-17 Problem 18.

19. In Problem 18, if the hole in the base of the tank has a 3-in. I.D., and there are no friction losses, what flow rate could be expected?

20. What delivery in gal/min would be required to shoot a water jet approximately 25 ft in the air from a vertical tube with a 1.5-in. I.D.?

CHAPTER 4

HOW FLUIDS FLOW

4-1 INTRODUCTION

When a fluid flows, its volume is displaced as it changes location from one place to another in the system. The time it takes the fluid to change location determines the flow rate, and this, in turn, determines how much power the fluid transmits through the system. For example, doubling the displacement and flow rate would double the fluid power transmitted through the system—assuming, of course, that the pressure remains constant.

In addition to Bernoulli's theorem which gives us the basic energy equation, there are a number of laws that govern the flow of fluids. These serve as the basis for several equations used to make system calculations. From the principle of *conservation of mass* which says, in effect, that the mass of the fluid remains constant, we have the continuity equation. The *law of conservation of energy* which says, in effect, that energy can neither be created nor destroyed, allows us to convert energy from one form to another and develop several equations that govern flow. Finally, the concept of *momentum* which describes the change in the velocity of the mass of fluid, is the basis for equations governing friction.

Here, theory and practice are brought together. Because the flow of fluids is complex, the laws and equations that govern flow through fluid power systems must be confirmed through experimentation. Thus, many of the formulas that have their basis in theory are modified slightly with flow and friction factors to describe conditions as they actually exist in practice.

4-2 APPLICATIONS OF THE DISPLACEMENT FORMULA

While displacement can be calculated from the bore diameter and stroke length, it is more common to use tabled values to solve this formula. Table 4-1 lists the bore area diameter, with the area of the blank end of the cylinder down the left margin

56

TABLE 4-1 DISPLACEMENT OF CYLINDERS IN CUBIC INCHES

Bore area		Stroke (in.)										
(in.)	(in.²)	3	6	12	18	24	30	36	42	48	54	60
1	0.785	2.36	4.71	9.42	14.13	18.84	23.55	28.26	32.97	37.68	42.39	47.1
1⅛	0.994	2.98	5.96	11.93	17.89	23.86	29.82	35.78	41.75	47.71	53.68	59.64
1½	1.77	5.31	10.62	21.24	31.86	42.48	53.10	63.72	74.34	84.96	95.58	106.2
2	3.14	9.42	18.84	37.68	56.52	75.36	94.2	113.0	131.9	150.7	169.6	188.4
2½	4.91	14.73	29.46	58.92	88.38	117.8	147.3	176.8	206.2	235.7	265.1	294.6
3	7.07	21.21	42.42	84.84	127.3	169.7	212.1	254.5	296.9	339.4	381.8	424.2
3½	9.62	28.86	57.72	115.4	173.2	230.9	288.6	346.3	404.0	461.8	519.5	577.2
4	12.56	37.68	75.36	150.7	226.1	301.4	376.8	452.16	527.5	602.9	678.2	753.6
4½	15.90	47.7	95.4	190.8	286.2	381.6	477.0	572.4	667.8	763.2	858.6	954.0
5	19.63	58.89	117.8	235.6	353.3	471.1	588.9	706.7	824.5	942.2	1060	1178
6	28.26	84.78	169.6	339.1	508.7	678.2	847.8	1017	1187	1356	1526	1696

and the stroke in inches across the top. The numbers in the Table represent the cubic inch displacement of the cylinder. The Table can be used to determine the working volume of the cylinder on the extension stroke. If the displacement on the return stroke is required, the displacement of the rod must be subtracted from the displacement on the extension stroke.

Example 4-1

A ram with a 4-in. bore diameter cylinder and 2-in. rod (Fig. 4-1) has a stroke of 24 in. What would be the displacement of the cylinder through the extension and retraction strokes?

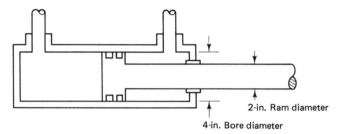

2-in. Ram diameter

4-in. Bore diameter   **Figure 4-1**   Example 1.

Solution   On extension, the displacement can be read by finding the place in Table 4-1 where the bore diameter and stroke length intersect. Finding the 4-in. bore diameter down the left margin and the 24-in. stroke across the top, the displacement is given as 301.4 in.³

On the return stroke the displacement of the ram would have to be subtracted from this value. In this case Table 4-1 can be used again. Locating the intersection of a 2-in. bore and a 24-in. stroke, the displacement is found to be 75.36 in.³ Subtracting this value from the bore displacement of 301.4 in.³ results in a value of 226.04 in.³ on the return stroke. Thus the combined displacement of the extension and retraction strokes would be

$$V_t = V_e + V_r$$

and

$$V_t = 301.4 \text{ in.}^3 + 226.04 \text{ in.}^3 = 527.4 \text{ in.}^3$$

where V_t = total volume (in.³)

V_e = extension volume (in.³)

V_r = retraction volume (in.³)

4-3 APPLICATIONS OF THE CONTINUITY EQUATION

It is common practice in sizing components to solve the continuity equation with nomographs. The continuity equation is solved with the conversion constants inserted so that solutions are given directly in the desired units. For example, Fig. 4-2 solves the continuity equation in English units:

$$\text{Area (in.}^2) = \frac{\text{Gal/min} \times 0.3208}{\text{Velocity (ft/sec)}}$$

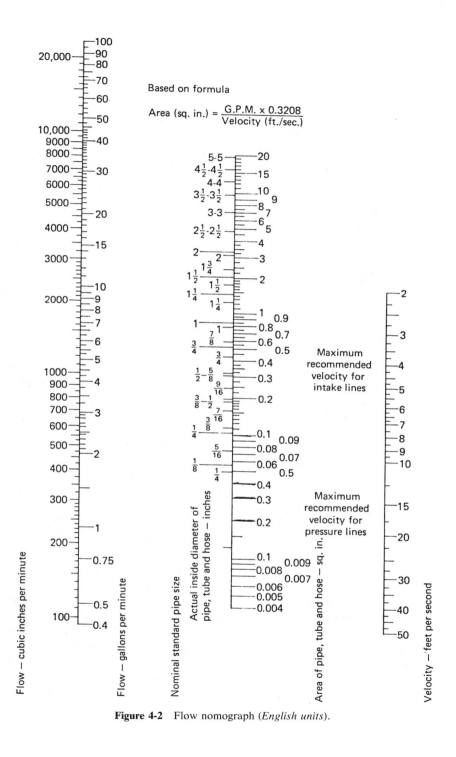

Based on formula

$$\text{Area (sq. in.)} = \frac{\text{G.P.M.} \times 0.3208}{\text{Velocity (ft./sec.)}}$$

Flow — cubic inches per minute

Flow — gallons per minute

Nominal standard pipe size

Actual inside diameter of pipe, tube and hose — inches

Maximum recommended velocity for intake lines

Maximum recommended velocity for pressure lines

Area of pipe, tube and hose — sq. in.

Velocity — feet per second

Figure 4-2 Flow nomograph (*English units*).

All that is required is to determine which two of the three variables are given by the problem and then use a straightedge across these two values to read the third quantity directly.

Example 2

Compute the velocity in a 1-in. I.D. suction line of a hydraulic system through which 10 gal/min is being pumped.

Solution Using Fig. 4-2, and placing a straightedge across the values for the 10 gal/min flow rate in the left column and the 1-in. actual I.D. pipe size in the center column, the velocity in the suction line is read directly as 4.1 ft/sec from the right column. Values thus determined will be approximately equal to the computed values.

Example 3

What size standard pipe would be required to deliver 40 gal/min if the velocity in the line was not to exceed 7 ft/sec?

Solution Placing a straightedge across the values for the 40 gal/min in the left column of Fig. 4-2, and the velocity of 7 ft/sec in the right, the actual I.D. of the pipe is given at 1½ in., which is slightly less than a 1½ in. standard pipe.

4-4 VISCOSITY

The viscosity of a fluid determines how easily it will flow and is required to solve most flow problems. Viscosity is a measure of the internal resistance of a fluid to shear and is related to the internal friction of the fluid itself. Thick fluids flow more slowly than thin fluids, because they have more internal friction. Viscosity numbers are assigned to describe the relative differences in the ability of a fluid to flow in comparison with other fluids. Higher numbers are assigned to thicker fluids, lower numbers to thinner fluids.

The oil film between the moving and stationary plates may be thought of as a series of fluid layers separating the two parts. Oil adheres to both surfaces. The velocity of the oil at the stationary plate is zero and the velocity of the oil at the moving plate equals the velocity of that surface. Between the two surfaces, the velocity of the oil varies on a straight line between zero at the stationary surface, and the speed of the moving plate at its surface. What happens is the moving plate causes the adjacent layers of oil between the two plates to shear and the force necessary to move the plate is what we call viscosity (Fig. 4-3).

There are really two ways to describe the viscosity of a fluid: one is the *absolute* or dynamic viscosity designated by μ (Greek letter mu), and the other is the kinematic viscosity designated by ν (Greek letter nu). Absolute viscosity equals the force required to move a flat surface with an area of one unit at a velocity of one unit, when the two are separated by an oil film one unit thick. Absolute viscosity is illustrated in Fig. 4-4. Kinematic viscosity, which is used frequently in hydraulic calculations, is simply equal to the absolute viscosity divided by the mass density of the fluid. That is,

$$\nu = \frac{\mu}{\rho} \tag{4-1}$$

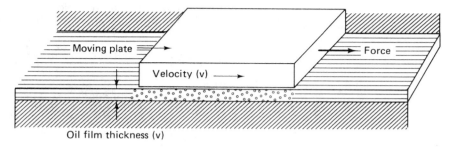

Figure 4-3 Viscous friction.

Unit velocity

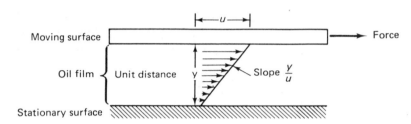

Figure 4-4 Absolute viscosity.

Because viscosity is calculated in all three systems of units: English (ft, slug, sec), traditional metric (cm, gm, sec) and SI metric (m, kg, sec), it is sometimes difficult to remember which viscosity is being described. Figure 4-5 can be used as a reference to keep them separate. Because the reyn and poise are large units of absolute viscosity, the micro-reyn (one millionth of a reyn or 10^{-6} reyn) and centipoise (one hundredth of a poise or 10^{-2} poise) are used to simplify calculations. To follow through from absolute to kinematic viscosity, the equivalent kinematic unit of viscosity to the centipoise (cP) is the centistoke (cSt).

	Absolute viscosity	Kinematic viscosity
English units	lb-sec/ft² lb-sec/in² (reyn) Saybolt Seconds Universal (SSU)	ft²/sec (no name) in²/sec (Newts)
Traditional metric units	dyne·sec/cm² (poise, P) poise × 10^{-2} = cP	cm²/sec (Stokes, St) Stokes × 10^{-2} = cSt
SI metric units	N·s/m² (Pascal-second)	m²/sec (no name)

Figure 4-5 Viscosity units.

When absolute viscosity is converted to kinematic viscosity, using Eq. 4-1, care must be exercised to assure that the units are correct. When μ is given in reyns,

$$\nu(\text{in.}^2/\text{sec}) = \frac{\mu\left(\dfrac{\text{lbf-sec}}{\text{in.}^2}\right)}{\rho\left(\dfrac{\text{slugs}}{\text{in.}^3}\right)} = \frac{\mu\left(\dfrac{\text{lbf-sec}}{\text{in.}^2}\right)\left(\dfrac{12\ \text{in.}}{\text{ft}}\right)}{\rho\left(\dfrac{\text{lbf-sec}^2}{\text{ft}} \times \dfrac{1}{\text{in.}^3}\right)}$$

and

$$\nu(\text{Newts}) = \frac{12\,\mu(\text{reyns})}{\rho(\text{slugs})}$$

When μ is given in poise,

$$\nu(\text{cm}^2/\text{sec}) = \frac{\mu\left(\dfrac{\text{dyne-sec}}{\text{cm}^2}\right)}{\rho\left(\dfrac{\text{gm}}{\text{cm}^3}\right)} = \frac{\mu\left(\dfrac{\text{dyne-sec}}{\text{cm}^2}\right)}{\rho\left(\dfrac{\text{dyne-sec}^2}{\text{cm}}\right)\left(\dfrac{1}{\text{cm}^3}\right)}$$

and

$$\nu(\text{Stokes}) = \frac{\mu(\text{poise})}{\rho(\text{gm/cm}^3)}$$

When μ is given N $\cdot$ s/m^2,

$$\nu(\text{m}^2/\text{sec}) = \frac{\mu\left(\dfrac{\text{N} \cdot \text{s}}{\text{m}^2}\right)}{\rho\left(\dfrac{\text{kg}}{\text{m}^3}\right)} = \frac{\mu\left(\dfrac{\text{N} \cdot \text{s}}{\text{m}^2}\right)}{\rho\left(\dfrac{\text{N} \cdot \text{s}^2}{\text{m}}\right)\left(\dfrac{1}{\text{m}^3}\right)}$$

and

$$\nu(\text{no name}) = \frac{\mu(\text{Pascal-second})}{\rho(\text{kg/m}^3)}$$

The conversion from kinematic viscosity in cSt to Newts is given by

$$(\text{Newts}) = (\text{cSt}) \times 0.001552 \qquad (4\text{-}2)$$

and from cSt to the SI metric system (no name) by

$$(\text{m}^2/\text{sec}) = (\text{cSt}) \times 10^{-6} \qquad (4\text{-}3)$$

The kinematic viscosity in cSt and absolute viscosity in cP can be determined by ASTM test procedure D 445 which measures the time required for a fixed amount of oil to flow through a calibrated capillary instrument using gravity flow at one of the two constant temperatures. The equipment and procedure are shown in Figure 4-6. The time is measured in seconds and then multiplied by the calibration constant for the viscometer to obtain the kinematic viscosity of the oil sample in cSt. The absolute viscosity is derived by multiplying the kinematic viscosity by the density of the oil.

One of the more common units of absolute viscosity is the Saybolt Second Universal, abbreviated SSU, and it is measured using ASTM test procedure D-88 with equipment like that shown in Figure 4-7. The resistance of the fluid to flow is

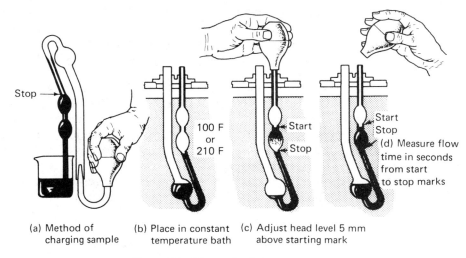

Stop →

100 F
or
210 F

→ Start

→ Stop

Start
Stop

(d) Measure flow
time in seconds
from start
to stop marks

(a) Method of
charging sample

(b) Place in constant
temperature bath

(c) Adjust head level 5 mm
above starting mark

Figure 4-6 Kinematic viscometer.

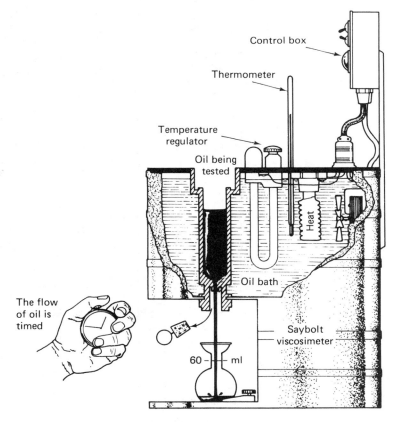

Control box

Thermometer

Temperature
regulator

Oil being
tested

Heat

Oil bath

The flow
of oil is
timed

Saybolt
viscosimeter

60 — ml

Figure 4-7 Saybolt viscometer.

measured as the time that it takes a 60 ml sample of oil to drain through a standard orifice at a constant temperature of 100°F (37.7°C) or 210°F (98.9°C). The elapsed time is the SSU viscosity for the fluid at the given temperature. For thicker fluids, the same test is repeated using a larger orifice to derive the Saybolt Seconds Furol (SSF) viscosity.

The conversion from kinematic viscosity in cSt to the equivalent viscosity in SSU is read from charts in ASTM Procedure D 2161. Basic conversion values are given for 100°F. For temperatures other than 100°F, kinematic viscosities are converted to SSU viscosities by using temperature correction factors. A number of other conversions are made among the various measures of viscosity using the conversion tables given in Appendix C.

Approximate conversions between the four measures of viscosity most commonly used are given by Fig. 4-8. By placing a straightedge horizontally across

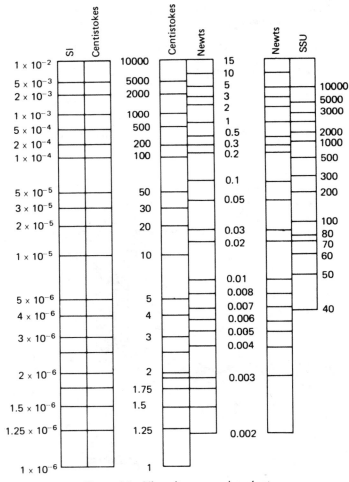

Figure 4-8 Viscosity conversion chart.

How Fluids Flow   Chap. 4

the chart, relative comparisons can be made between them. For example, if it is known that a hydraulic pump requires a fluid with an absolute viscosity in the range of 100 to 3000 SSU, an equivalent oil with a kinematic viscosity of just over 20 to 100 centistokes would be required.

Temperature affects viscosity inversely. Hydraulic fluids thin out at high temperatures and become thick at low temperatures. How much the viscosity changes with temperature can be determined for the range of expected temperatures encountered by the particular application. The most critical component affected is usually the pump. If the viscosity number is too high, the oil will be too thick and the pump will cavitate or break. If the viscosity number is too low, excessive leakage and wear will occur. Most pump manufacturers specify a range of viscosity no higher than 4000 SSU, including cold starts, or lower than 65 SSU at machine operating temperatures.

4-5 NONCOMPRESSIBLE FLOW IN PIPES

For practical purposes, the flow of noncompressible fluids in pipes is considered to be *laminar* or *turbulent*. Laminar flow is governed by the viscous nature of the fluid and the fluid flows in a streamline. When the flow is turbulent, the flow is mixed up and disorganized.

In circular pipes where the flow is governed by the viscous nature of the fluid, layers of unit thickness called laminae assume the shape of thin shell concentric tubes sliding one over another successively (Fig. 4-9). Viscous friction between the layers dissipates the potential energy and returns it to the fluid as heat. Here we are assuming that steady conditions prevail. That is, that the velocity and volume of fluid flowing past a fixed point in the conduit are constant. The distribution of these thin shell tubes is parabolic, and they telescope as they flow in a circular conduit. Near the center of the stream, the velocity of the thin shell tubes is greater, even though each has a velocity that is constant along its length. At the center of the stream, then, the velocity of the fluid is at the maximum; whereas at the inside wall of the conduit the velocity is zero, and this is where the greatest shear and internal friction occur.

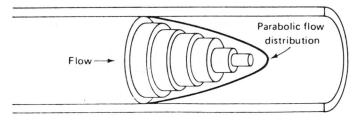

Figure 4-9 Viscous flow.

Turbulent flow, on the other hand, is characterized in a disorganized manner where the smooth flow of the streamline is disrupted (Fig. 4-10). Agitation within the flowing fluid stream is such that it is difficult to determine the flow path of

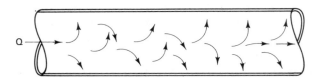

Figure 4-10 Turbulent flow.

individual particles, and experimental rather than theoretical equations are used to solve flow problems.

Between laminar and turbulent flow is a span of critical velocities where the flow is neither smooth and laminar nor disrupted and turbulent. Below the critical velocity the flow is laminar, and above the critical velocity the flow is turbulent.

4-6 REYNOLDS NUMBER

Osborne Reynolds[1,2] (1842–1912) discovered that viscous flow in pipes was related to the dimensionless ratio N_R (called Reynolds number) of the product of the fluid velocity (v), the diameter of the pipe (D), and the density of the fluid (ρ), to the absolute viscosity (μ) of the fluid. Reynolds number is expressed as

$$N_R = \frac{vD\rho}{\mu} \tag{4-4}$$

If the kinematic viscosity rather than the absolute viscosity is used,

$$N_R = \frac{vD}{\nu} \tag{4-5}$$

To be dimensionless, the units of the factors in the Reynolds equation must cancel. If the velocity is given in ft/sec, the diameter of the pipe (D) in inches, the mass of the fluid (ρ) in slugs/in³ and the absolute viscosity (μ) in reyns,

$$N_R = \frac{v(\text{ft/sec})D(\text{in.})(\text{slugs/in.}^3)}{\mu(\text{reyns})}$$

and

$$N_R = \frac{v(\text{ft/sec})D(\text{in.})(\text{lbf-sec}^2/\text{ft-in.}^3)}{\mu(\text{lbf-sec/in.}^2)}$$

Where the kinematic viscosity in Newts rather than the absolute viscosity is given

$$\nu(\text{Newts}) = \frac{12\mu(\text{reyns})}{\rho(\text{slugs/in.}^3)}$$

$$N_R = \frac{v(\text{ft/sec})\ D(\text{in.})(12\ \text{in./ft})}{\nu(\text{in.}^2/\text{sec})}$$

[1] For a detailed accounting of Osborne Reynolds' experiment, see G.A. Tokaty, *A History and Philosophy of Fluidmechanics.* Henley-on-Thames, Oxfordshire: G.T. Foulis and Co. Ltd., 1971

[2] Or see Reynolds' original paper, "An Experimental Investigation of the Circumstances Which Determine Whether the Motion of Water Shall Be Direct of Sinuous, and the Law of Resistance in Parallel Channels," *Philosophical Transactions of the Royal Society,* 174, Part III, 935 (1883), or *Scientific Papers,* London: Cambridge University Press, 1900–1903, Vol. II, pp. 51–105.

and

$$N_R = \frac{12(v)(D)}{\nu} \qquad (4\text{-}6)$$

Reynolds apparatus (Fig. 4-11) passed water through horizontal tubes of different diameters from 1/4 to 2 inches at increasing velocities to find the velocity at which a dye bled into the stream would show a transition from laminar to turbulent flow. This work, which started about 1880, and work by other investigators which replicated similar experiments through 1910, indicated that the upper and lower critical limits for the change from laminar to turbulent flow lie between N_R of 2000 and 4000. Below 2000, the flow is laminar, whatever its previous state, and this is considered to be the lower critical limit. Above 4000, the flow becomes or remains turbulent. The range of critical velocities, then, is considered to be $2000 \leq N_R \leq 4000$.

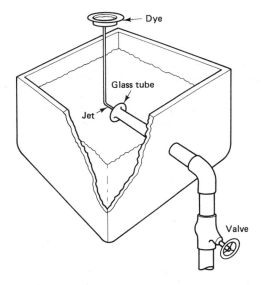

Figure 4-11 Reynolds apparatus.

Example 3

Oil with a kinematic viscosity of 0.05 Newts is flowing in a 1-in. pipe at the rate of 100 gal/min (Fig. 4-12). Is the flow laminar or turbulent?

Solution The Reynolds number will be computed from Eq. 4-6 (with compatible units)

$$N_R = \frac{12(v)(D)}{\nu}$$

First, the velocity (v) is computed from

$$v = \frac{Q}{(A)(3.12)}$$

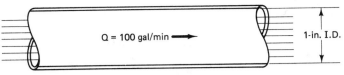

Q = 100 gal/min →

1-in. I.D.

Fluid viscosity = 0.05 Newts

Figure 4-12 Example 3.

and

$$v = \frac{100}{(0.785)(3.12)} = 40.83 \text{ ft/sec}$$

Solving Eq. 4-6 for N_R

$$N_R = \frac{(12)(40.83)(1)}{(0.05)} = 9799.2$$

and the flow is turbulent.

Example 4

An oil with a specific gravity of 0.85 and an absolute viscosity of 27 cP is flowing in a 1-in. I.D. pipe. Compute the range of critical velocities.

Solution   There are two ways to approach the problem. One is to solve it directly in absolute viscosity units given, using Eq. 4-4; and the other is to convert the absolute viscosity to kinematic viscosity in English units and solve Eq. 4-6 for the velocity in ft/sec. Both solutions will be given here. Reynolds Eq. 4-4 is solved for the velocity.

$$N_R = \frac{vD\rho}{\mu}$$

and

$$v = \frac{N_R \mu}{D\rho}$$

At the upper end of the critical velocity range $N_R = 4000$, and at the lower end of the critical velocity range $N_R = 2000$. Also notice that in cm, gm, sec, system of units, the density and Sg are numerically equal. And since the absolute viscosity is given in cP, this must be converted to poise using $cP \times 10^{-2} = P$. Solving Eq. 4-4 for the velocity when $N_R = 4000$,

$$v = \frac{(4000)(0.27\ P)}{(1\ \text{in.})(0.85\ \text{gm/cm}^3)}$$

$$v = \frac{(4000)\left(0.27\ \dfrac{\text{dyne-sec}}{\text{cm}^2}\right)}{\left(1\ \text{in.} \times 2.54\ \dfrac{\text{cm}}{\text{in.}}\right)\left(0.85\ \dfrac{\text{dyne-sec}^2}{\text{cm}} \times \dfrac{1}{\text{cm}^3}\right)}$$

and

$$v = 500.23 \text{ cm/sec}$$

How Fluids Flow   Chap. 4

Velocity ⟶

Fluid specific gravity 0.85
absolute viscosity 27 cP

1-in. I.D.

Figure 4-13   Example 4.

or

$$v = \frac{\left(500 \ \frac{cm}{sec}\right)}{\left(2.54 \ \frac{cm}{in.} \times 12 \ \frac{in.}{ft}\right)} = 16.4 \ \text{ft/sec}$$

Following through then to solve for the lower critical velocity,

$$v = \frac{(2000)\left(0.27 \ \frac{\text{dyne-sec}}{cm^2}\right)}{\left(1 \ \text{in.} \times 2.54 \ \frac{cm}{in.}\right)\left(0.85 \ \frac{\text{dyne-sec}^2}{cm} \times \frac{1}{cm^3}\right)}$$

and

$$v = 250.12 \ \text{cm/sec}$$

or

$$v = \frac{\left(250.12 \ \frac{cm}{sec}\right)}{\left(2.54 \ \frac{cm}{in.} \times 12 \ \frac{in.}{ft}\right)} = 8.2 \ \text{ft/sec}$$

When English units are used to solve the problem, Reynolds Eq. 4-6 (with compatible units) is solved for the velocity,

$$N_R = \frac{12vD}{\nu}$$

and

$$v = \frac{N_R \nu}{12D}$$

Absolute viscosity in cP is converted first to kinematic viscosity in cSt and then to Newts so the units in the Reynolds formula are compatible:

$$\nu = \frac{\mu}{\rho} = \frac{27}{0.85} = 31.76 \ \text{cSt}$$

Notice here that the value of the Sg was used for the density. Converting cSt to Newts using Eq. 4-2,

$$\nu(\text{Newts}) = 0.001552 \times \nu(\text{cSt})$$

and

$$\nu = 0.001552 \times 31.76 = 0.49 \text{ Newts}$$

Substituting in the rearranged Reynolds formula and solving for the upper critical velocity limit,

$$v = \frac{(4000)(0.049)}{(12)(1)} = 16.4 \text{ ft/sec}$$

Finally, substituting and solving for the lower critical velocity

$$v = \frac{(2000)(0.049)}{(12)(1)} = 8.2 \text{ ft/sec}$$

Thus, in both cases the critical velocity for the fluid is from 8.2 ft/sec to 16.4 ft/sec.

4-7 SUMMARY

A fluid power system transmits power by displacing the fluid from one place to another in the system. Displacement measured over time becomes flow rate. For a given flow rate, the continuity equation, $Q = A v$, relates the cross section area of the conductor to the velocity of the fluid. For the continuity equation to be valid, the flow must be steady. That is, the velocity and volume of fluid flowing through any cross section area of the system must remain constant over time. This condition would allow the velocity of the fluid to change as the cross section of the conductor increases or decreases, but would not allow surging that increases or decreases the mass flow rate. The continuity equation is widely used to solve for the conductor size when the required flow rate for the system is given. And in practice, the velocity in pressure lines is limited to 10 ft/sec to keep flow losses low.

Noncompressible flow in pipes is affected by the velocity in another way. At low velocities, the flow is laminar and the fluid flows in a smooth streamline. Above a certain velocity, the flow becomes disorganized, unpredictable, and turbulent. Osborne Reynolds found that laminar and turbulent flow were related to a dimensionless ratio which became known as Reynolds number, N_R. At values below $N_R = 2000$, the flow is laminar, whereas above $N_R = 4000$ the flow is turbulent. Between $N_R = 2000$ and $N_R = 4000$ the flow is neither laminar nor turbulent and describes what is known as the critical range. Most fluid power applications involve flows above the critical range.

STUDY QUESTIONS AND PROBLEMS

1. What are the extension and retraction displacements of a double-acting hydraulic cylinder that has a 2-in. bore, a 1-in. rod, and an 18-in. stroke?
2. What is the displacement of a single-acting ram with a 4-in. bore and a 30-in. stroke?
3. What is the displacement of a 5-cylinder piston motor with a bore diameter of 2.0 in. and a stroke of 3.0 in.?

4. A pump delivers 10 gal/min through a standard 3/4-in. pipe. What is the flow velocity?

5. What is the smallest intake line that could be used for a pump rated at 25 gal/min if the inlet velocity is limited to 4 ft/sec?

6. What flow rate will a standard 2-in. pipe deliver if the velocity is limited to 10 ft/sec?

7. A double-end rod cylinder with a 2-in. bore and 1-in. dia. rod cycles through a 12-in. stroke at 80 cycles per minute. What flow rate is required from the pump?

8. An automatic log splitter with a 5-in. bore cylinder and a 1.5-in. rod cycles through a 30-in. stroke 6 times each minute. If the flow velocity is limited to 5 ft/sec, what inside diameter hose would be required at the blank end of the cylinder?

9. What nominal standard pipe size would be required for an intake line to deliver 30 gal/min if the velocity is limited to 3 ft/sec?

10. If the maximum recommended velocity were 30 ft/sec, what would be the approximate flow rate that would ensue from a 7/16-in. tube?

11. What would be the actual I.D. of a pipe that would deliver 7.5 gal/min at a velocity of 3 ft/sec? What is the cross section area of the pipe and what would be the nearest nominal pipe size?

12. The area of the cylinder in Fig. 4-1 is based on the formula

$$\text{Area (in.}^2) = \frac{\text{Flow rate (gal/min)} \times \text{Constant (0.3208)}}{\text{Velocity (ft/sec)}}$$

If the constant were to be expanded to 7 digits past the decimal point rather than the 4 given in the figure, what would the last 3 digits be?

13. A hydraulic oil with a Sg. of 0.85 has an absolute viscosity of μ = 425 poise. Convert this value to kinematic viscosity in centistokes.

14. Convert the kinematic viscosity of 1.5 Newts to cSt and SI metric units.

15. An oil has a kinematic viscosity of 160 SSU. Consult Appendix C and list the equivalent viscosity conversions in cSt, ft²/sec and SI metric (m²/s).

16. Using Fig. 4-8, convert a kinematic viscosity of 100 SSU to Newts, cSt, and SI metric units (m²/s).

17. Oil with a Sg. of 0.86 and an absolute viscosity of 12.5×10^{-5} lbf-sec/ft² is flowing through a 2-in. I.D. pipe at 65 gal/min. Is the flow laminar or turbulent?

18. Oil with a kinematic viscosity of 0.07 Newts is flowing through a 2-in. I.D. pipe at 500 gal/min. Is the flow laminar or turbulent?

19. Compute the *critical velocity* range for a fluid with a viscosity of 0.045 Newts and a specific gravity of 0.91, flowing through a 1/2-in. I.D. tube.

20. Compute the *critical velocity* range of an oil with a Sg. of 0.90 and an absolute viscosity of 27 cP flowing in a 1-in. I.D. pipe.

CHAPTER *5*

FRICTION LOSSES IN HYDRAULIC SYSTEMS

5-1 INTRODUCTION

When fluid is pumped through a fluid power system, a certain amount of the energy in the fluid is lost to friction. *Major losses* occur as the fluid flows through pipes, hoses, and tubing, while *minor losses* occur at valves, fittings, bends, enlargements, contractions, and orifices. Major losses are calculated for a given length of pipe. Minor losses, on the other hand, first must be converted to losses through an equivalent length of straight pipe using various experimental friction factors. To arrive at the total loss for a circuit, the major and minor flow losses are combined and substituted in one of the formulas that determine the pressure and horsepower losses associated with pumping the fluid.

The friction generated from the flow of fluid through pipes brings to mind the notion that the fluid rubs the boundary surface as if a bullet were being pushed through the barrel of a gun, generating heat losses at the boundary. This is not what happens. What occurs is a rubbing action between the fluid particles themselves as they move through the pipe. When the flow is laminar, fluid layers move one past another, like a telescope, generating friction that can be related to the viscosity of the fluid. At the outside boundary, the fluid velocity is zero; at the center of the fluid stream, the velocity is maximum; and through the cross section of the stream, the friction gradient is linear.[1]

The transition from laminar to turbulent flow introduces another important cause for friction and that is the cross flow and intermingling of particles as the total mass moves through the length of the pipe. The effect of this intermingling of flowing particles is to void the relationship between the relative roughness of the boundary surface and the Reynolds number, which changes the method by which the friction factor is calculated. Again, while the friction is not between the flow-

---

[1] This is strictly true only for Newtonian fluids.

ing fluid and the wall itself, *the rate of shear* and heat generated are, in fact, greatest near the wall, and here is where most of the energy transfer occurs.

5-2 DARCY-WEISBACH AND HAGEN-POISEUILLE FORMULAS

The basic equation that governs viscous noncompressible flow in pipes is

$$h_f = f \left(\frac{L}{D}\right)\left(\frac{v^2}{2g}\right) \tag{5-1}$$

where h_f is the head loss required to pump the fluid, f is a dimensionless friction factor, L is the length of the pipe, D is the internal diameter of the pipe, v is the velocity of the fluid, and g is the acceleration due to gravity. This is known as the Darcy-Weisbach formula for viscous flow.

The basic formula says, in effect, that the head loss associated with pumping a fluid through a system is a function of the velocity head $(v^2/2g)$, the length to diameter ratio of the pipe (L/D), and a friction factor (f) that is derived experimentally.

Example 1

Oil with a Sg of 0.91 is flowing through a 2-in. I.D. pipe at the rate of 200 gal/min (Fig. 5-1). If the friction factor is 0.05, what would be the pressure drop through 800 ft. of pipe?

Solution   First, the velocity of the fluid is determined from the continuity equation.

$$v = \frac{Q}{(A)(3.12)}$$

and

$$v = \frac{(200)}{(3.14)(3.12)} = 20.4 \text{ ft/sec}$$

Substituting in the Darcy-Weisbach formula to solve for head loss

$$h_f = f \left(\frac{L}{D}\right)\left(\frac{v^2}{2g}\right)$$

$$h_f = (0.05) \frac{(800 \text{ ft})(20.4 \text{ ft/sec})^2}{\left(\frac{2 \text{ in.} \times 1 \text{ ft}}{12 \text{ in.}}\right)(2 \times 32.2 \text{ ft/sec})}$$

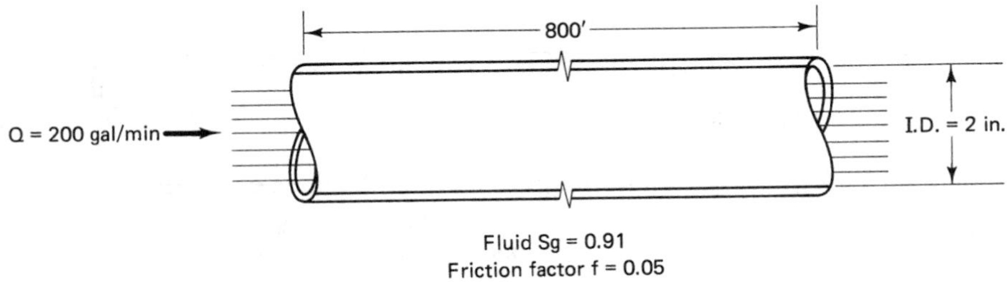

Q = 200 gal/min

800'

I.D. = 2 in.

Fluid Sg = 0.91
Friction factor f = 0.05

Figure 5-1   Example 1.

and

$$h_f = 1551 \text{ ft}$$

Converting the head loss h_f to pressure loss h_{psif}

$$h_{psif} = (0.433)(0.91)(1551 \text{ ft}) = 611 \text{ lbf/in.}^2$$

The value of the friction factor in the Darcy-Weisbach formula is largely determined by whether the flow is laminar or turbulent. For laminar flow, the friction factor has been determined experimentally to be

$$f = \frac{64}{N_R} \tag{5-2}$$

When this value is substituted in the Darcy-Weisbach formula for head loss due to friction, the resulting equation is one form of the Hagen-Poiseuille formula for losses due to friction when laminar flow prevails. That is,

$$h_f = \frac{64}{N_R} \frac{Lv^2}{D2g}$$

and

$$h_f = \frac{32 L v^2}{N_R D g} \tag{5-3}$$

If the value for N_R is substituted in Eq. 5-3

$$N_R = \frac{vD}{\nu}$$

$$h_f = \left(\frac{\nu}{vD}\right)\left(\frac{32 L v^2}{Dg}\right)$$

and

$$h_f = \frac{32 \nu L v}{D^2 g} \tag{5-4}$$

This is the more common form of the Hagen-Poiseuille formula.

Example 2

Oil with a Sg of 0.85 and a viscosity of 0.08 Newts is flowing in a 1-in. I.D. pipe 400 ft long, at the rate of 15 gal/min (Fig. 5-2). Determine first that the flow is laminar, and then compute the pressure drop in lbf/in.^{2}.

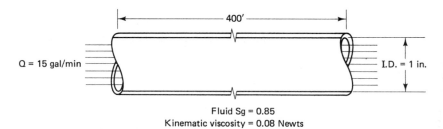

Fluid Sg = 0.85
Kinematic viscosity = 0.08 Newts

Figure 5-2 Example 2.

Friction Losses in Hydraulic Systems   Chap. 5

Solution The flow is considered to be laminar if the Reynolds number is less than 2000.

First, the velocity is determined to be

$$v = \frac{Q}{(A)(3.12)}$$

and

$$v = \frac{(15)}{(0.785)(3.12)} = 6.12 \text{ ft/sec}$$

Substituting in the Reynolds formula,

$$N_R = \frac{12vD}{\nu}$$

and

$$N_R = \frac{(12)(6.12)(1)}{(0.08)} = 918$$

so the flow is laminar. Substituting in the Hagen-Poiseuille formula for pressure drop when laminar flow prevails,

$$h_f = \frac{32\nu L v}{D^2 g}$$

$$h_f = \frac{(32)(0.08 \text{ in.}^2/\text{sec})(400 \text{ ft})(6.12 \text{ ft/sec})}{(1\text{-in.})^2 (32.2 \text{ ft/sec}^2)}$$

and

$$h_f = 194.6 \text{ ft}$$

Finally, solving for the pressure loss where

$$h_{psif} = (0.433)(S_g)(h_f)$$

$$h_{psif} = (0.433)(0.85)(194.6 \text{ ft}) = 71.6 \text{ lbf/in.}^2$$

Notice that in the solution of the Hagen-Poiseuille formula that the units must be compatible such that h_f is in units of ft or m.

5-3 f-FACTOR FOR TURBULENT FLOW

During laminar flow, the friction factor is relatively independent of the surface condition of the inside diameter of the pipe. When the flow is completely turbulent, the f-factor is read from Fig. 5-3 by locating the place of intersection of the Reynolds number and the relative roughness of the conductor surface, and then reading the friction value from the left or right margins. The relative roughness of the pipe wall is computed as the dimensionless ratio of the absolute roughness ε (Greek letter epsilon) to the inside diameter of the pipe D. That is,

$$\text{Relative roughness} = \frac{\varepsilon}{D} \qquad (5\text{-}5)$$

where the absolute roughness equals the average projection of the surface imperfections on the inside wall of the pipe. The absolute roughness of pipes made from

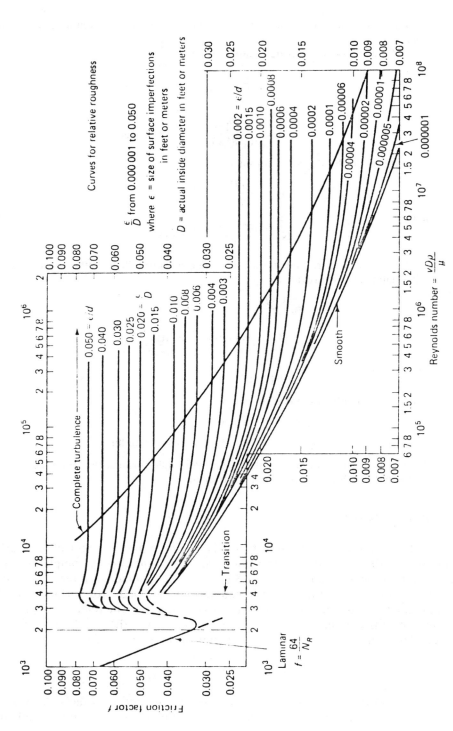

Figure 5-3 Friction factor vs. Reynolds number for turbulent flow (*Reprinted by permission of McGraw-Hill Book Company*).

TABLE 5-1 ABSOLUTE ROUGHNESS OF NEW COMMERCIALLY AVAILABLE PIPE AND TUBING*

	ε/D in Feet	ε/D in Meters
drawn tubing	0.000 005	0.000 001 5
commercial steel pipe	0.000 15	0.000 046
asphalted cast iron	0.0004	0.000 12
galvanized iron	0.0006	0.0002
cast iron	0.00085	0.000 26
concrete	0.001 to 0.01	0.000 30 to 0.0030
riveted steel	0.003 to 0.03	0.000 91 to 0.0091

* From *Fluid Mechanics*, J. A. Sullivan.

different materials is not the same. For example, drawn steel tubing is smooth compared to some types of commercial steel pipe. Table 5-1 lists the absolute roughness values for several types of commercially available pipe and tubing. It must be remembered that values shown are for new surface conditions and do not account for deterioration from such causes as condensation and rust at the surface after they have been in use for some time. The procedure for solving typical problems is as follows:

1. Determine the fluid velocity through the pipe.
2. Compute the Reynolds number to confirm that the flow is completely turbulent.
3. Find the absolute roughness of the pipe from Table 5-1.
4. Compute the relative roughness from the absolute roughness and the inside diameter of the pipe.
5. Locate the f-factor used in the Darcy-Weisbach formula for turbulent flow.
6. Finally, solve the Darcy-Weisbach formula for pressure to determine the head loss or pressure drop.

Example 3

Oil with a *Sg* of 0.85 and a viscosity of 0.30 Newts is flowing through a 2-in. I.D. wrought iron pipe at the rate of 600 gal/min (Fig. 5-4). What is the pressure drop in 500 ft.?

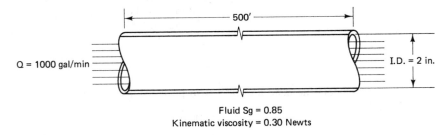

Fluid Sg = 0.85
Kinematic viscosity = 0.30 Newts

Figure 5-4 Example 3.

Solution 1. The velocity is determined from

$$v = \frac{Q}{(A)(3.12)}$$

and

$$v = \frac{600 \text{ gal/min}}{(3.14)(3.12)} = 61.24 \text{ ft/sec}$$

2. The Reynolds number is determined using Eq. 4-6:

$$N_R = \frac{12vD}{\nu}$$

and

$$N_R = \frac{(12 \text{ in./ft})(61.24 \text{ ft/sec})(2 \text{ in.})}{(0.30 \text{ in.}^2/\text{sec})} = 4899 = 4.9 \times 10^3$$

3. From Table 5-1, the absolute roughness of wrought iron pipe is given as 0.0002 ft.

4. The relative roughness factor is determined from Eq. 5-5:

$$\text{Relative Roughness} = \frac{\varepsilon}{D}$$

and

$$\text{Relative Roughness} = \frac{(0.0002 \text{ ft})}{(2 \text{ in.})(1 \text{ ft/12 in.})} = 0.0012$$

5. The friction factor f is determined by locating the Reynolds number ($N_R = 4.9 \times 10^3$) across the top of Fig. 5-3 and the relative roughness curve close to ($\varepsilon/D = 0.0012$) near the right margin. The curve is followed to the left until it intersects with the Reynolds number, and looking to the left of this intersection, the friction factor is read as approximately 0.040.

6. Finally, solving the Darcy-Weisbach formula for pressure drop when turbulent flow prevails,

$$h_f = f\left(\frac{L}{D}\right)\left(\frac{v^2}{2g}\right)$$

$$h_f = (0.040)\frac{(500 \text{ ft})(61.24 \text{ ft/sec})^2}{\left(\frac{2 \text{ in.} \times 1 \text{ ft}}{12 \text{ in.}}\right)(2 \times 32.2 \text{ ft/sec}^2)}$$

and

$$h_f = 6988 \text{ ft}$$

Converting the head loss h_f in ft to pressure loss h_{psif} in lbf/in.^{2}

$$h_{psif} = (0.433)(Sg)(h_f)$$

$$h_{psif} = (0.433)(0.85)(6988 \text{ ft}) = 2572 \text{ lbf/in.}^2$$

5-4 C-COEFFICIENTS

Some manufacturers prefer to express the flow characteristics of components, particularly flow control values, in terms of C-coefficients. They are also used to describe friction losses associated with orifices and short tubes.

From Torricelli's theorem, the theoretical velocity of a free stream emitted horizontally from the base of a fluid source such as a water tank is

$$v = \sqrt{2gh}$$

where the losses due to friction are assumed to be 0. If these losses are incorporated as a discharge coefficient C_d of the velocity,[2]

$$v = C_d\sqrt{2gh}$$

And since the velocity through the orifice equals the flow rate Q divided by the cross section area of the opening,

$$v = \frac{Q}{A}$$

The flow rate through an orifice or valve can be computed from

$$Q = AC_d\sqrt{2g\,\Delta h} \tag{5-6}$$

where Δh is the head loss across the orifice or valve associated with the flow rate Q.

In actual applications, the velocity as well as the discharge coefficient C_d are determined for each orifice and valve within a specified range of flow rates and temperatures. Values for C_d range from 0.5 to 1.0. Several orifices and their discharge coefficients are given in Fig. 5-5.

In English units where the orifice area is given in in.^{2}, Δh is in ft and g equals 32.2 ft/sec^2, the flow rate Q in gal/min is

$$Q = A(\text{in.}^2)\left(\frac{1\text{ gal}}{231\text{ in.}^3}\right)C_d\sqrt{2 \times 32.2\text{ (ft/sec}^2)\,\Delta h(\text{ft})}$$

$$Q = A(\text{in.}^2)\left(\frac{1\text{ gal}}{231\text{ in.}^3}\right)C_d\left(\frac{8.025}{1}\right)\left(\frac{\text{ft}}{\text{sec}}\right)\left(\frac{12\text{ in.}}{1\text{ ft}}\right)\left(\frac{60\text{ sec}}{1\text{ min}}\right)\sqrt{\Delta h}$$

and

$$Q = 25.01 AC_d\sqrt{\Delta h} \tag{5-7}$$

Where the pressure drop (lbf/in.^{2}) rather than the head loss is given

$$h = \frac{p \times 2.31}{Sg}$$

---

[2] This is an oversimplification of the concept. Actually, the discharge coefficient $C_d = C_v C_c$, where C_v is the coefficient of the velocity through the orifice and equals the actual velocity through the orifice divided by the theoretical velocity—that is, $C_v = v_a/v_t$. And the contraction coefficient C_c equals the cross section area of the jet A_j where it narrows a distance half the diameter of the stream after discharge at a place called vena contracta, divided by the actual cross section of the orifice opening A_o—that is $C_c = A_j/A_o$.

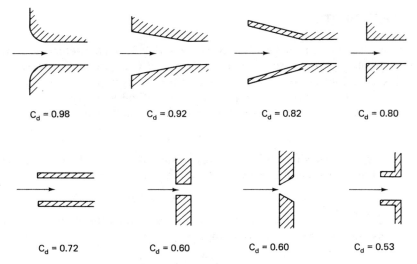

$C_d = 0.98$ $C_d = 0.92$ $C_d = 0.82$ $C_d = 0.80$

$C_d = 0.72$ $C_d = 0.60$ $C_d = 0.60$ $C_d = 0.53$

Figure 5-5   Orifice shapes and discharge coefficients C_d.

and

$$Q = 38.06 A C_d \sqrt{\frac{\Delta p}{Sg}} \tag{5-8}$$

In SI metric units where the orifice area is given in mm^2, Δh is in m, g equals 9.81 m/s², and the flow rate Q in ℓ/min is

$$Q = A(mm^2) \left(\frac{\ell \text{ liter}}{10^6 \text{ mm}^3}\right) C_d \sqrt{2 \times 9.81 \text{ m/s}^2 \, \Delta h(m)}$$

$$Q = A(mm^2) \left(\frac{\ell \text{ liter}}{10^6 \text{ mm}^2}\right) C_d \left(\frac{4.43}{1}\right)\left(\frac{m}{s}\right)\left(\frac{10^3 \text{ mm}}{m}\right)\left(\frac{60 \text{ sec}}{1 \text{ min}}\right) \sqrt{\Delta h}$$

and

$$Q = 2658 \times 10^2 A C_d \sqrt{\Delta h} \tag{5-9}$$

Where the pressure drop (Pascals) rather than the head loss is given

$$h = \frac{1.02 \times 10^{-4} p}{Sg}$$

and

$$Q = (2658 \times 10^2) A C_d (1.01 \times 10^{-2}) \sqrt{\frac{\Delta p}{Sg}}$$

and

$$Q = 2685 A C_d \sqrt{\frac{\Delta p}{Sg}} \tag{5-10}$$

Example 4

The pressure drop Δp across an orifice with a diameter of 2 in. is 120 lbf/in.^{2} (Fig. 5-6). If the Sg of the fluid is 0.85 and the flow rate is 1000 gal/min, compute the discharge coefficient for the orifice.

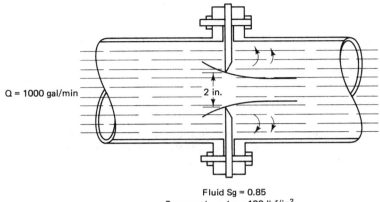

Q = 1000 gal/min

2 in.

Fluid Sg = 0.85
Pressure drop Δp = 120 lbf/in.^{2}

Figure 5-6 Example 4.

Solution The cross section area of a 2-in. I.D. orifice is 3.14 in^2. And the discharge coefficient is computed by solving Eq. 5-8 for C_d:

$$C_d = \frac{1}{38.06} \times \frac{Q}{A} \sqrt{\frac{Sg}{\Delta p}}$$

$$C_d = \frac{1}{38.06} \times \frac{1000}{3.14} \sqrt{\frac{0.85}{120}}$$

and

$$C_d = 8.37 \times 0.084 = 0.70$$

5-5 K-VALUES

Minor losses occur as the fluid undergoes sudden expansions or contractions, or as the fluid flows through pipe fittings, valves, and bends. These losses are associated with the Bernoulli equation and defined as the number of velocity heads lost due to friction. When turbulent flow prevails, the value of this factor is determined experimentally and then assigned a K-value for that fitting or pipe configuration. Head loss is thus computed from

$$h_{ff} = K \left(\frac{v^2}{2g} \right) \tag{5-11}$$

This formula does not hold true when viscous flow prevails.

Sudden enlargement and reduction in the cross section of the pipe result in losses because the fluid must change directions abruptly, causing increased turbulence in the form of eddies near where the two are joined. These eddies that accompany sudden enlargement can be shown both theoretically and experimentally to generate a K-value[3] such that

$$K = \left(1 - \frac{D_1^2}{D_2^2}\right)^2 \qquad (5\text{-}12)$$

where D_1 is the inside diameter of the smaller pipe and D_2 is the inside diameter of the larger pipe. Sudden reduction in the cross section of the pipe produces a K-value such that

$$K = 0.5 \left(1 - \frac{D_1^2}{D_2^2}\right) \qquad (5\text{-}13)$$

where D_1 and D_2 are the inside diameters of the smaller and larger pipes, respectively. Figures 5-7 and 5-8 illustrate these configurations and list K-values associated with various diameter ratios.

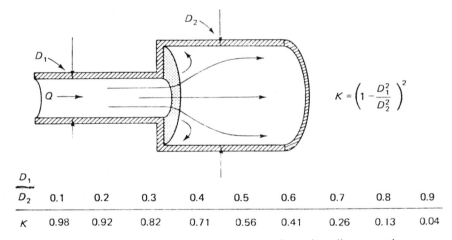

$\frac{D_1}{D_2}$	0.1	0.2	0.3	0.4	0.5	0.6	0.7	0.8	0.9
K	0.98	0.92	0.82	0.71	0.56	0.41	0.26	0.13	0.04

Figure 5-7 K-values for sudden enlargements for various diameter ratios.

K-values for pipe fittings, valves, and bends are determined empirically for each configuration by manufacturers and independent researchers and have been found to be relatively independent of size. The value of K can be computed from

$$K = f_t \left(\frac{L}{D}\right) \qquad (5\text{-}14)$$

where f_t equals the friction factor in the completely turbulent range and L/D is the ratio of the length of the fitting to its inside diameter. Figure 5-9 illustrates several

---

[3] *Flow of Fluids Through Values, Fittings and Pipe.* Technical Paper No. 410, Chicago: Crane Company, 1976, p. 2–11.

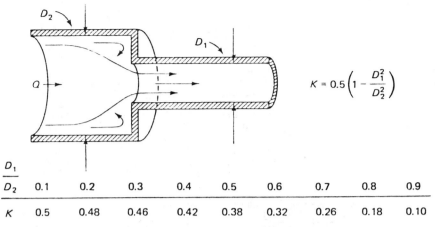

$$K = 0.5\left(1 - \frac{D_1^2}{D_2^2}\right)$$

$\dfrac{D_1}{D_2}$	0.1	0.2	0.3	0.4	0.5	0.6	0.7	0.8	0.9
K	0.5	0.48	0.46	0.42	0.38	0.32	0.26	0.18	0.10

Figure 5-8 K-values for sudden reductions for various diameter ratios.

of the more common configurations, equivalent lengths, and typical K-values associated with their length-to-diameter ratios.[4]

Losses due to expansion and contraction also occur as fluid enters and exits from pipes—for example, where they connect to the base of liquid tanks. Fig. 5-9 also illustrates several standard entrance and exit configurations and the K-values associated with each.

Following is an example that will illustrate how these losses occur in a hydraulic piping system.

Example 5

Hydraulic oil with a Sg of 0.91 and a kinetic viscosity of 0.05 Newts is flowing at 50 gal/min through a hydraulic circuit that contains an inward and outward projection, a globe valve, four 90° standard elbows, a sudden enlargement, and a sudden contraction. If the fittings and tubing have an I.D. of 1 in., what is the total pressure loss through the nine fittings (Fig. 5-10)?

Solution   First, the velocity is determined from

$$v = \frac{Q}{A \times 3.12} = \frac{50}{0.785 \times 3.12} = 20.4 \text{ ft/sec}$$

Second, determine that the flow is completely turbulent. Since the kinetic viscosity is in Newts, the Reynolds number is determined from Eq. 4-6

$$N_R = \frac{12vD}{\nu} = \frac{(12)(20.4)(1)}{(0.05)} = 4896$$

and the flow is turbulent.

[4] Ibid., pp. 26–29.

Valves—fittings	$\dfrac{L}{D}$	K-value
swing check valve	135	2.50
globe valve	340	10.00
gate valve		
(full open)	13	0.19
1/4 closed	35	1.15
1/2 closed	160	5.60
3/4 closed	900	24.00
cock valve	18	.26
close pattern return bend	50	2.20
standard tee	60	1.80
standard 90° elbow	30	0.90
standard 45° elbow	16	0.42

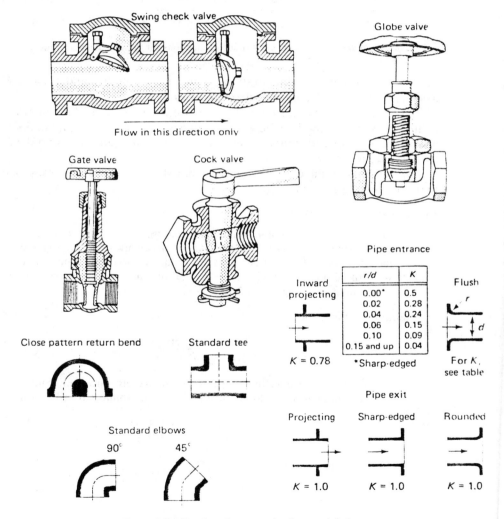

Figure 5-9 *K*-values for several valves and fittings.

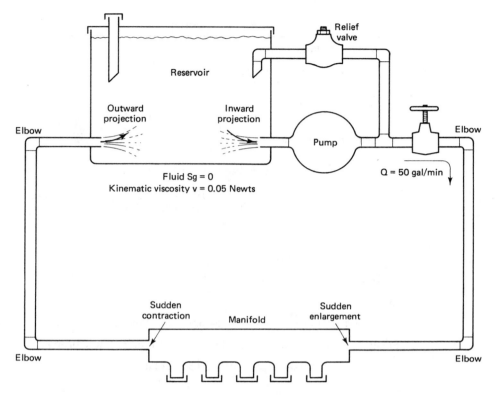

Figure 5-10   Example 5.

The *K*-values are taken from Fig. 5-7 and Fig. 5-9 and substituted into the respective formulas to solve for head losses through the fittings as follows:

Fitting	Formula	Friction loss
1. Inward Projection	$h_i = K \left(\dfrac{v^2}{2g}\right) = (0.78) \left(\dfrac{20.4^2}{2 \times 32.2}\right)$	= 5.04
2. Outward Projection	$h_o = K \left(\dfrac{v^2}{2g}\right) = (1.0) \left(\dfrac{20.4^2}{2 \times 32.2}\right)$	= 6.46
3. Globe Valve	$h_g = K \left(\dfrac{v^2}{2g}\right) = (10.0) \left(\dfrac{20.4^2}{2 \times 32.2}\right)$	= 64.62
4. 4 Std. 90° Elbows	$h_e = K \left(\dfrac{v^2}{2g}\right) = (4)(0.90) \left(\dfrac{20.4^2}{2 \times 32.2}\right)$	= 23.26

Fitting	Formula	Friction loss
5. Sudden Enlargement		

$$h_n = K \left(\frac{v^2}{2g}\right)$$

$$h_n = \left(1 - \frac{D_1^2}{D_2^2}\right)^2 \left(\frac{v^2}{2g}\right) = \left(1 - \frac{1^2}{2^2}\right)^2 \left(\frac{20.4^2}{2 \times 32.2}\right) \qquad = 3.63$$

6. Sudden Contraction

$$h_c = K \left(\frac{v^2}{2g}\right)$$

$$h_c = 0.5 \left(1 - \frac{D_1^2}{D_2^2}\right)\left(\frac{v^2}{2g}\right) = (0.5) \left(1 - \frac{1^2}{2^2}\right)\left(\frac{20.4^2}{2 \times 32.2}\right) = 2.42$$

Summing the minor losses

Total $h_{ff} = h_i + h_o + h_g + h_e + h_n + h_c$

and

Total $h_{ff} = (5.04) + (6.46) + (64.62) + (23.26) + (3.63) + (2.42) = 105.43$ ft.

Finally, converting the total head loss h_{ff} to a total pressure drop p_{psif} for the circuit

Total $h_{psif} = (0.433)(Sg)(\text{Total } h_{ff})$

and

Total $h_{psif} = (0.433)(0.91)(105.43) = 41.5$ lbf/in.^{2}

5-6 EQUIVALENT LENGTH

It is evident that the head or pressure loss through a sudden enlargement, contraction, fitting, or valve is equivalent to the loss through some length of straight pipe. For example, the loss through a globe valve may be the same as the loss through several feet or meters of the same size straight pipe. In notation,

$$h_{ff} = h_f$$

where h_{ff} is the head loss through restriction of some configuration and h_f is the loss through an equivalent length of straight pipe. Expanding these from Eqs. 5-11 and 5-1, we obtain

$$h_{ff} = K \left(\frac{v^2}{2g}\right) = f \left(\frac{Lv^2}{D2g}\right) = h_f$$

from which

$$L = D \left(\frac{K}{f}\right) \qquad\qquad (5\text{-}15)$$

where L and D are in the same units and K and f are dimensionless. In the English system of units, if D is the diameter of the conductor given in in., and L is to be computed in ft,

$$L = \frac{D}{12}\left(\frac{K}{f}\right) \qquad (5\text{-}16)$$

Example 6

Oil with a Sg of 0.85 and a kinematic viscosity of 0.10 Newts is flowing through a test rig consisting of a 1-in. I.D. cast iron open check valve at the rate of 150 gal/min. Determine the equivalent length of the valve (Fig. 5-11).

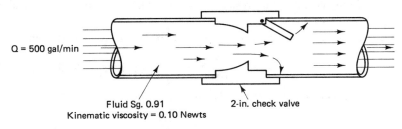

Fluid Sg. 0.91
Kinematic viscosity = 0.10 Newts

Figure 5-11 Example 6.

Solution First, the velocity is determined from

$$v = \frac{Q}{A \times 3.12} = \frac{150}{0.785 \times 3.12} = 61.24 \text{ ft/sec}$$

Second, the Reynolds number is computed from

$$N_R = \frac{12vD}{\nu} = \frac{(12)(61.24)(1)}{(0.10)} = 7349 = 7.349 \times 10^3$$

and the flow is completely turbulent.

Third, from Table 5-1 the absolute roughness of cast iron is given as 0.00085 ft, and the relative roughness is

$$\text{Relative Roughness} = \frac{\varepsilon}{D} = \frac{(0.00085)}{(0.0833)} = 0.010$$

Substituting the Reynolds number of 7.349×10^3 and a relative roughness of 0.010 in Fig. 5-3, the friction factor is read from the margin as approximately 0.045. Finally, solving Eq. 5-16 to determine the equivalent length,

$$L_e = \frac{D}{12}\left(\frac{K}{f}\right)$$

and

$$L_e = \frac{(1)(2.5)}{(12)(0.045)} = 4.6 \text{ ft}$$

5-7 CIRCUIT CALCULATIONS

Calculating the equivalent length and pressure drop for a hydraulic circuit involves determining the flow rate through the circuit plumbing, computing the Reynolds number, determining the friction factors and equivalent lengths for the pipe and fittings, and then combining the series of equivalent lengths into a total equivalent length. A final calculation determines either total pressure or fluid horsepower loss resulting from the friction of the total equivalent length. Other factors such as the velocity at the suction line become evident, and pipes can be resized as required to limit velocities at critical places in the circuit.

Example 7

Hydraulic oil with a kinematic viscosity of 80 cSt and a Sg of 0.88 is flowing through the system in Fig. 5-12, at the rate of 100 gpm. The pipe and fittings are uncoated cast iron. The valves are cast iron. The discharge pressure at the pump is 800 psi. Compute the total pressure drop and fluid horsepower loss resulting from friction losses in the system.

Solution   The velocity (v_1) in the 2-in. pipe is computed from

$$v_1 = \frac{Q}{(A_1)(3.12)} = \frac{100}{(3.14)(3.12)} = 10.20 \text{ ft/sec}$$

The velocity (v_2) in the 1-in. pipe is

$$v_2 = \frac{100}{(0.785)(3.12)} = 40.83 \text{ ft/sec}$$

The Reynolds number is computed for each velocity to determine if the flow is laminar or turbulent. For the 2-in. pipe, the Reynolds number is computed from

$$N_{R(1)} = \frac{12vD}{\nu}$$

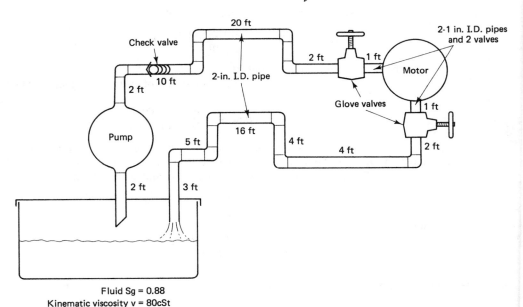

Figure 5-12   Example 7.

where the kinematic viscosity is in Newts. Converting the viscosity given in cSt to Newts (Eq. 4-2)

$$\nu(\text{Newts}) = 0.001552 \ (\text{cSt})$$

and

$$\nu = (0.001552)(80) = 0.124 \ \text{Newts}$$

Substituting in the Reynolds formula

$$N_{R(1)} = \frac{(12)(10.20)(2)}{0.124} = 1974$$

and the flow is laminar in the 2-in. pipe section. Again, substituting in the Reynolds formula to solve for $N_{R(2)}$

$$N_{R(2)} = \frac{(12)(40.83)(1)}{0.124} = 3951$$

and the flow is turbulent in the 1-in. pipe section.

The first equivalent length to be computed is through the 2-in. section of pipe and fittings that consist of 70 ft of straight pipe, 11 elbows, and 1 check valve. The K-factors for the fittings are

$$\begin{array}{lr} 1 \ \text{Check Valve (K = 2.5)} & 2.5 \\ 11 \ \text{Elbows (K = 0.9)} & \underline{9.9} \\ & \text{Total } 12.4 \end{array}$$

Since the flow is laminar, the friction factor (f) is computed from (Eq. 5-2)

$$f = \frac{64}{N_R} = \frac{64}{1974} = 0.032$$

The equivalent length for the fittings in the 2-in. pipe, then, is

$$L_E = \frac{D_1}{12} \left(\frac{K}{f_1} \right)$$

and

$$L_E = \frac{(2)(12.4)}{(12)(0.032)} = 64.58 \ \text{ft}$$

and the total equivalent length of the 2-in. pipe is (70 + 64.58) 134.58 ft.

Since the flow is laminar, the pressure drop through the 2-in. pipe section is computed from the Hagen-Poiseuille Formula (Eq. 5-1)

$$h_f = f \left(\frac{L}{D_1} \right) \left(\frac{v^2}{2g} \right)$$

where $f = 0.032$

$$h_f = (0.032) \ \frac{(134.58 \ \text{ft})(10.2 \ \text{ft/sec})^2}{\left(\dfrac{2 \ \text{in.} \times 1 \ \text{ft}}{12 \ \text{in.}} \right) (2 \times 32.2 \ \text{ft/sec}^2)}$$

and
$$h_f = 41.7 \text{ ft}$$

Converting the head loss in the 2-in. section and fittings h_f to pressure loss h_{psif}

$$h_{psif} = (0.433)(0.88)(41.7 \text{ ft}) = 15.9 \text{ lbf/in.}^2$$

The equivalent length of the 1-in. section of pipe and fittings consists of 2 ft of straight pipe and 2 globe valves. The K-factor for the globe valves is

$$2 \text{ Globe Valves } (K = 10) = 20$$

Since the flow is turbulent, the friction factor must be obtained from the graph in Fig. 5-3. Using an absolute roughness for cast iron of 0.00085 for the two valves, the relative roughness is computed from Eq. 5-5:

$$\text{Relative Roughness} = \frac{\varepsilon}{D}$$

and

$$\text{Relative Roughness} = \frac{(0.00085)}{(1/12)} = 0.010$$

Using Fig. 5-3 and locating the Reynolds number ($N_{R(2)} = 3951$) and the relative roughness (0.010) for both valves, the friction factor for the valves is approximately 0.05. Computing the equivalent length for both valves

$$L_{E(2)} = \frac{D_2}{12} \left(\frac{K}{f_2} \right)$$

$$L_{E(2)} = \frac{(1)(20)}{(12)(0.05)} = 33.33 \text{ ft}$$

and the total equivalent length of the 1-in. pipe is (2 + 33.33 = 35.33 ft).

Since the flow is turbulent, the pressure drop through the 1-in. section of pipe is computed from Eq. 5-1 where the value of f can be taken as 0.05.

$$h_f = f \left(\frac{L}{D_2} \right) \left(\frac{v^2}{2g} \right)$$

$$h_f = (0.05) \frac{(35.33 \text{ ft})(40.8 \text{ ft/sec})^2}{\left(\frac{1 \text{ in.} \times 1 \text{ ft}}{12 \text{ in.}} \right) (2 \times 32.2 \text{ ft/sec}^2)}$$

and

$$h_f = 547.9 \text{ ft}$$

Converting the head loss in the 1-in. section and fittings h_f to pressure loss h_{psif}

$$h_{psif} = (0.433)(0.88)(547.9 \text{ ft}) = 208.8 \text{ lbf/in.}^2$$

Notice that a slight error is introduced by adding the equivalent length of the two globe valves with the 2 ft of straight pipe in computing the pressure drop through the 1-in. section since the relative roughness and resulting friction factors for the fittings and the straight sections of pipe are not equal (0.05 vs. 0.042). This is not so in the case of the 2-in. pipe and fittings, however, where the flow is laminar and the friction is relatively independent of the surface conditions.

The total pressure drop for the system from both sections of pipe and fittings is

2-in. fittings and section	15.9 lbf/in.²	
1-in. fittings and section	208.8 lbf/in.²	
	224.7 lbf/in.²	

and the fluid horsepower loss is

$$\text{Fhp} = \frac{p \times Q}{1714}$$

$$\text{Fhp} = \frac{(224.7)(100)}{1714} = 13.1 \text{ hp}$$

It is also noticed that a velocity of 10.2 fps at the intake line is greater than the 4–5 fps usually recommended to keep the flow well within the laminar range. In addition, the 1-in. fluid lines connected to the pump would be larger in practice to reduce velocities and accompanying pressure and fluid horsepower losses associated with using undersized conductors.

5-8 SUMMARY

Resistance to flow is generated by friction within the fluid. Head and pressure losses due to friction are computed using the Darcy-Weisbach formula. When the flow is laminar and Reynolds numbers are below $N_R = 2000$, the friction factor is relatively independent of the surface conditions of the conductor and has been found experimentally to be

$$f = \frac{64}{N_R}$$

When this value is substituted in the Darcy-Weisbach formula, the result is the Hagen-Poiseuille formula for head loss when laminar flow prevails.

When the flow is turbulent—Reynolds numbers $N_R = 4000$ and above—the f-factor for the Darcy-Weisbach formula is influenced by the viscosity of the fluid, the ratio of the length to the diameter of the conductor, and the absolute roughness of the conductor wall surface.

When friction losses through short restrictions such as orifices, valves, and fittings are encountered, C-coefficients, K-factors, and equivalent length values are used to compute losses directly, or losses associated with an equivalent length of pipe that would account for the same loss. Discharge coefficients C_d are commonly used in combination with Torricelli's theorem to compute the flow and head loss associated with an orifice or valve. K-values, on the other hand, are defined as the number of velocity heads lost to friction, and as a factor are multiplied by the velocity head $v^2/2g$ to arrive at the head loss associated with friction through the fittings.

Finally, circuit calculations are made by summing the major head loss through the length of pipe or tubing, and then the minor losses through the fittings, valves, sudden enlargements, orifices, and components such as filters and heat exchangers.

STUDY QUESTIONS AND PROBLEMS

1. A hydraulic oil with a Sg of 0.85 is piped from the pump to the motor through 30 ft. of 1/2-in. I.D. hose. If the flow rate is 20 gal/min and the friction factor is 0.040, what would be the pressure drop along the length of the hose?

2. A service truck transfers hydraulic oil with a kinematic viscosity of 200 cSt through a 2-in. I.D. hose 100 ft. long to a bulldozer at a flow rate of 50 gal/min. What is the pressure drop between the service truck and the bulldozer if the fluid Sg is 0.85?

3. What diameter pipe would be required to deliver a hydraulic fluid with a Sg of 0.86, 100 ft. through drawn steel tubing if the velocity is limited to 10 ft/sec, the friction factor is 0.05, and the pressure drop to 25 lbf/in.^{2}?

4. Hydraulic oil with a Sg of 0.85 is transferred between two tanks 300 m apart through a 6.5 cm pipe at 1,000 l/min. If the f-factor is given as 0.04, what is the pressure drop between the two tanks?

5. Hydraulic oil with a Sg of 0.90 is pumped 25 ft through a 1-in. I.D. off-line filter loop. If the oil has a kinematic viscosity of 0.1 Newts and the head loss is 5 lbf/in.^{2}, what flow rate could be expected through the filter?

6. An oil with a Sg of 0.95 and kinematic viscosity of 50 cSt is pumped through a section of 1.5-in. I.D. drawn steel tubing at the flow rate of 100 gal/min. Determine the absolute roughness, relative roughness, and friction factor for the tubing.

7. The pressure drop associated with hydraulic fluid with a Sg of 0.93 flowing across a 1-in. I.D. sharp edge orifice is 20 lbf/in.^{2}. If the discharge coefficient is 0.60, what flow rate in gal/min could be expected through the orifice?

8. The flow rate of a hydraulic fluid with a Sg of 1.08 through a 2-in. I.D. orifice is 50 gal/min. If the pressure drop across the orifice is 5 lbf/in.^{2}, what is the discharge coefficient?

9. What pressure drop in lbf/in.^{2} could be expected across a 1.5-in. I.D. orifice with a $C_d = 0.70$ through which a hydraulic fluid with a Sg of 0.86 is flowing at the rate of 80 gal/min?

10. What I.D. orifice with a $C_d = 0.60$ would be required to allow a flow rate of 25 gal/min with a pressure loss of 10 lbf/in.^{2}, if the hydraulic fluid has a Sg of 1.0?

11. Hydraulic oil with a Sg of 0.92 and a kinematic viscosity of 0.01 Newts flows first through a sudden reduction, and then through a sudden enlargement. If the larger pipe has a 3-in. I.D. and the smaller pipe has a 1-in. I.D., what pressure loss could be expected through these two fittings, if the flow rate is 60 gal/min?

12. Hydraulic fluid with a Sg of 0.95 and a kinematic viscosity of 0.02 Newts is flowing through a 3/4 in. I.D. line at 35 gal/min. If a 3/4 closed gate valve is inserted in the line, what pressure drop could be expected?

13. A 1-in. I.D. hydraulic circuit contains four standard 90° elbows and a swing check valve in series. If hydraulic fluid with a Sg of 0.87 and a kinematic viscosity of 0.03 Newts is flowing at the rate of 30 gal/min, what is the loss through these five fittings?

14. Hydraulic fluid is flowing horizontally 100 m through a 3-cm I.D. distribution manifold at 250 l/min. Compute the pressure loss when $N_R = 2000$.

15. A 1/2 closed gate valve is used as a temporary flow control in a 1/2-in. tube to limit the flow to 28 gal/min. If the fluid has a Sg of 1.08 and a kinematic viscosity of 30 cSt, what would be the equivalent length of the valve if the f-factor is assumed to be 0.05?

16. What is the equivalent length of 10 cast iron 3/4-in. I.D. standard 90° elbows placed in a hydraulic circuit which carries 48 gal/min of a hydraulic fluid with a Sg of 0.86 and a kinematic viscosity of 0.04 Newts?

17. Oil with a Sg of 0.85 and an absolute viscosity of 7×10^{-6} N · s/m² is pumped horizontally through a 5- cm I.D. commercial steel pipe at 400 l/min. What is the pressure drop associated with the friction in a 200 m section?

18. Fig. 5-13 illustrates an off-line heat exchanger circuit consisting of a tee, three 90° elbows, a check valve, a filter ($\Delta p = 5$ lbf/in.²), a control valve ($C_d = 0.75$), two globe valves, a heat exchanger ($\Delta p = 40$ lbf/in.²) and 50 ft of 1-in. drawn steel tubing with an I.D. equaling 0.902 in. If the fluid has a $Sg = 0.95$ and a $\nu = 0.035$ Newts, calculate the maximum velocity, flow rate, and total pressure drop in the circuit if the flow is kept at the upper limit of the laminar flow range.

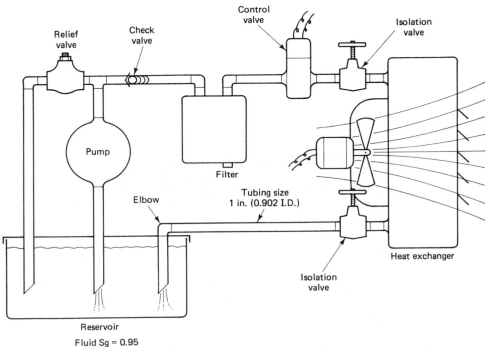

Figure 5-13 Problem 18.

19. The relief valve portion of the circuit in Fig. 5-13 consists of a tee, relief valve ($C_d = 0.5$) valve, 90° elbow, and 20 ft. of 1-in. drawn steel tubing with an I.D. = 0.902 in. If the flow rate remains in the laminar range, what would be the pressure drop through this portion of the circuit?

20. In Problem 18, assuming the pressure drop across the filter and heat exchanger remained the same, what would be the total pressure loss if the flow rate were doubled?

CHAPTER 6

HYDRAULIC FLUIDS

6-1 INTRODUCTION

The primary purpose of the hydraulic fluid is to transmit power. Fluid is picked up by the pump from the reservoir, fed through the control valves to cylinders and motors where the power is expended, and then returned to the reservoir where it is cooled and settled before starting the cycle again. How well a fluid transmits power is determined by how easily it is pumped, how stiff it is, and a number of other service-related properties that determine how suitable it is to a particular application system and environment. About 80% of the fluids used are petroleum-base hydraulic oils.

Specific properties that affect the performance of a hydraulic fluid include its viscosity, specific gravity or API gravity, viscosity index, pour point, neutralization number, flash point, fire point, auto ignition temperature, antiwear properties, resistance to oxidation and rust, and defoaming and detergent dispersing properties. These properties determine not only present suitability, but resistance to deterioration during the time the fluid will be used in the system.

Filters are used to keep the fluid clean, and maintenance personnel change them periodically to keep fluid contamination levels to an acceptable level. At set intervals, samples are taken from the machines and tested to determine if the fluid should be cleaned, changed, or returned to a distributor for complete reclamation. Where a machine failure has been experienced, fluid testing determines where the fault lies. Recently developed in-line techniques permit monitoring the condition of the fluid in the machine during the working cycle.[1]

[1] E.C. Fitch. An Encyclopedia of Fluid Contamination Control. (Stillwater, OK: Fluid Power Research Center, Oklahoma State University, 1980). Part 2.6.

6-2 TYPES OF FLUIDS

A number of liquids, including water, have been used as hydraulic fluids. Industry uses five basic types:

1. Petroleum-base hydraulic oils.
2. Synthetic fluids.
3. Water glycol fluids.
4. Oil-in-water and water-in-oil emulsions.
5. High Water Content Fluids (HWCF).

About 80% of the 250–275 million gallons of oil sold each year is petroleum-base oil, with the remaining 20% being the other four types which are fire resistant. Industries that use fire-resistant fluids include forging and extrusion, coal mining, chemical-petroleum power, die casting, foundries, fabrication, injection molding and the primary metals, and some marine applications.

Water was the first hydraulic fluid but, because of several undesirable characteristics, it was replaced by petroleum oil. Water froze, evaporated, corroded the system and was a poor lubricant. Petroleum oil is used because it is relatively inexpensive and has several desirable characteristics. To be effective in a hydraulic system, a fluid must

1. Transmit fluid power efficiently
2. Lubricate the system
3. Dissipate heat generated by the system
4. Not deteriorate seals and parts of the system
5. Remain stable over time with changes in temperature and operating conditions.

If there is a danger of fire, then fire resistance must be added to this set of characteristics, and that means one of the fire-resistant fluids must be selected.

Petroleum-base hydraulic oils are refined from selected crude oil and formulated with additives to prevent wear, rust, oxidation, and foaming. They are the least expensive of the hydraulic oils and are used where fire and high temperatures are not a problem. Hydraulic oil has the natural ability to transmit fluid power efficiently, has good lubrication qualities, dissipates heat reasonably well under normal operating conditions, and is compatible with most sealing materials. It is also long-lasting and stable so long as the temperature is below 150°F (65.5°C), and it is kept free of contaminants by filtration. The condition is checked periodically by taking a sample and testing it. This determines not only the condition of the oil, but the condition of machine components such as pumps and motors which deposit wear debris in the oil that can cause the system to fail.

Fire-resistant fluids are of four basic types:

1. synthetics,
2. water glycols, and

3. water-in-oil emulsions with 35–40% water (called invert emulsions because oil surrounds the water droplets, and

4. high water content fluids (90% or more water).

Typically, a hydraulic fire starts when a hose bursts and sprays fluid against a hot surface. Straight synthetics prevent ignition of the fluid because of the fire-resistant qualities of the fluid. Water glycol and water-in-oil emulsions prevent ignition when the water in the fluid turns to steam and snuffs out the fire. Water-glycol and water-in-oil emulsions are not totally fire resistant, however, and they will burn if the temperature is sufficiently high over an extended period of time. High water content fluids will not burn because the water content to additive ratio is too high for the additive to ignite.

Oil-in-water emulsions are so called because the oil in each molecule is surrounded by water. This makes the oil the discontinuous phase and the water the continuous phase. Oil-in-water emulsions use the continuous water phase for cooling, such as in cutting oil and coolants. **Water-in-oil** emulsions, on the other hand, are mixtures of oil and water with the oil surrounding the water droplets. Thus, the oil is the continuous phase and usually, though not always, this gives superior lubricating quality. Water-in-oil emulsions are called invert emulsions.

High water content fluids (HWCF) receive their name from the American National Standards Institute (ANSI) Standard ANSI B93-5N *Practice for the Use of Fire Resistant Fluids in Industrial Hydraulic Fluid Power Systems*. They are often referred to as high water base fluids.

Soluble oils were first available as cutting fluids and coolants. In *neat* form (straight) and higher concentrations, they were also recommended as gear lubricants and hydraulic fluids on the same machine. This arrangement had the advantage of allowing the coolant and lubricant to be mixed, even though the coolant contained very high levels of contaminants from the machining or grinding operation, and in noncritical applications this was acceptable.

The HWCF fluids, however, are different. They were developed in the 70s, partly in response to the OPEC situation of 1973, but more importantly, from the potential advantage in cost savings. The cost of HWCF fluids is about 20% that of petroleum-base hydraulic fluids, and 5% that of straight phosphate esters. The magnitude of the savings is dramatic when one considers that the automotive industry, which is the largest user, adds 60 million gallons of make-up oil annually.

Typically, HWCF fluids are 95% water and 5% additive, or 90% water and 10% additive. The additive portion consists of viscosity improvers (water is too thin), anti-friction additives, rust inhibitors, and bacteria growth and sludge inhibitors. If a fluid contains water, it will grow microbes and this is a continuing problem with HWCF fluids, particularly as the fluid ages. Biocides and fungicides are used to control bacteria growth.

There are a number of HWCF fluids available, each with its own chemical composition and performance characteristics. *Micro-emulsions* are formulations with emulsified particles less than one micron dispersed in a solution which is nearly clear in concentrate form, but turns opaque when used in a hydraulic system. Micro-emulsions have a high film strength and good wear resistance.

Synthetic solubles contain nonpetroleum water soluble lubricants in an aqueous solution. The lubrication molecules are polarized (charged) and thus attracted to the metal parts in the system where they form a plating that serves as a boundary lubrication layer between moving parts in the system. *Thickened fluids* is a name given to solutions where the additive forms a mechanical matrix that gives the water solution the viscosity and lubricity of a petroleum-based fluid. The mechanical matrix is unlike conventional petroleum-base thickeners and viscosity improvers which use long chain polymer linkages that break irreversibly under high shear conditions. Instead, the synthetic properties of the mechanical matrix reform without apparent fatigue. *Neo-synthetic* or *macro-emulsions* are no-oil chemical compositions that form synthetic lubrication droplets in the 1-10 micron range. The solution is opaque and resembles the old emulsified mineral oils, even though the macro-emulsions are completely synthetic. Macro synthetic emulsions are also polarized to plate component surfaces with a high viscosity layer of lubricant, and some macro-emulsions are recommended for multiple use: gear box lubricant in neat form, and cutting fluid and hydraulic fluid in mixed form. Micro-dispersions are no-oil solutions with very fine nonsoluble particles dispersed in the fluid by surface active ingredients. Multiple function micro-dispersions also contain what are called "second ingredients" to plate surfaces, serve as thickeners, and act as wetting agents to speed heat transfer. All HWCF fluids contain rust and corrosion inhibitors, as well as biocides and defoaming agents. High water base fluids are not affected adversely by fine filtration and the additives do not separate out.

6-3 FLUID APPLICATIONS

A major task facing fluid power mechanics and technicians is to select the proper fluid for a specific application. The question to be answered is this: which fluid goes in which machine, given its characteristics and the conditions under which it must operate?

Under ordinary circumstances, and where fire is not a problem, one of the petroleum-base hydraulic oils is the obvious choice. If fire is a hazard, then one of the fire-resistant fluids must be chosen. The first step in the process is to consult the equipment manufacturer who has assembled a great amount of information based upon experience and test data. It should also be remembered that the manufacturer's guarantee may require the use of specified hydraulic fluids. Before selecting a fluid to fill a new machine, or before adding fluid to a machine that is in operation, consult the machine manual. Also be sure you know what kind of oil is in the machine, because some of them do not mix and can cause trouble. Oil-synthetic blends, for example, can attack the seals used with petroleum-base hydraulic fluids.

A petroleum-base hydraulic fluid is the least expensive of the hydraulic oils, with the exception of the oil-in-water emulsions (water is the continuous phase) and the recently developed HWCF fluids composed of water with chemical additives. Phosphate ester fluids cost about seven times more than hydraulic fluid oil. Phosphate esters are true synthetics, however, and have the advantage of continued service without the possibility of separation of the continuous phase from the emulsion, or periodic replenishment of additives which evaporate with continued service, thus changing the viscosity and wear characteristics of the fluid.

Hydraulic fluid is added to replenish the loss from leaks, and this can be very expensive if the system is not leak free. For example, one estimate places the combined leakage at more than one million barrels of oil hydraulic fluid each year in the United States alone. In coal mining, which uses fire-resistant fluids, the estimate is one gallon of hydraulic fluid lost for each 17 tons of coal mined in the United States. The most skeptical estimates from the industry believe fully 85% of all hydraulic fluid put in the machinery leaks out. The ongoing cost of adding fluid to the system is important, then, when the loss due to leaks is considered. The risk of contamination also increases each time the reservoir is opened to add make-up fluid.

Straight phosphate esters and oil synthetic blends which have a phosphate ester base are suitable for all applications, have excellent lubricating and anti-wear qualities, and for practical purposes are equal to premium grade petroleum-base hydraulic oil. They are available in a wide range of viscosities which suit them to uses in plants and outside, from high-speed precision machine tools which require low viscosity fluids, to sub-zero aircraft and mobile equipment applications. Make-up fluid is added directly without regard for changing the viscosity or chemical make-up of the fluid.

Water glycol fluids are designed for use in hydraulic systems working in areas with a source of heat, ignition, or where there is a potential fire hazard. They are a mixture of water, ethylene glycol, and a high viscosity lubricating and thickening agent, with additives to prevent corrosion.

Water glycol fluids can be used in gear pumps at full capacity without premature failure. Vane pumps can be operated at pressures to 2000 lbf/in.^{2} (13.8 MPa) and speeds to 1200 rpm. Axial piston pumps can be operated at pressures of 3500 lbf/in.^{2} (24 MPa) at 1200 rpm, but if higher speeds are required (1750 rpm), the pressure must be reduced to attain equivalent service. Most water glycol fluids are not recommended for use in radial piston pumps. Water glycol fluids are limited to use in low to medium pressure and noncritical applications because their water content limits their lubricating and resistance-to-wear qualities. Water glycol fluids do attack zinc, cadmium, and aluminum, however, and these metals must be removed from the system to prevent deterioration and contamination of the system. No special hoses, seals, or packing materials are required when changing over from hydraulic oil, although neoprene, Buna N, and viton are the most common compounds used for seals, and nylon, butyl, or neoprene are used most commonly for hoses. The water content in water glycol fluids excludes the use of leather, cork, untreated cotton, and cellulose packing. Conventional paints are unsuitable for use with water glycol fluids and unless compatible paints are selected initially, questionable paints should be removed and the surface left unpainted. Fluid filters made of fine mesh metal screens and waterproofed cotton fiber give satisfactory service, but active clay or absorbent filters will remove ingredients in the solution and so must be avoided. Water glycol fluids will dissolve certain pipe sealing compounds and dopes the same as phosphate ester fluids. Only compatible pipe sealing compounds and dopes can be used, and those with questionable contents should be discarded. Water glycol fluids have low toxicity and are not irritants to the skin.

The HWCF fluids are relatively inexpensive and have excellent fire-resistant qualities. The lubrication quality depends upon which fluid additive is used as does the viscosity stability under severe use and at elevated temperatures. Because 90% of the fluid is water, HWCF fluids have excellent heat transfer qualities, but the temperature must be kept within the range of 40°F to 120°F (4.4°C − 48.9°C) to prevent freezing and reduce evaporation, both of which cause the fluid condition and concentration to change. Because the viscosity of most HWCF fluids is lower than petroleum oil and straight synthetics, they tend to scour the system and carry foreign particles in suspension. Fine filtration is recommended. The higher water content requires close monitoring of the oil additive package to keep the chemical balance correct and control the microbe level. Experience gained in the automotive and mining industries indicates that centrally located reservoirs mounted above the pump inlet give the best results. This arrangement prevents pump cavitation and allows the fluid to be added, conditioned, and monitored at one location, rather than at several machines.

Figure 6-1 illustrates an overhead tank power unit with a single motor and pump. This allows the pump to receive fluid without restriction at greater than atmospheric pressure. HWCF fluid pump specifications call for no less than full atmospheric pressure. The inlet filter cartridge is serviced through the top of the reservoir without draining the fluid.

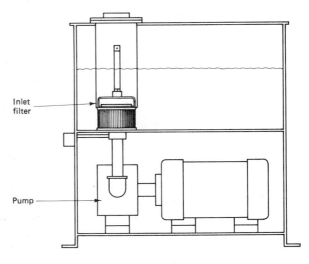

Figure 6-1 Overhead reservoir and inlet filter for HWCF fluid application.

6-4 SERVICE-RELATED PROPERTIES

Properties that affect the performance of fluids include specific gravity, viscosity, viscosity index, bulk modulus, pour point, neutralization number, flash point, fire point, auto ignition temperature, anti-wear properties, resistance to oxidation and rust, defoaming, and detergent dispersant properties. The objective is to select a fluid that has the properties for a specific application and then have them remain stable with continued use for the recommended time between changes. This could be as long as 1500 hours (or three years) in some cases. HWCF fluids require a higher standard of maintenance to keep the chemical balance correct, and the level of biocides and fungicides high enough to control microbe growth.

In operation, high viscosity fluids cause sluggish operation, transmit power less efficiently, and generate excessive heat in the fluid. This is due to the higher internal friction in the fluid, which has the effect of increasing its static inertia. In extreme cases with high viscosity at low temperatures, pumps and motors will break if the machine is operated. Low viscosity oils flow more easily because of lower internal friction, but slippage in the pump and motor increase, allowing the fluid to bypass the close fitting surfaces of components. External leakage also increases visibly through components that have an exposed moving part such as cylinder rod seals, but so too does internal leakage increase through control valves and past cylinder pistons.

The selection of the proper viscosity, then, must be made with the normal operating temperature (usually taken at the pump) of the machine in mind. This will allow the fluid to transmit the power most efficiently, while still lubricating and cooling the machine without excessive leakage. For most applications, this will place the viscosity in the range of 100 SSU to 200 SSU (approx. 20.6 cSt to 43.2 cSt), depending upon the pump being used. Gear pumps are more tolerant of a wide range of viscosity (50 SSU to 1000 SSU), even though the preferred viscosity would be on the order of 200 SSU given the larger fit and clearances of this pump design. Piston pumps and motors, on the other hand, are less tolerant of a wide range of viscosity (50 SSU to 300 SSU), with a preferred viscosity of about 150 SSU because of the close fit and clearances in the pump and motor design. The manufacturer supplies this information for the pump, and the equipment manufacturer provides it for the complete machine.

This viscosity of a number of hydraulic fluids and petroleum-base products is given in Fig. 6-2. A good reference oil to consider is SAE 10W engine crank-case oil. Notice that the viscosity of SAE 10W engine oil traces a diagonal line across the graph, and in the temperature range of 100°F to 210°F (37.7° to 98.9°C) has a range of viscosity of just under 22 SSU to just over 45 SSU. SAE 10W engine oil is not a recommended hydraulic fluid, but it is useful as a reference oil against which to compare the viscosity of a number of different hydraulic fluids to get a feel for the change in viscosity with changes in temperature. For example, most petroleum-base hydraulic oils have a similar viscosity pattern, as do the water glycols, phosphate esters, and water-in-oil emulsions, but not HWCF fluids. Plain water at 100°F has a viscosity of 30 SSU, and so do several of the HWCF formulations. This requires that the system be designed around the fluid to accommodate

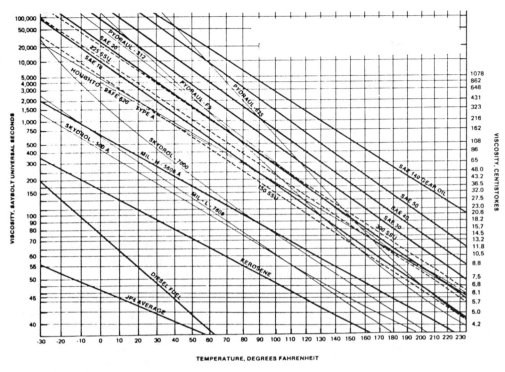

Figure 6-2 Viscosity for a number of hydraulic fluids and petroleum base products (*courtesy of Sun Oil Co.*).

the viscosity. Thickeners and viscosity improvers can be added, but this design limitation must be recognized at the outset.

The change in the viscosity as the temperature of the fluid increases is described by the viscosity index. It measures the stability between two temperature extremes. As the temperature increases, the fluid becomes thinner and the viscosity decreases. When the viscosity index scale was established in 1929, it measured the degree to which the viscosity of an oil would increase when it was cooled from a temperature of 210°F (98.9°C) to 100°F (37.8°C). A Pennsylvania crude paraffin base fraction was designated to have a VI of 100. A coastal crude naptha base fraction was designated to have a value of 0, indicating that it changes the most with changes in temperature. Other oils were compared to these two reference oils and assigned values between 100 and 0, corresponding to their relative change in viscosity with respect to changes in temperature. Since then, oils have been developed that exceed the VI of 100.

The proper viscosity index of a fluid for a specific application is determined from the fluid temperature change requirements of the system. Production machinery, for example, keeps the oil within a narrow temperature range, and a low viscosity index would be suitable. Mobile hydraulic equipment, however, operating where the temperature of the oil may vary from below freezing to near 160°F requires a fluid with a very stable viscosity with respect to temperature change, and would call for an oil with a VI of at least 100.

Viscosity index is computed by using Saybolt Second Universal (SSU) designations for the reference oils (one with a VI of 100 and the other with a VI of 0) and for the oil for which the viscosity index is to be determined. That formula is

$$VI = \frac{(L - U)}{(L - H)} \times 100 \tag{6-1}$$

where L is the SSU viscosity of a reference oil at 100°F with a viscosity index of 0 that has the same viscosity at 210°F as the oil to be calculated; H is the SSU viscosity of a reference oil at 100°F with a viscosity index of 100 that has the same viscosity at 210°F as the oil to be calculated; and U is the viscosity at 100°F of the oil whose viscosity index is to be calculated. The table, composed of SSU values at 100°F, lists viscosity index values vertically down both margins and SSU values at 210°F horizontally across the top. To determine the viscosity index, the SSU viscosity values at 210°F and 100°F are located, respectively, across the top of the table and down the appropriate column, and the third value, viscosity index, is then read from the corresponding vertical margin.

Example 1

Determine the viscosity index of an oil with a SSU viscosity of 105 at 210°F and 1376 at 100°F.

Solution   The SSU viscosity of 90 sec at 210°F given for the oil whose viscosity index is to be determined is found by reading across the top of Table 6-1. The SSU viscosity of the sample oil at 100°F is then found by reading down the column under 105. This value is 1376. The viscosity index is then found by tracing the horizontal line of values back to the left or right margins. The VI is 95. This high value indicates a narrow range of viscosity in the temperature interval between 100° and 210°F.

Comparison of oils of different viscosity indexes can be made by plotting viscosity versus temperature curves for several samples. Figure 6-3 shows a plot of four oils with the same SSU viscosity of 90 at 210°F, but with much different viscosities at 100°F. The corresponding viscosity index for each sample is listed. The procedure generates data that may be used as in Example 1 to determine the viscosity index.

Caution must be exercised in the interpretation and use of the viscosity index at elevated temperatures. Fluids with the same SSU viscosity at 100°F and the same viscosity index may have much different viscosities at higher temperatures. That is, the relationship between viscosity and temperature may not be linear, with the result that the actual viscosity index may, in fact, be much different from the computed value outside the temperature range of 100 to 210°F. The implication is that the viscosity index is strictly accurate only within this temperature range.

Bulk modulus of elasticity is a measure of the rigidity of the fluid. It gives an indication of how much the fluid "gives" or compresses under pressure. This is particularly important in positioning circuits where the fluid must act almost like a solid to maintain accuracy even with changes in load. Mathematically, bulk mod-

VI*

TABLE 6-1 VISCOSITY INDEX (*FROM* LUBRICATING ENGINEERS MANUAL, © *UNITED STATES STEEL CORPORATION*)

SUS 210 F	40	45	50	55	60	65	70	75	80	85	90	95	100	105	110	115	120	125	130	135	140	145	150	155	SUS 210 F
0	138	265	422	596	781	976	1182	1399	1627	1865	2115	2375	2646	2928	3220	3524	3838	4163	4498	4845	5202	5570	5959	6339	0
5	136	261	414	584	763	953	1153	1364	1585	1816	2059	2311	2573	2846	3129	3423	3727	4042	4366	4701	5046	5402	5768	6145	5
10	135	256	405	570	745	930	1124	1329	1543	1767	2002	2246	2500	2765	3038	3323	3616	3920	4233	4557	4890	5234	5587	5950	10
15	133	252	397	557	727	907	1095	1294	1502	1718	1946	2182	2427	2683	2947	3222	3505	3799	4101	4413	4734	5065	5406	5756	15
20	132	247	389	545	710	884	1066	1259	1460	1670	1889	2117	2355	2601	2856	3121	3394	3677	3968	4269	4578	4897	5225	5562	20
25	130	243	380	532	692	861	1038	1224	1418	1621	1833	2053	2282	2520	2765	3021	3284	3556	3836	4125	4423	4729	5044	5368	25
30	129	238	372	519	674	837	1009	1188	1376	1572	1776	1989	2209	2438	2674	2920	3173	3434	3703	3981	4267	4561	4863	5173	30
35	127	234	364	506	656	814	980	1153	1334	1523	1720	1924	2136	2356	2583	2819	3062	3313	3571	3837	4111	4392	4682	4979	35
40	126	230	355	493	638	791	951	1118	1293	1474	1663	1860	2063	2274	2492	2718	2951	3191	3438	3693	3955	4224	4501	4785	40
45	124	225	347	480	621	768	922	1083	1251	1425	1607	1795	1990	2193	2401	2618	2840	3070	3306	3549	3799	4056	4320	4590	45
50	123	221	339	468	603	745	893	1048	1209	1377	1551	1731	1918	2111	2311	2517	2729	2948	3173	3405	3643	3888	4139	4396	50
55	121	216	330	455	585	722	864	1013	1167	1328	1494	1667	1845	2029	2220	2416	2618	2827	3041	3261	3487	3719	3957	4202	55
60	119	212	322	442	568	699	835	978	1125	1279	1438	1602	1772	1948	2129	2316	2507	2705	2908	3117	3331	3551	3776	4007	60
65	118	207	314	429	550	676	806	943	1084	1230	1381	1538	1699	1866	2038	2215	2396	2584	2776	2937	3175	3383	3595	3813	65
70	116	203	305	416	532	653	777	908	1042	1181	1325	1473	1626	1788	1947	2114	2285	2462	2643	2829	3019	3215	3414	3619	70
75	115	199	297	403	514	630	749	873	1000	1132	1268	1409	1553	1703	1856	2014	2175	2341	2511	2685	2864	3046	3233	3425	75
80	113	194	288	391	497	606	720	837	958	1083	1212	1345	1480	1621	1765	1913	2064	2219	2378	2541	2708	2878	3052	3230	80
85	112	190	280	378	479	583	691	802	916	1035	1155	1280	1408	1539	1674	1812	1953	2098	2246	2397	2551	2710	2871	3036	85
90	110	185	272	365	461	560	662	767	875	986	1099	1216	1335	1457	1583	1711	1842	1976	2113	2253	2396	2542	2690	2842	90
95	109	181	263	352	443	537	633	732	833	937	1042	1151	1262	1376	1492	1611	1731	1855	1981	2109	2240	2373	2509	2647	95
100	107	176	255	339	426	514	604	697	791	888	986	1087	1189	1294	1401	1510	1620	1733	1848	1965	2084	2205	2328	2453	100
105	106	172	247	326	408	491	575	662	749	839	930	1023	1116	1212	1310	1409	1509	1612	1716	1821	1928	2037	2147	2259	105
110	104	167	238	314	390	468	546	627	707	790	873	958	1043	1131	1219	1309	1398	1490	1583	1677	1772	1869	1966	2064	110
115	103	163	230	301	372	445	517	592	666	741	817	894	970	1049	1128	1208	1287	1369	1451	1533	1616	1700	1785	1870	115
120	101	159	222	288	355	422	488	557	624	693	760	829	898	967	1037	1107	1176	1247	1318	1389	1460	1532	1604	1676	120
125	99	154	213	275	337	399	460	522	582	644	704	765	825	886	946	1007	1066	1126	1186	1245	1305	1364	1423	1482	125
130	98	150	205	262	319	375	431	486	540	595	647	701	752	804	855	906	955	1004	1053	1101	1149	1196	1242	1287	130
135	96	145	197	249	301	352	402	451	498	546	591	636	679	722	764	805	844	883	921	957	993	1027	1061	1093	135
140	95	141	188	236	284	329	373	416	457	497	534	572	606	640	673	704	733	761	788	813	806	859	880	899	140

→ Viscosity index

Vertical bold type—viscosity index. Horizontal bold type headings—SUS @ 210 F, tabled values are SUS @ 100F

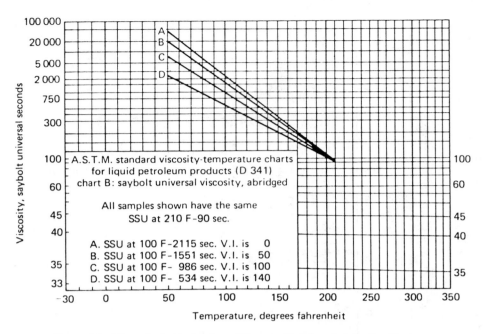

Figure 6-3 Viscosity index for four different oils (*from Lubrication Engineers Manual,* © *United States Corporation*).

ulus K is the reciprocal of compressibility, which equals the ratio of the change in volume to the change in pressure. Entrained air, for example, decreases the bulk modulus of hydraulic oil because it makes it spongy. The formula for bulk modulus is

$$K = \frac{\Delta p}{\Delta V/V} = \frac{p_2 - p_1}{\dfrac{V_1 - V_2}{V_1}} \tag{6-2}$$

and since $\Delta V/V$ is dimensionless, K is expressed in units of pressure. At room temperature and atmospheric pressure, water has a $K = 300,000$ lbf/in.² or, in SI units, approximately 2068 MPa. Hydraulic fluids will have a K of between 300,000 and 400,000 at room temperature in the pressure range of 1000 to 6000 lbf/in². Bulk modulus becomes important when the amount of air entrained in the fluid lowers the value to the place where the accuracy and safety of the hydraulic circuit are impaired.

Example 2

With an increase in load, fluid in an elevator hydraulic cylinder undergoes a change in pressure from 1000 lbf/in.² to 2000 lbf/in.². If the K of the fluid is 350,000 lbf/in.², how much can the fluid be expected to compress?

Solution Solving Eq. 6-2 for the change in volume,

$$\Delta V = \frac{V_1(p_2 - p_1)}{K} = \frac{1(2000 - 1000)}{350,000} = 0.0029$$

or about 1/4 of 1 percent. This means that within the pressure range given, the cylinder would be able to position the load to within that limitation if the same volume of fluid entered the cylinder.

Pour point is measured to indicate the lowest temperature at which an oil will flow. Hydraulic fluid specifications for low temperature applications, particularly those involving mobile equipment, use pour point as an indication of the ability of the oil to be pumped as the temperature drops. Diesel fuel oils also fall in this classification, since diesel power is used to power much of the equipment.

Pour point is determined by ASTM procedure D 97, which measures the observed movement of an oil sample held horizontally in a test tube. The pour point is taken as 5°F above the temperature of the solid point where the oil sample shows no indication of movement under controlled conditions. As a general rule, the pour point should be 15° to 20°F below the lowest temperature of the system during start-up to be sure that the pump will not cavitate and become damaged. A typical hydraulic fluid with a viscosity of 10 cSt at 100°F (60 SSU) will have a pour point of −50°F or below. Where temperatures are too low to pump the oil, it may be heated using emersion elements so long as the heat applied does not exceed 15 watts per in.^{2} of heating element.

The neutralization number is the designation that reflects to what degree the hydraulic fluid is acid or alkaline. Petroleum-base fluids that have been in service for some time have a tendency to become acid, and this causes deterioration of the fluid, bearings, component parts, and seals. It is also very important to control the alkalinity of water glycol fluids to assure continued good performance, and any time the water content is adjusted, the fluid alkalinity must be checked after a run-in period of 24 hours. The *acid number* refers to the number of mg of potassium hydroxide necessary to neutralize a 1-gm sample of hydraulic fluid when the ASTM D 974 test procedure is used. Neutralization is achieved by an observable color change in the solution. Conversely, the *base number* is the number of mg of acid (alcoholic potassium hydroxide) that is required to titrate 1 gm of a strong base solution. Hydraulic fluids are fortified with additives to reduce the tendency to become acid and keep the neutralization number below 0.1 during normal service. The procedure to carry out ASTM procedure D 974 is shown in Fig. 6-4. At the point where the color change indicates neutralization has been reached, the neutralization number is calculated from

$$\text{Neutralization number} = \frac{\text{Total ml of titrating solution} \times 5.61}{\text{weight of the sample used}} \qquad (6\text{-}3)$$

The flammability of a hydraulic fluid is described from the flash point, fire point, and auto ignition temperatures. The flash point is the temperature at which a test flame passed over the surface will ignite the gases generated as the fluid is heated and its temperature increases. The fire point is reached when the test flame will ignite the fluid and keep it lit for five seconds. The auto ignition temperature is reached when the sample will self-ignite and combustion continues. These tests indicate how hazardous the fluid will be in the presence of hot metals,

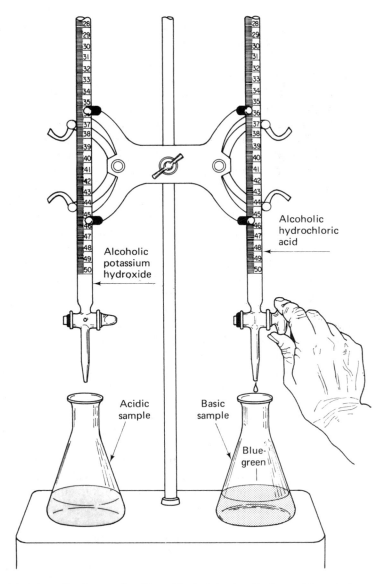

Figure 6-4 Neutralization number (*from The Lubrication Engineers Manual,* © *United States Steel Corporation*).

open flames, or elevated temperatures. Typical applications include the coal mines, ships, aircraft, and space craft.

Antiwear additives in hydraulic fluids reduce wear caused by friction between moving parts such as shafts and bearings, piston rings and cylinder walls, and pump vanes and bodies. Wear is the permanent displacement of surface materials between two surfaces. Those parts that lose material change in dimension, and in some cases the clearance changes. Tests that measure antiwear properties evaluate the change in dimension of parts losing materials or their change in weight. When the loss is too small to measure the dimensional or weight loss, parts may be radiated, and the loss measured as the amount of radiation that is present in the lubricant after the wear test.

The two standard procedures for testing antiwear properties of hydraulic fluids are

1. Four-Ball Wear Test (ASTM D 2266).
2. Vickers Hydraulic Pump Wear and Oxidation Test (ASTM D 2271 and D 2882).

The Four-Ball Wear Test exerts a vertical force through a rotating steel ball (Fig. 6-5), and measures the coefficient of friction and amount of material displaced between three stationary steel balls (or three discs in a holding cup) and the fourth ball, which is rotated at constant speed. The contact surfaces between the test are immersed in the hydraulic fluid during the test, and the relationship that defines friction is

$$f = 2 \sqrt{2} \left(\frac{F}{L}\right)\left(\frac{r}{s}\right) \tag{6-4}$$

Where f is the coefficient of friction, a number without dimension between 0 and 1, F is the force in gm acting on the torque arm, L is the load applied vertically

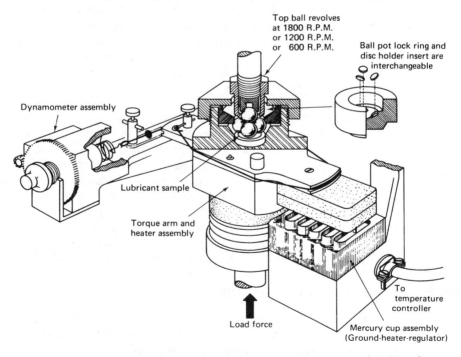

Figure 6-5 Four-ball wear tester (*from Lubrication Engineers Manual,* © *United States Steel Corporation*).

Sec. 6-4   Service-Related Properties

through the rotating steel ball against the three test balls, r is the length of the torque arm in cm, and s is the diameter of the three test balls in cm. Typical test results for the coefficient of friction range from 0.10 to 0.75 mm at the termination of the test, which usually lasts one hour. Scar diameters are measured with a calibrated low-power microscope.

The Vickers Pump Tests, which are standards in the industry, measure the wear that occurs in a vane-type pump operating as part of a hydraulic system under preset conditions of pressure, temperature, and time. Wear, viscosity, color, neutralization number, and other properties are recorded as the pump operates. Pump parts affected by wear (cam ring and vanes) and the filter screen are examined for changes in weight at the termination of the test.

Oxidation is a reaction between the fluid and oxygen which results in the formation of acid and sludge. Oxidation inhibitors reduce the catalytic effect of adjacent metals that promote the reaction. The oxidation reaction rate increases at elevated temperatures, particularly with petroleum-base hydraulic oils. For example, the rate approximately doubles for every 18°F (10°C), and it is estimated that the life of the oil is halved for each 15°F rise in temperature. Tests that measure the oxidation stability of hydraulic oils increase the temperature of the fluid to approximately 200°F in the presence of oxygen for a specified number of hours or until the neutralization number reaches 2.0.

Rusting is the corrosion of the ferrous parts of the machinery in the presence of water in the hydraulic fluid. Moisture enters the system as condensation, through leaks in the reservoir, or through exposed parts such as cylinder rods and shaft bearings. Rusting is not the same as oxidation. Oxidation produces a sludge and acid condition, while rusting, which occurs at the metal surface, produces flaking. Rust inhibitors plate the ferrous surfaces, forming a thin protective coating on the metal that prevents the reaction from occurring. The standard test for rusting (ASTM D 665) exposes a polished steel rod to a 10 percent mixture of water in mineral oil, stirred and held at 140°F for 24 hours. The presence of rust indicates failure of the test, with 5 percent or more indicating severe failure.

Foaming is the result of entrainment of air in oil. Most oils normally contain air in solution, some as much as 10% by volume. Air in solution is not usually harmful, although it promotes oxidation. Air entrainment is caused by improper oil levels in the reservoir or severe agitation. Inhibited oils with foaming depressants promote the release or "breaking out" of the air from the fluid rather than preventing entrapment initially. Air also can be separated out by passing the fluid through a sieve screen in the reservoir. Tests for resistance to foaming bubble air through a preheated sample to promote foaming and then measure the time it takes for the foam that forms to release and settle out the air.

6-5 FLUID STORAGE AND HANDLING

Knowing where the fluid is used gives an indication of the condition and service records of hydraulic machinery. The fluid handling and storage system maintains an inventory of fluids used in the machinery, and keeps accurate records of types, purchase dates, and usage. This provides the necessary records to know not only how much fluid is being used overall, but where, when, why, and how much fluid is used for each piece of machinery. As a total usage, it also lets purchasing summarize associated costs and place new orders with suppliers against future needs.

Hydraulic fluid is stored centrally in a clean, dry, well lit, controlled storage area where maintenance personnel can pick up the necessary supply for machines needing make-up fluid, or for use in changing fluid. It is important that the person in charge of the storage area be certain that it is kept clean and clear of spillage and that the containers used for storage and transport of the oil are clearly marked to match labels placed on machines so there will be no mistake about which fluid is in which container and in which machinery it should be used. Because there will be some spillage and drips, provisions must be made to keep the floor clean and dry. This can be done by placing containers under drips and by using an absorbent floor compound to dry up spills. Dust will also cling to the oil on drums and containers, and so these must be wiped periodically with clean, lint-free rags to prevent dirt and grime build-up.

Mobile lubrication trucks with storage and dispensing capability are used to service aircraft and mobile equipment fleets.

It is estimated by a recent study that there is a 50/50 chance that the make-up fluid added to a machine will be dirtier than the tolerance level of the machine. This means that about half the time the fluid added to the machine contains dirt, water, or other contaminants when it is drawn from storage, or that these were introduced when it is added to the machine. This can occur in a number of ways that can be avoided. Remember that the storage area may contain used fluid which is to be reclaimed as well as new fluid used for changes and make-up. These should be clearly marked and stored separately. Placing drums of new fluid upright allows for dirt and perhaps water to accumulate on the top and enter the drum when it is opened. It also makes for a convenient storage area which accumulates an assortment of cans, containers, and other items that have dirty bottoms. Storing drums of new fluid in racks on their sides eliminates this problem. Should water and dirt enter the drum of oil, it will be on the bottom, and at its lower edge, and care must be taken to be sure the first bit drawn is clear of water and dirt. Turning the drum slightly so that the faucet valve is not at the lowest point in the drum will reduce this possibility. When the drum is apparently empty, it can be turned slightly to release the last contents at the bottom and a careful check made to be sure this oil also is clean.

Oil is added to machinery using funnels with strainers or a dispensing pump. The container and funnel must be kept clean, and if this condition cannot be maintained, then prepackaged, sealed can containers in quart, gallon, or five-gallon sizes should be used and dispensed through a strainer with a clean spout. The method used will depend upon the cost of the fluid and how much fluid is added, but the objective remains the same: to be sure that the new oil remains clean when it is added to the reservoir. Before the cap or breather to the reservoir is removed, the area around it should be wiped with a clean cloth to prevent dirt entering the system. Some authorities suggest the use of air breathers which will remove dirt and contaminant particles from the air entering the system, down to the limits of the oil filter installed on the system. Presently they will pass airborne particles two or three times that size.

6-6 FILTRATION

The preventive maintenance program services the fluid filtration system consisting of strainers, filters, and general cleanliness of the machine to keep the fluid clean. This requires cleaning strainers, changing filters, keeping the machine clean, sampling the fluid from time to time to determine its condition, and cleaning the fluid and changing unusable fluid.

Strainers and filters remove particulate matter and silt from the hydraulic fluid. Technically, strainers direct the fluid in a straight line through the element which is usually made up of one or more fine mesh screens attached to a sheet metal core, and thereby trap particles mechanically. Filters, on the other hand,

(a)

Figure 6-6 (a) Typical suction line strainer equipped with ceramic magnets (*courtesy of Schroeder Brothers Corporation*); (b) Return line filter (*courtesy of Arlon International, Inc.*).

direct the fluid in a tortuous path through one or more layers of a porous element to remove smaller particles and silt. Because filters are more complex than simple strainers in their construction, they have higher pressure losses. For this reason, they are usually placed in the return line to the reservoir rather than in suction lines which might cause the pump to cavitate as they become loaded. Magnets and other active materials are often used to increase the effectiveness of strainers and filters to remove metal contaminants from the fluid. Even bronze and other nonmetallic particles are attracted in the magnetic field. Indicators are frequently located on the filter housing to warn the operator and maintenance personnel that they should be changed. A typical machine-mounted strainer and filter are shown in Fig. 6-6.

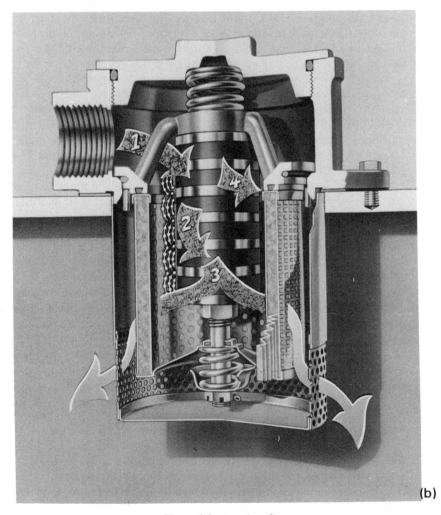

(b)

Figure 6-6 (*continued*)

Strainers remove particles from the fluid down to about 40 micrometers, which is the lower limit of visibility of the naked eye, but filters are capable of removing particles down to one micrometer. Filter maintenance has a marked effect on the life of components since by actual count more than 90% of the contaminants are smaller than 10 micrometers. The relative size of micronic particles magnified 500 times is shown in Fig. 6-7.

Machine-mounted filters can be located in a number of places including high pressure lines, the suction line, a separate filtration circuit, or in the return line before the fluid re-enters the reservoir. Locating a course "full flow" pressure line filter just after the pump, for example, would protect components from large particle wear debris contamination downstream. A "last chance" filter in the pilot-operated circuit of a close tolerance valve would also protect the valve and assure its operation at critical times. However, care must be exercised in the maintenance of filter elements since loading will cause circuit failure, by-passing unfiltered oil, and even bursting of the filter element, releasing trapped contamination to migrate through system components. The type of filtration circuit used then is influenced by the cost of a system failure (for example, before an expensive pump or motor), system pressures, system tolerance for contamination, the degree of filtration desired, and by accessibility for servicing. So long as a portion of the oil is circulated through the filter at all times, where the filter is located is often more important than the type of circuit used because a filter that is not serviced as scheduled is not effective in controlling fluid contamination. Current practice places the main filter in an easily serviced location with a saturation level indicator in a noticeable place so that the operator and maintenance personnel can see its condition. Some filters have "tell-tale" sensors that send warning signals which can be connected to lights, audible devices, or machine disabling switches. Filter location in a circuit and the various types of "last chance" filters are shown in Figs. 6-8 and 6-9.

The condition of the fluid is determined from samples taken from the machine. A number of procedures for sampling and testing are in use. Taking the sample from the machine without introducing additional contamination is very important. Samples can be drawn from the reservoir or from a tap placed in the pressure line. The system should be hot and running at working temperature. Drawing fluid from the system by cracking a joint is not recommended, since this will almost certainly introduce additional contamination with the sample, as well as cause unnecessary disturbance of the plumbing. Placing a special petcock in the system to bleed fluid through a pressure-reducing orifice and coiled tube from the reservoir reduces this possibility and is a better method of drawing the sample. Sample petcocks should be covered, capped when not in use, wiped clean with a lint-free cloth and flushed before drawing the sample.

The cleanliness of hydraulic fluid is described by the number and size of foreign particles it contains, and the standard accepted by the ISO (International Standards Organization) and BSI (British Standards Institute) is ISO/TC 131/SC-6 (1974). The code range number that is assigned to a particular hydraulic fluid to describe its contamination level reflects the number of particles per unit volume greater than five micrometres and 15 micrometres in size. The number of smaller

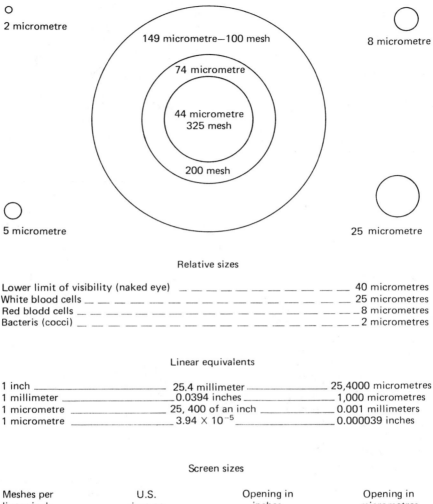

Relative sizes

Lower limit of visibility (naked eye) _ _ _ _ _ _ _ _ _ _ _ _ 40 micrometres
White blood cells _ _ _ _ _ _ _ _ _ _ _ _ _ _ _ _ _ 25 micrometres
Red blodd cells _ _ _ _ _ _ _ _ _ _ _ _ _ _ _ _ _ 8 micrometres
Bacteris (cocci) _ _ _ _ _ _ _ _ _ _ _ _ _ _ _ _ _ 2 micrometres

Linear equivalents

1 inch _____ 25.4 millimeter _____ 25,4000 micrometres
1 millimeter _____ 0.0394 inches _____ 1,000 micrometres
1 micrometre _____ 25, 400 of an inch _____ 0.001 millimeters
1 micrometre _____ 3.94 × 10⁻⁵ _____ 0.000039 inches

Screen sizes

Meshes per linear inch	U.S. sieve no.	Opening in inches	Opening in micrometres
52.36	50	0.0117	297
72.45	70	0.0083	210
101.01	100	0.0059	149
142.86	140	0.0041	105
200.00	200	0.0029	74
270.26	270	0.0021	53
323.00	325	0.0017	44
		0.00039	10
		0.000019	.5

Figure 6-7 Relative size of micrometer particles multiplied 500 times.

size particles gives an indication of the silting condition of the fluid, whereas the large size gives an indication of the number of large particles that promote wear in hydraulic components. Table 6-2 is used to assign code range numbers to actual

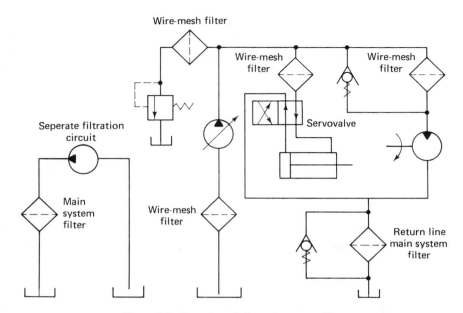

Figure 6-8 Location of the main system filter.

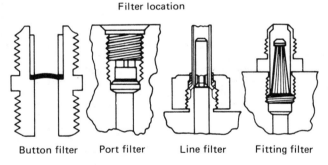

Button filter   Port filter   Line filter   Fitting filter

"Last chance" wire mesh filters

Figure 6-9 Various types of last chance wire mesh filters.

contaminant particle counts in parts per milliliter. For example, a code designation of ISO 17/14 indicates there are approximately 1300 particles of 5-micrometre size and approximately 160 particles of 15-micrometre size per milliliter of fluid. Typical permissible cleanliness levels for fluids from a number of industries are aerospace (ISO 15/12), commercial airplanes (ISO 16/13), mobile heavy equipment (ISO 18/15), and machine tools (ISO 15/12).

Filter performance is measured by the extent to which silt and particulate matter can be removed from the fluid. Filters are proportional in their cleansing action—that is, they filter out only a part of the dirt and silt above a specified size that enters with the fluid at the inlet, rather than filter out all particles and silt above a certain size. Several terms are used by the industry in an effort to standardize ratings and make equal comparisons between filters. These include *abso-*

TABLE 6-2 RANGE CODE NUMBERS
AND CONTAMINATION LEVELS

Maximum no. of particles per mi. represented by range no.	Range no.
10,000,000	30
5,000,000	29
2,500,000	28
1,300,000	27
640,000	26
320,000	25
160,000	24
80,000	23
40,000	22
20,000	21
10,000	20
5,000	19
2,500	18
1,300	17
640	16
320	15
160	14
80	13
40	12
20	11
10	10
5	9
2.5	8
1.3	7
0.64	6
0.32	5
0.16	4
0.08	3
0.04	2
0.02	1
0.01	0.9
0.005	0.8
0.0025	0.7

lute filtration rating, *mean* filtration rating, *nominal* filtration rating, and *filtration ratio*. The unit of measurement used in these ratings is the micrometer, which is approximately 0.000039 in.

The *absolute* filtration rating is given as the micrometer size of the largest hard particle that will pass through the filter under specified test conditions. This indicates the size of the largest opening through the filter element. When a filter is rated at 3μ absolute, this means that it will remove 100% of those particles larger than 3μ and about 99% of those particles larger than 1μ. The *mean* filtration rating is given as the μ size of the average pore in the filter. The *nominal* filtration rating is an arbitrary value given by the filter manufacturer and is not well standardized by the industry. Because of the variance in definition, use of the term is not widely accepted by standards organizations.

The *filtration ratio* is the number of particles greater than the micrometer rating of the filter in the influent fluid (on the upstream side of the filter), to the number of particles greater than the same micrometer size in the effluent fluid (on the downstream side of the filter). The higher the filtration ratio, the more particles above that size that are captured by the filter. A filtration ratio of two means that half the total of all particles above the given micrometer size are captured by the filter. A ratio of one means that the filter is incapable of capturing any particles above the designated size.

Two standard filter designations are given by Greek symbols, alpha (α) and beta (β). Alpha ratings are used to designate coarse filters used for prefiltering and for lubrication systems. Beta ratings are used for fine hydraulic filters. The filtration ratio is derived from ISO Standard 4572 Multipass Filter Performance Test,[2] which measures and compares the contamination level of a test fluid that is injected with AC fine test dust as the specified contaminant upstream of the filter to the level of contamination measured downstream of the filter. The test is stopped when the pressure drop across the filter reaches a predetermined limit—for example 20 lbf/in.[2]—which indicates that the filter has been saturated. A rating of $B_{10} = 12$ means that about 92% of the particles larger than 10 micrometers were captured by the filter when the sample was taken. Obviously, as the filter becomes loaded, the filtration ratio is reduced. The beta ratio is taken, then, as the final reading when the saturation of the filter is reached.

The actual derivation of the beta ratio is

$$\text{Beta ratio} = \frac{\text{Particle count upstream}}{\text{Particle count downstream}} \qquad (6\text{-}5)$$

Example 3

What is the beta ratio for a filter which has 36,815 particles greater than a given size (μ) in the influent fluid, and 6347 particles greater than the same size in the effluent fluid?

Solution   Computing the beta ratio from Eq. 6-5,

$$\text{Beta ratio} = \frac{36,815}{6347} = 5.8$$

The beta efficiency derivation is

$$\text{Beta efficiency} = \frac{\text{Particle count upstream} - \text{Particle count downstream}}{\text{Particle count upstream}} \qquad (6\text{-}6)$$

Example 4

Compute the beta efficiency for Example 3.

[2] Allied United States Standards are ANSI Std. B93.31 and NFPA Std. T3-10.8.

Solution Computing the beta efficiency from Eq. 6-6

$$\text{Beta efficiency} = \frac{36,815 - 6347}{36,815} = 82.76\%$$

The efficiencies for a number of selected beta ratio values are given in Table 6-3.

There are a number of other tests to rate the performance of filter elements which are used to establish structural integrity. These include

1. Fabrication Integrity (A.N.S.I. Std. B93.22, and ISO Std. 2942)
2. Material Compatibility (A.N.S.I. Std. B93.23 and I.S.O. 2943)
3. End Load Test (A.N.S.I. Std. B93.21 and I.S.O. Std. 3723)
4. Flow Fatigue (A.N.S.I. Std. B93.24 and I.S.O. Std 3724)
5. Burst Test (A.N.S.I. Std. B93.25 and I.S.O. Std. 2941)

The condition of the fluid is determined not only by its cleanliness, but by its clarity, color, odor, water content, viscosity, gravity, and neutralization number.

Foreign particle concentration is determined using spectrographic analysis, ferrographic analysis, and particle counting methods. Fluid samples can be analyzed at the establishment which owns the laboratory equipment, or sent to one of the several laboratories which are maintained by independent businesses, research centers, and by the major oil companies which supply hydraulic fluid.

Spectrochemical analysis has been used for some time to measure the concentration of various elements in an oil sample. Both emission and absorption techniques are used to identify the type and concentration of the various elements which constitute the wear debris generated by the system. Ferrographic analysis

TABLE 6-3
EFFICIENCIES FOR
SELECTED BETA
RATINGS

Beta ratio	Efficiency
1.0	0
1.14	12.38
1.5	33.33
2.0	50.00
2.4	58.33
3.0	66.66
4.0	75.00
5.8	82.76
16.0	93.75
17.4	94.25
32.0	96.875
52.2	98.084
100.0	99.0
173.0	99.42

separates wear debris from a fluid sample with a strong magnetic field. Iron as well as other nonmagnetic elements are attracted by the field. Samples are viewed and photographed under a microscope to identify the type and density of material in the wear debris. The test is both quantitative and qualitative in nature. Automatic particle counters size and count particles entrained in a fluid sample that is passed through a photocell light sensor which measures the amount of light blocked by particles being carried by the sample. A particle size analyzer and test set-up for laboratory work are illustrated in Fig. 6-10.

Field test kits are also available to make on-site analyses of fluid samples. These are commonly called patch tests and compare contamination counts of the test fluid sample under a microscope with those of reference fluids or slides. The kit shown in Fig. 6-11 is used for all hydraulic fluids but water glycols and synthetics. In a typical patch test a standard volume of the test liquid is drawn through the membrane filter disc, and an examination of the disc is made under an 80× microscope. The results are compared with a set of reference photos supplied with the kit. The test provides a quick and inexpensive method to observe contamination levels in the fluid. To obtain more exact information about the contamination type and concentration, laboratory analysis is required. A number of the many tests conducted which are used to determine the condition of hydraulic fluids are listed in Appendix F.

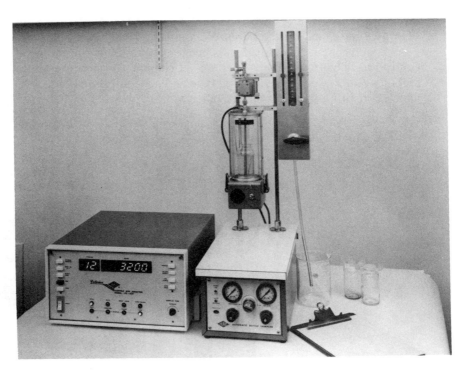

Figure 6-10 Particle size counter test set-up (*courtesy HIAC Instruments Division*).

Figure 6-11 Test kit for hydraulic fluids (*courtesy of Gulf Oil Corporation*).

Contaminated hydraulic fluid may be reclaimed by filtering it at the machine where it is used, or by removing it altogether and having it recycled by the supplier. Where there is a machine failure, or history of component failure, changing to new fluid should also include cycling a flushing fluid through the system to wash out the remainder of the used fluid. This is also necessary when changing from petroleum-based fluids to several of the fire-resistant fluids. Opening and cleaning the reservoir with an access plate is also scheduled with periodic maintenance to remove silt and metallic particles. Filters should be changed after a preset time—for example, after 50 hours of operation with the new fluid—to be sure that the fresh fluid is not carrying unacceptable levels of contamination. Some preventative maintenance programs recently have recommended filtering the new fluid *before* it is put in the machine to assure its cleanliness. Thereafter, the filter is changed using the recommended preventative maintenance schedule, or as its condition indicates.

6-7 SUMMARY

While the major function of the hydraulic fluid is to transmit power throughout the system, to be effective it also must lubricate, dissipate heat, not deteriorate seals, and remain stable over time with changes in temperature and operating conditions. The most important property of the fluid is its viscosity, which increases with pressure and decreases with temperature. The hydraulic fluid must be thin enough to transmit power efficiently, but should not thin out at working temperatures to the extent that it cannot lubricate moving parts with a cushion of oil sufficient to prevent excessive wear and leakage through cylinder rod and shaft seals.

The two worst conditions for a fluid to accommodate are dirt and heat. Large particles of dirt up to 40 micrometers wear the system severely, whereas silt in the 5–10 micrometer range promotes wear by lapping. Heat thins out the fluid causing inefficiency in the transmission of power, wear because of insufficient oil cushion between close fitting parts, and leakage. It also causes deterioration by oxidation.

Fire-resistant fluids are used in hazardous operations such as mining, shipping, and steel mills, where hot surfaces and open flames would cause ignition of petroleum-base oils should a hose break spraying hot fluid. Special care must be exercised to be sure that system seals are compatible with the fire-resistant fluid being added to the system.

The cleanliness of hydraulic fluid is described by the number and size of foreign particles it contains. Filtration removes large particles and silt from the fluid, which promotes wear. A proportion of the dirt and silt are removed by each pass of the fluid through the filter. The multipass filter test measures the filter efficiency by determining the ratio of particles greater than the given micrometer size upstream, to the number of particles greater than the same micrometer size measured downstream. Detailed analysis of the fluid is used to determine the number, size, and composition of the foreign particles it contains, as well as other properties such as specific gravity and neutralization number. Field "patch test" kits also are available to make on-site tests.

STUDY QUESTIONS AND PROBLEMS

1. Name four hydraulic fluids and describe their composition.
2. What are some of the characteristics of an effective hydraulic fluid?
3. What are the two most damaging conditions affecting a hydraulic fluid (conditions that will adversely affect the operation of hydraulic machinery)?
4. What effect does viscosity have on the operation of hydraulic machinery?
5. What is meant by viscosity index and how is it used?
6. Name five operational service-related properties of a hydraulic fluid and describe each.
7. What are the differences among the following: flash point, fire point, and auto ignition temperature?
8. What is the difference between oxidation and foaming in hydraulic oil?
9. How is the cleanliness of a hydraulic fluid measured?
10. Name three terms used to rate the effectiveness of filters.
11. How is the patch test used to analyze hydraulic fluid? What are its limitations?
12. What is meant by bulk modulus of elasticity, and what service property does it measure?
13. What is meant by filtration ratio?
14. What happens when a filter becomes loaded with contaminant?

15. Compute the viscosity index for an oil that has an SSU of 1000 at 100°F and 80 at 210°F.

16. An oil has a viscosity index of 100. If the viscosity is 150 at 210°F, what would be the expected viscosity at 100°F?

17. As a fluid becomes more compressible (spongy), does the bulk modulus increase or decrease?

18. A compression test causes a fluid to reduce in volume one percent when the pressure rises from 1000 to 5000 psi. What is its bulk modulus?

19. It is noticed that air ingested through a pump inlet leak causes the stroke of a forklift cylinder with a 4-in. I.D. bore and 24-in. stroke to vary 1/4 in. at full stroke when the pressure changes from 2000 to 4000 lbf/in^2. What is the bulk modulus of the fluid?

20. What is meant by a beta ratio of $B_{10} = 15$?

21. What is the contamination level of a hydraulic fluid with a range code of 18/15?

22. What would be the range numbers for a hydraulic fluid with 5000 particles greater than 5 microns and 320 particles greater than 15 microns?

23. If by actual count at the final pressure reading there are 40,000 particles greater than 10 microns upstream of the filter and 7500 particles greater than the same size downstream of the filter, what is the *beta ratio*?

24. What is the *beta efficiency* for the filter in Problem 23?

25. What is the relationship between *beta ratio* and *beta efficiency*?

CHAPTER 7

PUMPS

7-1 INTRODUCTION

Hydraulic pumps raise the energy level of the fluid. Pumping action involves the transfer of fluid from the low energy supply in the reservoir to the closed hydraulic system where high energy fluid is used to accomplish work. It is then returned to the reservoir. The energy or work equivalent added to the fluid by the pump can be accounted for by the volume of fluid from the pump outlet, and the pressure at which it is discharged. That is,

$$\text{Total Energy} = \text{Pressure} \times \text{Volume}$$

$$\text{Total Energy} = \frac{\text{Force}}{\text{Area}} \times \text{Area} \times \text{Distance}$$

and

$$\text{Total Energy} = \text{Force} \times \text{Distance}$$

where the force is in lbf (N) and the distance is in ft (m). For the purpose of explanation, the total output volume of the pump may be visualized as the total input volume of an infinitely large cylinder. This volume equals the area of the cylinder bore multiplied by some stroke length.

Pumps receive energy from outside the hydraulic system. Industrial hydraulic applications, such as manufacturing plant machinery, are driven by electric motors at one of the alternating current synchronous speeds. Common speeds are 1,200 and 1,800 rpm. Mobile hydraulic applications, such as power steering units, backhoe pumps, and other construction equipment, are driven at varying speeds up to 3,600 rpm, depending on the size of the equipment, by power take-off arrangements directly from the internal combustion engine or the transmission. Air and space craft hydraulic systems employ high speed pumps driven by direct current electric motors to speeds of 12,000 rpm, where weight considerations are critical and substantial power must be generated from small self-contained units.

The output of the pump constitutes the total source of energy available within the hydraulic system. The total of the energy outputs from the system by components such as the drive motor and hydraulic cylinders equals this total energy available from the pump if losses are ignored. That is,

$$\text{System Energy Output} = \text{Pump Energy Output}$$

In practice, the total fluid energy output from the pump always exceeds the total energy available to the fluid power motors and cylinders because of friction and heat losses within the system. The proper determination of the correct pump for a given application essentially involves matching the pump to the required fluid power actuator giving consideration to pressure, flow rates, available drive speeds and power, efficiencies, cost, size, maintenance, and the operational characteristics of the pump and system. These include pump noise, vibration, natural frequency, and flow characteristics.

7-2 PUMPING THEORY

A fluid is pumped when its volume is displaced and transferred from one place to another. The pumping action can be positive, as when an exact amount of fluid is displaced for each pump revolution or cycle, or it can be displaced and transferred using the inertia of the fluid in motion.

When the pumping action displaces a specified amount of fluid per revolution, it is referred to as a positive displacement pump. Gear, vane, piston, and screw pumps are positive displacement pumps. Pumps using the inertia principle to propel the fluid are nonpositive displacement pumps and include both centrifugal and propeller pumps. Positive displacement pumps are required for medium and high pressures of 500 lbf/in.² and up, while nonpositive displacement pumps are widely used for low pressure high-volume applications up to 600 lbf/in². Recent developments in nonpositive displacement pumps have extended their pressure range to above 1000 lbf/in². Positive displacement pumps are used where the primary consideration is one of pressure and power output, whereas nonpositive displacement pumps are used where high volume output or transfer of the fluid is of primary importance. Because of their simple design and fewer number of moving parts, nonpositive displacement pumps cost less to install and operate with less maintenance.

Essentially, a pump consists of a low pressure hydraulic fluid source, a pumping chamber attached to a drive mechanism, and a high energy outlet. Energy is added to the fluid in the pumping chamber by the action of the mechanical drive. In positive displacement pumps, the high and low pressure sections of the pumping chamber are separated so that the fluid cannot leak back and return to the low pressure side. The pumping chamber in nonpositive displacement pumps is connected so that as pressure increases, fluid within the pumping chamber circulates.

Nonpositive displacement pumps make use of Newton's first law of motion to move the fluid against system resistance. This law states, in effect, that a body in motion tends to stay in motion. The action of the mechanical drive in the pumping chamber speeds up the fluid so that its velocity accounts for its ability to move against the resistance of the system. In centrifugal pumps, rotational inertia is imparted to the fluid, whereas in propeller pumps, the inertia imparted to the fluid is translational. For a given pump size, the volume of fluid pumped is dependent on the speed of the rotating member and the resistance or pressure the pump must overcome.

Positive displacement pumps cause the fluid to move by varying the physical size of the pumping chamber in which the fluid is sealed. The low pressure inlet side of the pumping chamber increases in volume, a vacuum forms, and atmospheric pressure causes the hydraulic fluid to enter. As fluid moves through the pump and to the high pressure outlet, its volume is reduced, causing it to be expelled. Positive displacement pumps alternately increase and then decrease the volume of the pumping chamber to accomplish the pumping action. Since the volume per cycle is fixed by the positive displacement characteristic of the pumping chamber, the volume of fluid pumped for a given pump size is dependent only on the number of cycles made by the pump. Leakage is a factor, but is considered as a loss in volumetric efficiency and treated as a separate problem.

The ratio of the output from a pump to the input to the pump is a measure describing overall efficiency of the unit, and it takes into account such factors as leakage or pump cavity flow and losses due to mechanical friction. It is computed from

$$\text{Overall Pump Efficiency } (e_o) = \frac{\text{Output}}{\text{Input}} \times 100 \quad \text{as a percent} \qquad (7\text{-}1)$$

Overall pump efficiency is thus computed from

Overall Pump Efficiency $(e_o) =$

$$\frac{\text{Pump Output Fluid Horsepower } (Fhp_o)}{\text{Pump Input Brake Horsepower } (Bhp_i)} \times 100 \qquad (7\text{-}2)$$

$$e_o = \frac{\dfrac{p \times Q}{1{,}714}}{\dfrac{T_o \times N}{5{,}252}} \times 100 \qquad (7\text{-}3)$$

and

$$e_o = \frac{p \times Q}{T_o \times N} \times 3.06 \times 100 \text{ as a percent} \qquad (7\text{-}4)$$

where the overall efficiency (e_o) is computed as a percent, the pressure (p) is in lbf/in.^{2}, the flow rate (Q) is in gal/min, the observed torque (T_o) is in lbf-ft, and the speed (N) is in rpm. If the torque is required in lbf-in., which is commonly the case,

$$e_o = \frac{p \times Q}{T_o \times N} \times 36.7 \times 100 \qquad (7\text{-}5)$$

Example 1

A pump operates at 2,500 lbf/in.^{2} and discharges 3 gal/min. It requires 5 Bhp to drive the pump. Compute the overall efficiency of the pump. If the pump is driven at 1,725 rpm, what is the input torque to the pump?

Solution   From Eq. 7-2 for overall pump efficiency

$$e_o = \frac{\dfrac{p \times Q}{1,714}}{Bhp_i} \times 100$$

and

$$e_o = \frac{2,500 \times 3}{1,714 \times 5} \times 100 = 87.5\%$$

The torque input to the pump at 1,725 is computed from the brake horsepower input formula

$$Bhp_i = \frac{T_o \times N}{5,252}$$

$$T_o = \frac{Bhp_i \times 5,252}{1,725} = 15.22 \text{ lbf-ft or } 182.7 \text{ lbf-in.}$$

Theoretical pump torque is expressed in terms of pump displacement and a specified operating pressure. That is,

$$\text{Theoretical Torque } (T_t) = \frac{\text{Displacement } (V_p) \times \text{Specified Pressure } (p)}{2\pi}$$

and

$$T_t = \frac{V_p \times p}{2\pi} \tag{7-6}$$

where the theoretical torque (T_t) is expressed in lbf-in., the displacement (V_p) is in in.^{3} per revolution and the pressure (p) is in lbf/in^2. Expressed in units

$$T_t = \frac{\text{in.}^3 \times \text{lbf}}{2\pi \times \text{in.}^2}$$

and

$$T_t = \frac{\text{lbf-in.}}{2\pi} \tag{7-7}$$

If torque in lbf-ft is required

$$T_t = \frac{V_p \times p}{24\pi} \tag{7-8}$$

In SI metric units, theoretical torque is expressed in $N \cdot m$ from the calculation

$$T_t = \frac{(m^3/\text{rev})(N/m^2)}{2\pi} = \frac{N \cdot m}{2\pi} \tag{7-9}$$

Where the displacement of the pump is given in liters/rev and the pressure is in kPa instead of Pa (N/m^2),

$$T_t = \frac{(\text{liters}/1000)(N/m^2\ 1000)}{2\pi} = \frac{\text{liters} \times kPa}{2\pi} \qquad (7\text{-}10)$$

which, conveniently, gives the same result since liters $\times\ 0.001 = m^3$, and kPa $\times\ 1000 = $ Pa.

Torque efficiency (e_t), then, is computed from the ratio of the output torque (T_o) to input torque (T_i).

$$e_t = \frac{T_o}{T_t} \times 100 \quad \text{as a percent} \qquad (7\text{-}11)$$

where T_o is the observed torque and T_t is the theoretical torque. Torque efficiency is the same as mechanical efficiency.

Leakage or pump cavity flow in positive displacement hydraulic pumps is computed in gal/min from the volumetric efficiency where

Pump Volumetric Efficiency (e_v)

$$= \frac{\text{Pump Volume Output } (Q_o)}{\text{Pump Displacement } (V_p) \times \text{Speed } (N)} \times 100$$

$$e_v = \frac{Q_o}{\dfrac{V_p \times N}{231}} \times 100$$

and

$$e_v = \frac{231 \times Q_o}{V_p \times N} \times 100 \text{ as a percent} \qquad (7\text{-}12)$$

where the observed output volume flow of the pump (Q_o) is measured in gal/min, the displacement of the pump (V_p) is measured in in.^{3}, the speed of the pump (N) is measured in rpm, and the constant 231 converts in.^{3} to gal.

In SI units, if the delivery of the pump (Q_o) and displacement (V_p) are measured in liters per minute and liters respectively, the volumetric efficiency is simply

$$e_v = \frac{Q_o}{V_p \times N} \times 100 = \frac{(\text{liters/min})}{(\text{liters/rev})(\text{rev/min})} \times 100 \quad \text{as a percent} \qquad (7\text{-}13)$$

Example 2

Compute the volumetric efficiency of a pump that has a positive displacement of 3.75 in.^{3} and delivers 52 gal/min of fluid while operating at 3300 rpm.

Solution   Volumetric efficiency is computed from Eq. 7-12

$$e_v = \frac{231 \times Q_o}{V_p \times N} \times 100$$

$$e_v = \frac{231 \times 52}{3.75 \times 3300} \times 100 = 97.07\%$$

Losses due to mechanical friction within the pump are usually not computed directly because it is not easy to construct a means of measurement. Essentially, it would require driving the pump dry and computing the torque through the range of operating speeds necessary to drive the pump, and then computing the frictional horsepower. Other friction factors related to the presence of the fluid, however, would not be available. For these reasons, friction and mechanical efficiency are usually computed from overall and volumetric efficiencies. Since the overall efficiency is the product of all contributing efficiencies, that is,

Pump Overall Efficiency (e_o) = Pump Volumetric Efficiency (e_v)

$\times$ Pump Mechanical Efficiency (e_m)

$$e_o = e_v \times e_m \qquad (7\text{-}14)$$

and

$$e_m = \frac{e_o}{e_v} \quad \text{as a percent} \qquad (7\text{-}15)$$

This accounts for all losses that occur from the point where the coupling connects the mechanical drive to the pumping chamber, to the discharge port of the pump where output pressure and flow rate are available. The pump drive is usually an electric motor coupled directly to the pump input shaft. Frictional horsepower losses are computed as a portion of the brake horsepower supplied at the pump drive shaft from

Pump Frictional Horsepower (Fhp_p)

= Pump Brake Horsepower (Bhp_i)

$[1 -$ Pump Mechanical Efficiency (e_m)]

and

$$Fhp_p = Bhp_i(1 - e_m) \qquad (7\text{-}16)$$

Example 3

A hydraulic pump is observed to have an overall efficiency of 87% and a volumetric efficiency of 94% while consuming 10 Bhp measured at the pump drive shaft. Compute the mechanical efficiency and the frictional horsepower loss from the system.

Solution The mechanical efficiency is computed from Eq. 7-15:

$$e_m = \frac{e_o}{e_v} \times 100$$

and

$$e_m = \frac{87}{94} \times 100 = 92.55\%$$

The frictional horsepower loss from the system is computed as a percent of the power input at the pump drive shaft, and is the product of the input horsepower and the loss in mechanical efficiency of the pump. From Eq. 7-16,

$$Fhp_p = Bhp_i(1 - e_m)$$

$$Fhp_p = 10(1 - 0.92)$$

and

$$Fhp_p = 10 \times 0.08 = 0.8 \text{ bhp}$$

From this example it is seen that friction losses account for about 8% of the input at the pump drive shaft.

7-3 PUMP TYPES

Nonpositive displacement hydraulic pumps are divided into centrifugal, axial propeller and mixed flow pumps. Positive displacement pumps are divided into gear, vane, screw, and piston pumps. Positive displacement pumps are further classified as fixed or variable displacement and noncompensated or compensated design (see Fig. 7-1 for pump classification).

Rotary pumps transfer fluid from the low pressure inlet port to the high pressure outlet port through the motion of a rotating drive shaft and internal transfer mechanism. Gear, vane, screw, and some piston pumps are a part of this classification. Pumps are considered to be of the rotating type if the internal drive element transmits fluid by a rotating rather than a reciprocating motion.

Nonpositive Displacement Pumps: Centrifugal pumps are used to pump large volumes of fluids at relatively low pressures. Like other nonpositive displacement pumps, they share the common advantages of low production cost, simplicity of operation, high reliability, low maintenance factors, low noise level, and the ability to pump nearly all fluids without damage to internal parts. Because the inlet and outlet passages are connected hydraulically, centrifugal pumps are not self-priming and must be positioned below the level of the fluid or primed to start the pumping action. Most have large internal clearances, although some water

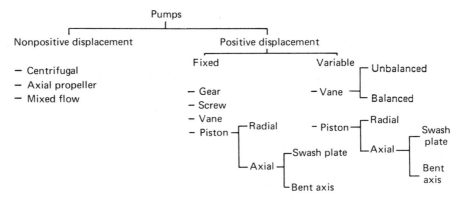

Figure 7-1 Pump classification.

circulating units use flexible impeller blades that reduce internal clearances to increase pumping efficiency.

Centrifugal pumps have their greatest application as supercharging pumps, as liquid transfer pumps and in low-pressure hydraulic applications requiring high fluid flow rates, such as in traverse feed mechanisms.

The parts of the centrifugal pump include an inlet port, a pumping chamber shaped as an involute, the drive shaft and impeller, and the outlet port (Fig. 7-2).

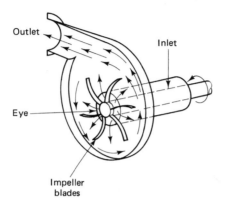

Figure 7-2 Centrifugal pump components.

In operation, rotation of the impeller imparts a centrifugal force to the fluid which slings it outward from the center of the impeller. The blades are curved opposite to the direction of rotation to increase the efficiency. Defuser blades are often attached stationary to the pump housing, and they redirect the fluid in such a way as to reduce the velocity and internal clearances thereby increasing the ability of the pump to develop pressure against external resistance to fluid flow (Fig. 7-3).

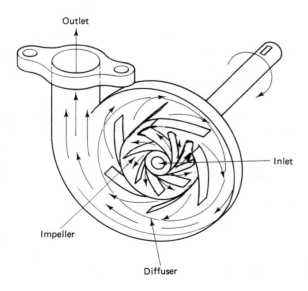

Figure 7-3 Defuser action.

While centrifugal pumps are used primarily for low-pressure applications, staging in series may be used to increase system pressure. Like other series circuits, the pressure is additive while the volume flow rate is the same through each of the pumps placed in series.

Propeller pumps are also nonpositive displacement pumps (Fig. 7-4). Fluid is swept along by the action of the close-fitting propeller in the pump housing. The fluid moves axially with respect to the direction of the propeller drive shaft. Propeller pumps are installed in water distribution pipes to raise the line pressure.

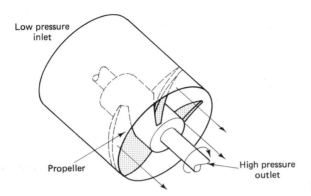

Figure 7-4 Propeller pump.

Fixed Delivery Positive Displacement Pumps: Positive displacement pumps transfer a constant amount of fluid for each cycle of operation. If this internal pump volume cannot be adjusted, the pump is considered to have a fixed displacement. If the internal volume of the pump is adjustable, the pump is considered to be a variable volume displacement pump. While the drive speed can also be used to change the volume flow rate for a positive displacement pump, this is a less desirable alternative than varying the internal displacement to change pump output.

External gear pumps are positive fixed displacement rotary pumps (Fig. 7-5). They are simple in design and comparatively inexpensive to manufacture and maintain. The pumping action occurs in gear pumps when the input drive causes one gear to turn the other. This action, in turn, causes fluid to be displaced from the inlet port to the outlet port in the following manner. The gear teeth seal as one rotates the other. As the teeth part on the suction side of the pump near the low pressure inlet port, they increase the volume of the inlet chamber causing a slight vacuum. The rotating gear teeth transfer fluid, forced in by atmospheric pressure and trapped in the gear teeth, around the outside periphery of the gears to the high pressure outlet chamber. Meshing of the teeth in the outlet chamber reduces the cavity volume by an amount equal to that displaced between the teeth as they mesh. This forces fluid from the outlet cavity and port at system pressure.

Spur gear pumps are an unbalanced design and cause side forces on the gears and supporting shaft, limiting both the pressure and the speed at which the pump can be operated. Balancing the external gear pump requires that passages be drilled either through the gears themselves, or in the end plates to equalize the

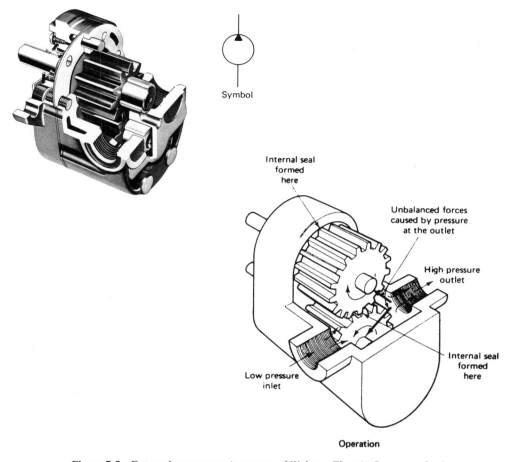

Symbol

Internal seal
formed
here

Unbalanced forces
caused by pressure
at the outlet

High pressure
outlet

Internal seal
formed
here

Low pressure
inlet

Operation

Figure 7-5 External gear pump (*courtesy of Webster Electric Company, Inc.*).

pressure in opposite directions on the gears (Fig. 7-6). Most gear pumps provide no means for balancing the out-of-balance forces on the gears and supporting shafts.

Internal gear pumps operate similarly to external gear pumps. The internal spur gear drives the outside ring gear which is set off center. Between the two gears on one side is a crescent-shaped spacer around which oil is carried. The inlet and outlet ports are located in the end plates between where the teeth mesh and the ends of the crescent-shaped spacer. In operation, oil is directed from the inlet to the outlet port in the following manner (Fig. 7-7). The internal gear drives the external ring gear and makes a fluid tight seal at the place where the teeth mesh. Rotation causes the teeth to unmesh near the inlet port, the cavity volume to increase, and suction to occur. Oil is trapped between the internal and external gear teeth on both sides of the crescent-shaped spacer and is carried from the inlet to the outlet cavity of the pump. Meshing of the gear teeth reduces the volume in

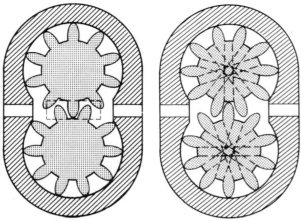

Drilled passages
equalize pressure

Figure 7-6 Balancing gear pumps
(*reprinted by permission of Machine
Design*).

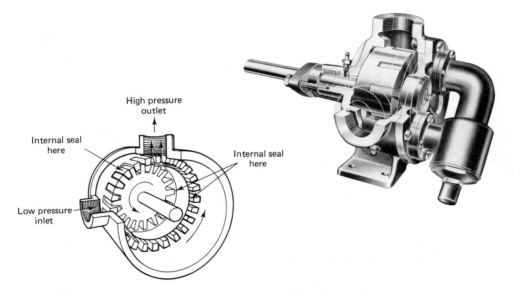

High pressure
outlet

Internal seal
here

Internal seal
here

Low pressure
inlet

Figure 7-7 Internal gear pump (*courtesy of Viking Pump Division, Houdaille
Industries, Inc.*).

the high pressure cavity near the outlet port and fluid exits from the outlet port.
Wear on internal gear pumps has a tendency to reduce the volumetric efficiency
more quickly than on external gear pumps. They are used mostly as lubrication
and charge pumps at pressures under 1000 lbf/in.^{2}, although staging increases
their pressure range to 4000 lbf/in^2.

Gear pumps may be operated singly, in tandem (Fig. 7-8), or mounted in
piggyback fashion to alter both pressure and volume characteristics. The speed

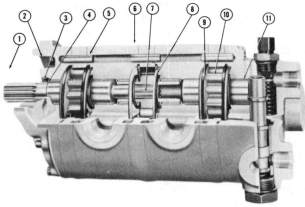

1. Shaft
2. Case screws
3. Double lip seal
4. Seal chamber
5. Aligning dowel pins
6. Aluminium casting

7. Splined shaft coupling
8. Pressure loaded seals
9. Pressure loaded end plates
10. Hardened gears
11. Needle bearings

Figure 7-8 Tandem gear pump (*courtesy of Webster Electric Company, Inc.*).

range for industrial external gear pumps is 1,500–2,500 rpm with pressures to 4,000 lbf/in.² available. Volume flow rates from less than 1 gpm to more than 100 gal/min and horsepower ratings to 100 bhp are available.

Screw pumps transfer fluid by displacing it axially through close-fitting chambers (Fig. 7-9). They are fixed displacement pumps. The only moving parts are the screws themselves. In a typical three rotor unit, one rotor transmits power

Figure 7-9 Screw pump (*courtesy of De Laval Turbine, Inc.*).

while the remaining two rotors act as idlers. The power rotor is the only driven element in the pump and is extended through the pump housing to connect to the pump drive. The idler rotors turn because of the action of the fluid transferred through the pump. They act as sealing elements and perform no work themselves.

Pumping action occurs when the meshing of the rotors seals and enfolds the fluid and then transmits it in the enclosures continuously in a uniform manner. The meshing of the screw rotors is a rolling action and this accounts for the quiet operation of these pumps. There are no internal adjustments for wear. Compensating movement within the pump accounts for its high reliability. Side thrusts, often associated with gear and vane pumps, are not present in screw pumps. Usually, only one main bearing is required at the pump housing to support the drive rotor and position the seal, which is isolated from direct exposure to the fluid. The unbalanced force within the screw pump is axial and on the center line of the drive rotor. This end thrust is supported by a thrust plate bearing surface near the drive end of the rotor shaft.

Several desirable characteristics are inherent in the design of screw pumps. They are quiet because the rolling action of the rotors eliminates the hydraulic whine, noise, and vibration traditionally associated with many positive displacement pumps. Nearly all fluids are compatible with the pump because only minimum lubrication properties are necessary. Water in oil emulsions and HWCF fluids are common examples which present a pumping problem because of minimum lubricity. Because the internal parts are few and rugged, these pumps are highly reliable. High speed operation is an additional advantage, and is controlled only by the ability of the fluid to fill the rotor cavities at the pump inlet. While commercial speeds of 3500 rpm are recommended, much higher speeds with special units are available. This feature allows direct coupling to drive motors. Screw pumps produce a nonpulsating flow. There is no churning, pump turbulence or pocketing of the fluid caused by rapid expansion and contraction of the pumping chamber, or by the reciprocating strokes of pump pistons. This accounts for the elimination of localized destructive forces sometimes encountered in other pump designs.

Screw pumps are used on a variety of production machinery to transfer fluid at medium to high pressures. Die casting machines, hydraulic presses, machine tools, and broaching machines, using volumes to 500 gal/min and pressures to 3000 psi, are common applications. For applications requiring high volumes, several screw pumps are sometimes mounted and operated in tandem.

Gerotor (generated rotor) pumps (Fig. 7-10) are a version of the internal gear pump. They have been used historically as lubricating pumps for internal combustion engines in the automotive industry and as low-pressure charging pumps, but recent advances in metal technology have extended their pressure range for use in fluid power in the pressure range of 1000–2000 lbf/in.^{2} per stage. Notice that the inner driving rotor has one less tooth than the outer driven rotor. This allows the internal volume of the pump to expand and contract by a total amount equal to the displaced tooth volume. The seal between the tip of the internal teeth and the driven rotor is caused by the driving action of the internal tooth. Gerotor

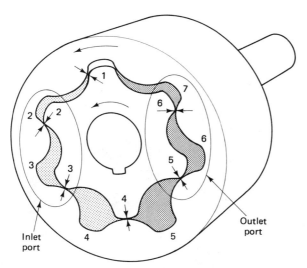

Figure 7-10 Gerotor pump.

pumps are made in displacements from 0.1 in.^{3} to 11.5 in.^{3} for flows of 0.5 to 60 gal/min.

Vane pumps are classed as fixed or variable displacement and unbalanced or balanced design. Four pump combinations are available:

1. Fixed displacement, unbalanced design.
2. Fixed displacement, balanced design.
3. Variable displacement, unbalanced design.
4. Variable displacement, balanced design.

The essential components of the vane pump include the inlet and outlet ports, driven rotor, sliding vanes, and stationary cam ring. The rotor, vanes, and cam ring, and sometimes the end wear plates, are replaceable as a cartridge unit (Fig. 7-11).

Pumping action occurs when the sliding vanes in the rotor alternately increase and decrease the crescent-shaped space between the rotor and the cam ring. The increasing volume causes suction at the inlet port and fluid to enter the low pressure inlet cavity. The inlet and outlet ports are isolated from each other by the spacing of the vanes. At the crossover point, the fluid volume is reduced in the outlet cavity at system pressure. The vanes in the rotor slots alternately extend and retract as the rotor turns through each half revolution, trapping fluid from the inlet port and transferring it to the outlet port.

Fixed displacement, unbalanced pumps permit pumping action on one side of the rotor only. This places a side thrust on the rotor and support bearings and requires robust support from the driving shaft and end bearings.

Pump vanes are located in the slotted rotor and follow the contour of the cam ring. Both centrifugal and the additional force from system pressure are

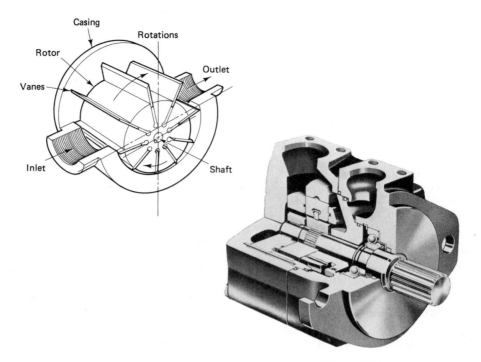

Figure 7-11 Vane pump components (*courtesy of Sperry Vickers*).

applied to the cross section of the vanes. During start-up, only centrifugal force accounts for locating and holding the vanes against the inner surface of the cam ring. A minimum start up speed is required to cause proper suction and pumping action.

Fixed displacement balanced design vane pumps locate two inlet ports and two outlet ports on opposite sides of the rotor (Fig. 7-12). The cam ring has an oval or elliptical shape. Both sides of the rotor have equal pressure exerted in opposite directions which has the effect of balancing out the forces. Lighter bearings are used with balanced pump designs rather than with unbalanced pump designs, since pump bearings are required to support only the external drive loads.

One method of hydraulically balancing the action of the pump employs an intravane that receives continuous outlet pressure applied to the projected cross section area of the intravane (Fig. 7-13). This assures a positive means of vane tracking by using system pressure to exert an outward force on the vane during operation and permits higher pressure without slippage.

Another design to balance vane pumps employs dual vanes in each rotor slot (Fig. 7-14). Pump vanes have tapered ends and a chamfered slot at their mating surfaces. The result is that system pressure is used not only to exert an outward force on the vanes, but to cause the vanes to seal against the rotor slot. Both of these features result in higher volumetric efficiencies. Having two rotor tips in

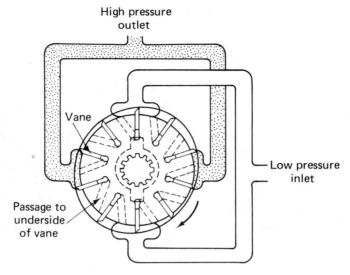

Figure 7-12 Fixed displacement pump design (*courtesy of Sperry Vickers*).

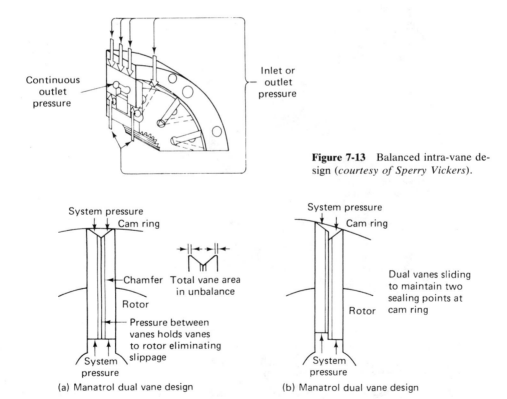

Figure 7-13 Balanced intra-vane design (*courtesy of Sperry Vickers*).

(a) Manatrol dual vane design

(b) Manatrol dual vane design

Figure 7-14 Dual vane design (*courtesy of Perker Hannifin*).

each rotor slot operating against the cam also reduces slippage at the ends of the vanes.

Figure 7-15 illustrates a recent concept in vane pump technology. Vanes in this rotary positive displacement pump are operated by pushrods, so the pump needs no springs or centrifugal or pressure loading that could increase friction. Opposing pump vanes are interconnected by pushrods and ride in a contoured rotor. As the rotor turns, the vanes follow the contour and are pushed back and forth by the pushrods. Thus, when one vane retracts, the opposite vane extends and pumps fluid through the outlet port. The simplified design features low rpm and high torque, and because the outer member rotates about the stator, various driving mechanisms including gear, belt, and chain drive can be utilized.

Fixed displacement piston pumps can be classified from the arrangement of the piston and drive mechanism and include axial, radial, and in-line crankshaft driven pumps. Axial piston pumps are further classified as in-line or bent axis drive piston pumps. Two in-line axial piston pump designs are common: the swash plate design and the check valve design.

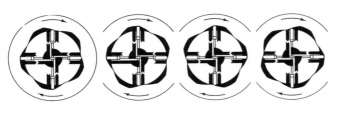

Figure 7-15 Compact vane pump design (*courtesy of Federal Brass Manufacturing Co., Inc.*).

Pumps   Chap. 7

In-line axial piston pumps operate at pressures to 3,000 lbf/in.[2] with high volumetric and overall efficiencies (above 96% and 85% respectively) because of the close fit between the reciprocating pistons and the cylinder bores. While speeds in the 3,000–4,000 rpm range are nominal for industry, some airborne applications are capable of speeds over 15,000 rpm.

Component parts of the swash plate design include the inlet and outlet ports, pump housing, valve plate, rotating group, shaft seal, bearings, and drive shaft (Fig. 7-16). The rotating group consists of the splined rotating cylinder block and spherical washer, cylinder block spring and force transmitting pins, pistons supported by shoes, swash plate, and retainer plate. Pumping action occurs as the rotating group causes the pistons to reciprocate as they follow the angle of the stationary swash plate. As the pistons retract in the cylinder bores, the rotating group passes over the kidney-shaped inlet port in the valve plate and fluid is inducted into the cylinder bore. Further rotation causes the piston to extend into the cylinder bore as the rotating group passes over the kidney-shaped outlet port in the valve plate and fluid is expelled.

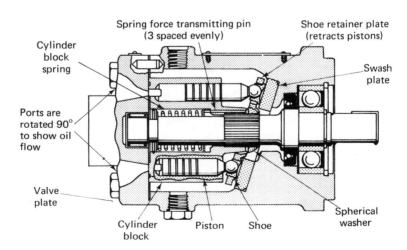

Figure 7-16 Fixed displacement swash plate piston pump (*courtesy of Sperry Vickers*).

Delivery from the pump (Q) is determined by the number of cylinders (n), the speed of the pump (N_t), the bore (d_b) and the stroke (S). Stroke, in turn, is dependent on the angle of the swash plate. For fixed displacement axial piston pumps, 30° is a common swash plate angle.

Example 4

What is the theoretical volume flow rate for a fixed displacement axial piston pump with a nine-bore cylinder group with a 0.5-inch bore and 0.75-inch stroke operating at 3000 rpm?

Solution

Volume Flow Rate (Q_t)

$$= \frac{\text{No. of Cyls. } (n) \times \text{Rpm } (N) \times \text{Stroke } (S) \times \text{Bore Area } (A)}{231}$$

$$Q_t = \frac{n \times N \times S \times A}{231}$$

$$Q_t = \frac{9 \times 3000 \times 0.75 \times 0.196}{231}$$

and

$$Q_t = 17.2 \text{ gal/min}$$

Fixed displacement bent axis pumps reciprocate the pistons in the rotating cylinder block through a bevel gear mechanism. The axis is tilted near the junction of the drive shaft and cylinder block. Displacement angles of 15°, 20°, 25°, 30°, and 40° are standard. These pumps operate at high efficiencies and have received widespread application in the aerospace industry. Units range in sizes from 2 to 45 lbf and flow rates of .25 to 150 gal/min at pressures to 5,000 lbf/in². The speed range is from 100 rpm to 12,000 rpm, depending on the size and

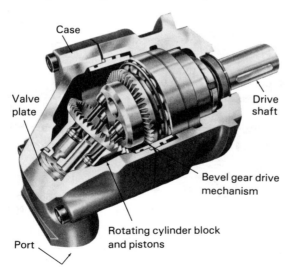

Case

Valve plate

Drive shaft

Bevel gear drive mechanism

Rotating cylinder block and pistons

Port

Figure 7-17 Fixed displacement bent axis pump (*courtesy of Volvo of America Corporation*).

application. Components include the rotating group, valve plate, antifriction bearings, and aluminum case (Fig. 7-17).

Fixed displacement radial piston pumps locate the pistons around the pump drive shaft at a right angle (Fig. 7-18). The pistons reciprocate in a rotating cylinder block. While many arrangements are used to accomplish the pumping and valving action, typically the pistons are attached to sliding shoes or rollers that follow an outer ring located off-center with respect to the rotating cylinder block. Rotation causes the pistons to reciprocate in the block. Appropriate valving during the inlet and discharge strokes is accomplished by having the cylinder block

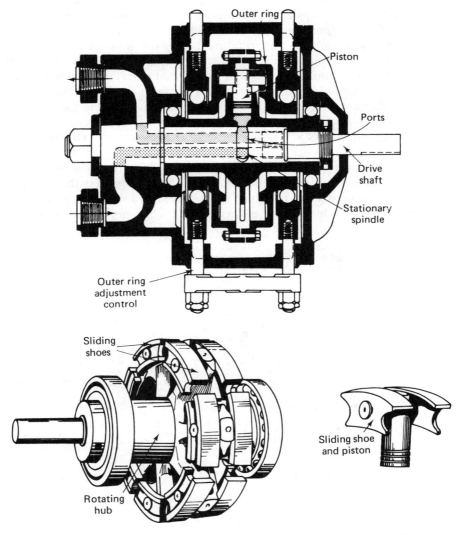

Figure 7-18 Fixed displacement radial piston pump (*courtesy of Mobil Oil Corporation*).

rotate on a fixed pintle (hollow shaft support) with internal inlet and discharge passages appropriately indexed so that when the pistons retract and extend in the cylinder bores, fluid is inducted and discharged through the inlet and outlet passages of the pintle and pump housing.

Variable Delivery Positive Displacement Pumps: Variable displacement vane pumps are usually of the unbalanced design (Fig. 7-19). The cam ring is round, permitting fluid to be inducted for one half-revolution and discharged for the other half-revolution. As in unbalanced fixed displacement vane pumps, this places a side thrust on the drive shaft and support bearings. The displacement in variable volume vane pumps is changed by moving the cam ring from the extreme eccentric off-center position where the displacement is maximum to the center position where the rotor and cam ring are concentric and the displacement is 0. Side thrusts on the pump shaft and bearings are proportional to both the degree of offset of the cam ring and pump speed that determine volume flow rate, and to system pressure. At the extreme offset position of the cam ring and maximum volume flow rate, system pressure is lowest, whereas in the concentric position, the volume flow rate is approximately 0 and the pressure is highest. Maximum side thrust occurs where the product of pressure and volume flow rate are the highest.

Variable volume vane pumps have widespread use in industry. Because they can infinitely vary the volume of the fluid pumped in the system, the need for unloading valves and plumbing often necessary for dumping system fluid is eliminated. Power loss and heat associated with unproductive pumping of fluid

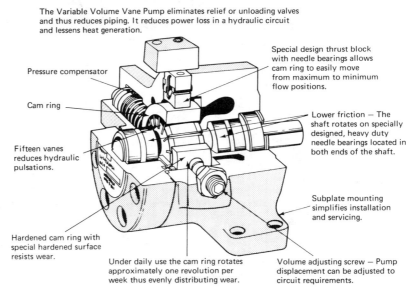

The Variable Volume Vane Pump eliminates relief or unloading valves and thus reduces piping. It reduces power loss in a hydraulic circuit and lessens heat generation.

Special design thrust block with needle bearings allows cam ring to easily move from maximum to minimum flow positions.

Pressure compensator

Cam ring

Lower friction — The shaft rotates on specially designed, heavy duty needle bearings located in both ends of the shaft.

Fifteen vanes reduces hydraulic pulsations.

Subplate mounting simplifies installation and servicing.

Hardened cam ring with special hardened surface resists wear.

Under daily use the cam ring rotates approximately one revolution per week thus evenly distributing wear.

Volume adjusting screw — Pump displacement can be adjusted to circuit requirements.

Figure 7-19 Variable displacement vane pump (*courtesy of Miller Fluid Power Division, Flick-Reedy Corporation*).

through relief valves is also reduced through pressure, flow, and load compensation.

Figure 7-20 illustrates a cutaway view of the pressure compensation mechanism for a vane pump of the unbalanced design. The maximum displacement setting is established by the stop-nut setting on the right side of the case which limits the travel of the cam ring. In the position shown, the cam pressure ring gives the pump maximum displacement. As the pressure builds between the pressure ring and the chambers between the cam pressure ring and the chambers that are formed between the vanes, it causes the cam ring to move to the left which compresses the governor spring and reduces the displacement of the pump. How

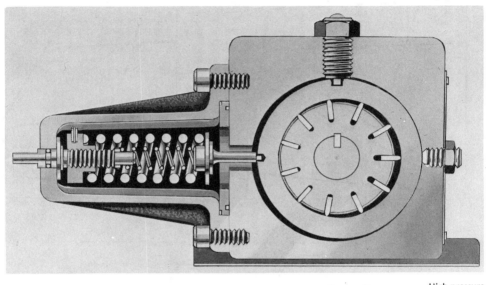

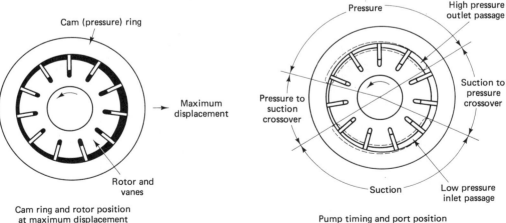

Figure 7-20 Cutaway of variable volume vane pump compensator (*courtesy of Racine Hydraulics Division, Dana Corporation*).

far the cam ring moves to the left is determined by the system pressure, spring rate, and setting which is controlled by the pressure adjustment screw at left. This is the pressure at which the cam ring centers, the flow is zero and the pressure reaches the governor spring setting, which is called the "deadhead" pressure.

The performance characteristics of a given pump are determined by the spring rate. Table 7-1 shows performance curves for the same pump with three

TABLE 7-1 PERFORMANCE CURVES FOR THREE VANE PUMP GOVERNOR SPRINGS (*COURTESY OF RACINE HYDRAULICS DIVISION, DANA CORPORATION*).

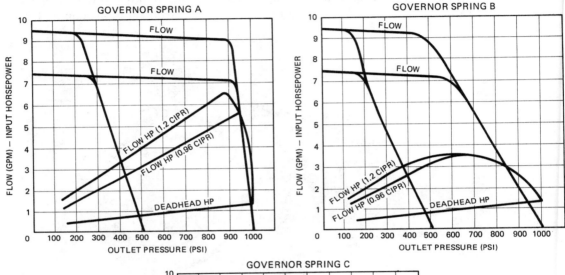

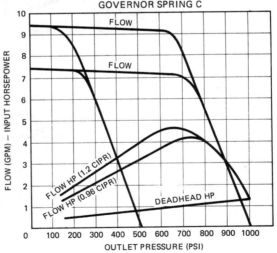

*NOTE – PERFORMANCE CHARACTERISTICS BASED ON:
· 1800 RPM SHAFT SPEED
· 150 SUS (32c St) OIL AT 120°F (49°C)
· 1.2 AND 0.96 CUBIC INCH DISPLACEMENT
  PRESSURE CAM RINGS

different springs. Spring A is designed to give maximum flow to a point as close to the maximum pressure setting of the pump as possible. High flow causes the horsepower required to drive the pump to increase as the pressure increases. Two straight lines are shown labeled Flow Hp for two different displacement pressure rings (1.2 CIPR and 0.96 CIPR). Governor spring B is a volume cut-off spring. It has a higher rate spring than A which allows the flow rate to taper off before the maximum pressure setting and pressure are reached. Governor spring C is a constant horsepower spring. It has an even higher rate spring than A or B, which causes the pressure to change such that the product of the pressure and the flow rate are more nearly constant. Notice that at pressures above 500 lbf/in.^{2},

Figure 7-21 Tandem mounting pressure compensated variable volume vane pumps (*courtesy of Racine Hydraulics Division, Dana Corporation*).

the flow lines (for both displacement cam rings) decreases steadily as the pressure increases to a maximum of 1000 lbf/in^2.

Tandem mounting allows two pumps to be driven from the same power source. The trailing pump can be the same size or smaller than the leading pump. Figure 7-21 shows two compensated vane pumps driven by a through shaft in the leading pump. The pressure compensating valves mounted to each of the pumps permits the operation of two independent systems from the same power unit (drive and reservoir), or the same system to operate at two independent pressures.

Piston displacement and volume flow rate in swash plate pump designs are varied by changing the angle of the swash plate, either mechanically or by the action of the yoke-actuating piston operating from fluid supplied through the pressure compensating valve (Fig. 7-22).

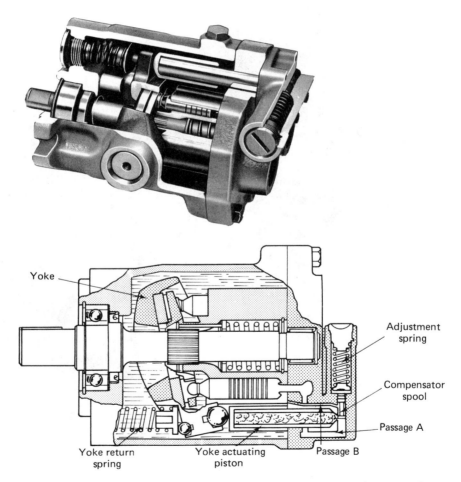

Figure 7-22 Variable displacement swash plate pump (*Courtesy of Sperry Vickers*).

Figure 7-23 illustrates another cutaway view of a variable volume swash plate pump equipped with a pressure compensator valve. The compensator control automatically reduces delivery when resistance is met to maintain the preset pressure in the system. When pressure in the delivery line is below the compensator setting, the control cylinder is connected to drain and the control spring positions the swash plate for full delivery. When pressure in the compensator line exceeds the compensator setting, the compensator valve is positioned to allow the pressure in the delivery line to enter the control cylinder behind the piston which, in turn, overcomes the control spring force and delivery until the pressure in the delivery line holds at the compensator setting. The pressure setting of the pump can be controlled automatically at the pump with the circuit shown, or it can be controlled at a remote location, for example at a control panel. A manual control valve option is also available.

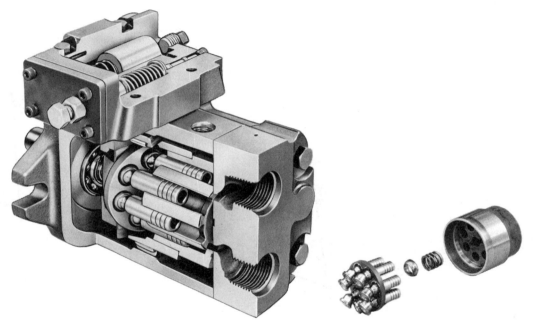

Figure 7-23 Variable volume swash plate pump with compensator valve and cylinder group detail (*Courtesy of Oilgear Company*).

Table 7-2 shows the performance characteristics of the variable displacement pressure compensated axial swash plate piston pump. This particular pump has a theoretical displacement of 4.41 in.^{3} and is rated at 32 gal/min and 3500 lbf/in.^{2} at 1800–2400 rev/min. A pump of this size weighs about 85 lbf and requires approximately 70 bhp at the input shaft. Volumetric efficiency is over 95% in the mid-range of pressure with overall efficiency about 85%.

Piston displacement and volume flow rate in radial piston pump designs are varied by changing the position of the cam ring with respect to the center line of

TABLE 7-2 PERFORMANCE CURVES FOR VARIABLE DISPLACEMENT PRESSURE COMPENSATED AXIAL SWASH PLATE PISTON PUMP (*COURTESY OF OILGEAR COMPANY*).

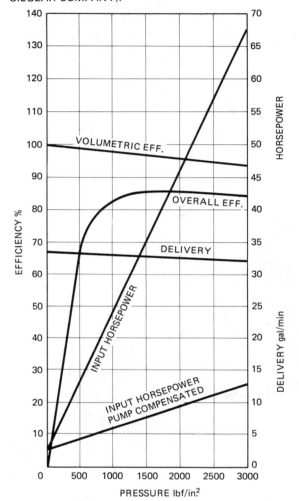

the supporting pintle (Fig. 7-24). This is accomplished mechanically, hydraulically, or electrically to cause varying conditions of flow. If the cam ring is moved over center, the action of the pump reverses the direction of fluid flow, even though the pump continues to rotate at constant speed in the same direction (Fig. 7-25).

Some industrial variable volume axial piston pumps vary the angle of the cam plate with an adjustment wheel to change the output volume (Fig. 7-26). A

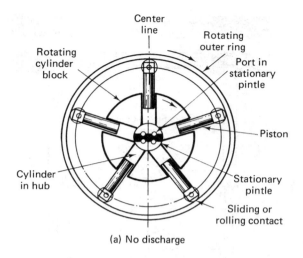

(a) No discharge

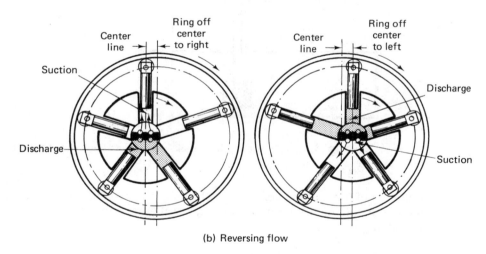

(b) Reversing flow

Figure 7-24 Variable discharge and direction from radial piston pump (*courtesy of Mobil Oil Corporation*).

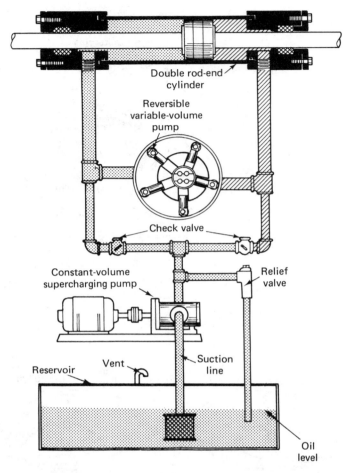

Figure 7-25 Reversible piston pump circuit (*courtesy of Mobil Oil Corporation*).

minimum of force at the control wheel is necessary to change the cam plate angle and pump volume using this system. Several other control input mechanisms are available.

7-4 PUMP PERFORMANCE

Pump performance characteristics are first analyzed independently of the rest of the hydraulic system and then as a part of the system. Both sets of data are valuable to the designer and technician. Analyzing the pump by itself gives an indication of its capabilities and performance based on the speed of operation and internal geometry and cost factors, whereas analyzing pump performance in the system essentially determines pump-system compatibility. In the first case, the

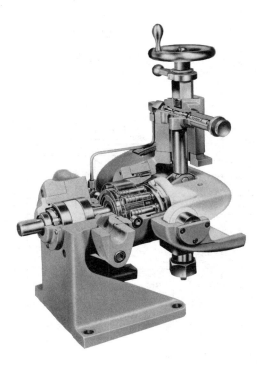

Figure 7-26 Variable displacement bent axis piston pump (*courtesy of Abex Corporation, Denison Division*).

system designer may observe performance curves to see if a specific pump has the pressure and volume flow rate to operate a given set of actuators. Or a technician may test a system pump independently to measure volumetric efficiency and pump wear after the pump has been in operation in a system. In the second instance, the system designer and technician may be computing the noise, vibration, frequency, cavitation, and flow characteristics of a specific pump to determine if the pump and existing system are compatible. While the two are necessarily complementary, in practice much of the hands-on work is completed independently.

Pump performance characteristics are interpreted from data in tabular form and then graphed. Independent variables (abscissa) are specified and dependent variables (ordinate) are read directly or computed. Specified independent variables usually include pressure and pump speed. Dependent variables read directly from the test apparatus or bench usually include pump delivery and power input in watts or bhp. Computed independent variables include horsepower output, horsepower loss, torque, volumetric efficiency, mechanical efficiency, and overall efficiency. In many instances pressure will be specified both as an independent and as a dependent variable based on the application and circumstance. Table 7-3 is a graphical representation of a typical fixed delivery pump's performance. Complete pump performance is sometimes put on one graph, or partial components of specific importance may be isolated and entered separately on one or more graphs. Tabular data are collected first, entered on the graphs, and then smooth curves are drawn to represent characteristics. The tabular entries are

Typical performance curves for Vickers Fixed
Displacement Aerospace Pumps
(Model PF-3918-30)

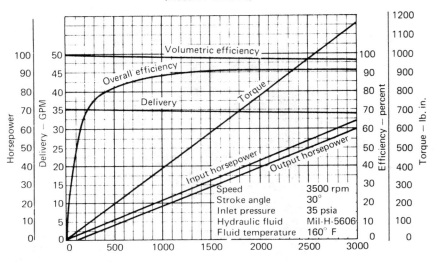

Pump outlet pressure — psi

most useful and easy to interpret when performance at a specific pressure or speed is desired (Table 7-4). Graphical representation has the greatest use when observing overall pump performance within the useable range of operation is the objective.

Pumps and reservoirs are sized to be compatible with the pressure, volume flow rate, and other operational characteristics associated with the system, such as heat generated causing a temperature rise. Minimum capacity requirements for the reservoir are twice the rated gal/min output of the pump to meet the volume requirements of cylinders or other actuators, plumbing, and accumulator if one is used. Whichever is the greater of these two requirements (pump or system) should be used to size the reservoir. Systems that subject the fluid to high temperatures should have oversized reservoirs and oil coolers to prevent overheating of the fluid.

TABLE 7-4 FIXED DELIVERY PUMP PERFORMANCE

Torque (T)	200.0	395.0	590.0	785.0	980.0	1175.0
Speed (N)	3500.0	3500.0	3500.0	3500.0	3500.0	3500.0
Input horsepower (Ihp)	11.1	21.9	32.7	43.6	54.4	65.2
Pressure (P)	500.0	1000.0	1500.0	2000.0	2500.0	3000.0
Delivery (Q)	35.5	35.1	35.0	34.9	34.5	34.2
Output horsepower (Ohp)	10.35	20.48	30.63	40.72	50.32	59.86
Volumetric efficiency (e_v)	99.0	98.5	98.0	97.5	97.0	96.5
Overall efficiency (e_o)	93.2	93.5	93.6	93.4	92.5	91.4

Example 5

A pump with a 5-gal/min capacity supplies fluid to an elevator cylinder with a 6-in. bore and 120-in. stroke. Compute the minimum size of the reservoir.

Solution The cylinder volume extended is computed from

$$\text{Cylinder Volume } (V_c) = \frac{\dfrac{\text{Area } (\pi d_c^2)}{4} \times \text{Stroke } (S)}{231}$$

$$V_c = \frac{\dfrac{3.14 \times 6^2 \times 120}{4}}{231}$$

and

$$V_c = 14.68 \text{ gal}$$

Since plumbing and fluid reserve have not been accounted for, a reservoir of at least 20 gal/min would be recommended. It also is observed that this is greater than twice the rated delivery of the pump ($2 \times 5 = 10$ gal/min).

7-5 PUMP SELECTION FACTORS

Pumps are selected using two basic criteria: (1) those relating to the actuator; and (2) those relating to the operation and efficiency of the pump itself. The basic parameter relating to the actuator is volume flow rate. Those relating to the operation and efficiency of the pump include system pressure, drive speed, cost, reliability, maintenance, and noise.

A system pressure should be selected that will provide existing and future actuators with the force necessary to move the load resistance. Higher system pressures allow for smaller system actuators and higher efficiencies, but are less tolerant of contaminants in the system and other lower quality components. Working pressure should also be compatible with the rest of the system should exchange of components be necessary.

Volume flow rates are selected to provide the necessary fluid and power to move specified load resistances required distances within available time limits.

That is,

$$Fhp = \frac{P \times Q}{1714}$$

must be equal to or greater than the anticipated output of the system.

The next parameter to be defined is drive speed. Standard speeds of 1,200 and 1,800 rpm are common. Higher speeds are associated with higher pump efficiencies, but noise, vibration, and cavitation are usually associated with high suction lifts and air entering the system. Accepted practice places the pump as close to the reservoir as possible and provides for cooling of the fluid (adequate reservoir size and design) as well as filtration. If HWCF fluids are used, the reservoir is placed above the pump inlet.

Another consideration is whether the pump will be of the nonpositive or positive displacement type, and, if positive, whether the displacement will be fixed or variable. System pressure is used to determine whether a nonpositive displacement pump should be used, whereas cost and circuitry are used to determine if the pump should have a fixed or variable displacement. Fixed displacement pumps require pressure relief and flow control valves to set the pressure and flow limits and prevent component rupture. Variable displacement pumps, on the other hand, may be pressure-, flow-, or load-compensated to limit system pressure and volume flow rate simultaneously.

Cost factors, system reliability, and required maintenance level are used to determine if the pump should be of the gear, vane, or piston type, although gear and vane types are usually associated with lower pressures. Some vane pumps have replaceable cartridges that allow rebuilding within minutes without removing the pump from the circuit. Gear pumps are usually less expensive and more tolerant of adverse conditions such as contaminants in the fluid, but have lower volumetric and overall efficiencies than vane pump counterparts. Piston pumps offer maximum volumetric and overall efficiency, as well as reliability, but cost is more than proportionally higher.

7-6 OTHER PUMP CHARACTERISTICS

Noise from pumps, measured on the decibel db(A) scale or sones scale, may be used over time to evaluate performance and service conditions. The relationship between sound levels on the sones and db(A) scale is seen in Table 7-5. Increases in noise usually indicate increased wear and imminent failure, unless a pipe or tube can be found vibrating against a machine or mounting. Sometimes the coupling is misaligned where the pump connects to the drive motor. While noise has been traditionally associated with pump wear and failure, recent redefinition of the term as unwanted or excessive sound recognizes the phenomenon as a form of environmental degradation. Department of Labor standards for noise levels in work areas also must be considered when specifying pumps, with or without enclosures, to meet occupational safety and health standards.

Pump vibration is influenced by both mechanical out of balance forces associated with the pump design and by the surges of fluid which exit the outlet port in the case of positive displacement pumps. The frequency of pump vibration for piston pumps is defined from the cycle rate (N) of the pump and the number of pistons (n) such that

$$\text{Pump Frequency } (f) = \text{Cycle Rate } (N) \times \text{No. of Pistons } (n)$$

and

$$f = N \times n \tag{7-17}$$

Pressure impulses or ripples in the system resulting from fluid surges from the pump have been shown to be equal to the pump frequency for pumps with an even number of pistons and to twice the pump frequency for pumps with an odd number of pistons.

TABLE 7-5 COMPARISON OF SONES VS. DB(A) SOUND
SCALES (*COURTESY OF SPERRY VICKERS*).

Sones	Loudness category
Below 30	Very quiet
30-40	Quiet
40-45	Medium quiet
45-50	Medium loud
50-60	Loud
Above 60	Very loud

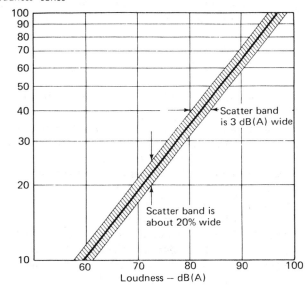

Pump cavitation occurs when suction lifts are excessive and local system pressure is below the vapor pressure of the fluid. This condition can be caused by inadequately charged or poorly designed inlet ports, as well as by mismatching the pump to the system. Bubbles that form in the low pressure region and travel through the pump are collapsed under high pressure and local high velocities and explosive forces causing eroding of pump parts. Similar phenomena result if air enters the system on the inlet side of the pump because of poor connections, low fluid level in the reservoir, or fluid foaming.

Pump wear under controlled simulated operation conditions can be determined from the loss of pump material over time. Two such tests used for vane pumps, ASTM 2271 and ASTM 2282, measure the amount of material displaced during the test run due to wear. Components of the pump cartridge, including the vanes, cam ring, and end plates, are weighed before and after test runs up to 100 hrs or more at constant pressure and temperature with a specified fluid. Wear is calculated from the weight loss of the parts and as a percentage from respective parts.

Other pump operational characteristics indicating progressive stages of wear include pump discharge pressure, internal fluid losses due to leakage past the close-fitting parts of the pumping elements, heat generated by the pump unit itself, and the appearance of contaminating metallic parts in filters and in the reservoir fluid. For positive fixed displacement pumps, a pressure drop at the outlet can also mean a drop in fluid discharge, and for a given pump speed, loss in volumetric efficiency is indicated. If an internal case drain is used, the flow volume from this source is also another indication of the same problem. In applications where the pump is of the positive displacement variable volume type, with either an internal or inaccessible case drain, excessive heat, and pump noise may be used as substitute measures, together with examination of the system fluid for metallic contaminants, to determine if pump disassembly and inspection are required.

7-7 SUMMARY

The hydraulic pump raises the energy level of the fluid by increasing the pressure, usually measured in ft of head, lbf/in.^{2}, kPa, or bars. Energy is added by the drive motor that turns or reciprocates the pump transferring fluid from the reservoir at a low energy level to the work cylinder or motor at a higher energy level. The energy level of the fluid is dissipated by the work performed against the load resistance, after which it returns to the reservoir at low pressure to be recycled through the system.

Pumping action involves transferring fluid from one location to another. That is, its volume is physically displaced by the movement of the internal parts of the pump. Positive displacement pumps accomplish this by having close-fitting parts increase and then decrease the size of the pumping chamber volume by an exact amount each time the pump cycles. The volume flow rate is then computed as a function of this displacement and the rate at which the pump cycles.

Nonpositive displacement pumps cause fluid to move by increasing its velocity within the pumping chamber. In centrifugal pumps, for example, a rotational inertia is imparted to the fluid, and the volume of fluid pumped is dependent upon the speed of the rotating member and the pressure resistance the pump must overcome at the discharge port.

Pumping efficiency can be computed by comparing the power available at the output of the pump to the power supplied at the input of the pump. The output usually is measured in terms of the pressure and volume flow rate at the discharge, whereas the input is computed in terms which describe the energy source—for example, in kilowatts for electric drive motors or horsepower for internal combustion engines.

Two components are commonly computed with pump overall efficiency: volumetric efficiency and mechanical efficiency. When positive displacement pumps are used, volumetric efficiency is computed by comparing the theoretical volume of fluid displaced as the pump rotates with the actual volume of fluid leaving the discharge port. Volumetric efficiencies between 85% and 95% are common. Mechanical efficiency is used to describe losses which result from the rubbing action of mechanical parts, fluid and other related causes. Typically it is computed from the overall and volumetric efficiencies. The relationship between the three is

Overall Efficiency = Volumetric Efficiency × Mechanical Efficiency

Pump performance is measured in terms of pressure, delivery, and cycle rate. Each or all three of these and others, such as the operating temperature of the fluid, may vary depending upon the application. Other factors which relate to performance and to the selection of an appropriate pump for a specific application include cost, conditions under which the unit must operate, reliability, expected life, and maintenance.

While the characteristics of the pump itself can be determined using bench tests, compatibility between the pump and a particular application must be determined by monitoring performance of the unit under actual operating conditions with related components in the system, such as the specified fluid, inlet and outlet pressure conditions, temperature fluctuations, the severity of environmental conditions, and pressure surging caused by loading conditions.

Pumps are selected, then, for a particular application using two basic criteria: those which relate to the operation of the pump itself and which can be monitored using bench tests, and those which relate to the operation of the system as a whole. Specifically, these include the size and configuration of the reservoir, the type of drive unit, speed and rotation, system plumbing, valving, cylinders and other actuators, presure variations, cost, reliability, maintenance, and noise.

Selection of an appropriate pumping unit typically follows the sequence:

1. Determine the flow rate.
2. Select an appropriate drive speed, direction, and unit.
3. Circuit conditions determine whether the pump will be of the positive or nonpositive type, and if positive, if the displacement will be fixed or variable. This will depend to a large extent upon the circuit selected.
4. Determine system pressure.
5. Select an appropriate reservoir and plumbing, including the use of a heat exchanger, if necessary.
6. Compute cost and other factors such as noise levels, vibration, wear characteristics, fluid filtration, and periodic maintenance schedule necessary to assure the longest possible life of the pump.

STUDY QUESTIONS AND PROBLEMS

1. List major differences between nonpositive displacement and positive displacement pumps.
2. How is the pumping action in nonpositive displacement pumps accomplished?

3. A pump operating at 2000 lbf/in.^{2} and delivering 5 gal/min requires 7 bhp to drive the input shaft. Compute the overall efficiency of the pump.

4. If the pump in Problem 3 is driven at 1200 rpm, what is the input torque to the pump shaft?

5. What is the overall efficiency of a pump driven by an 11 bhp power source that delivers fluid at 45 l/min at a pressure of 10.5 MPa?

6. A piston pump operating at 1200 rpm and 3000 lbf/in.^{2} delivers fluid at 10 gal/min. If the pump has an overall efficiency of 95%, what is the input torque from the driving unit?

7. A mobile application with the pump driven directly from the engine delivers 12 gal/min at a pressure of 2000 lbf/in^2. At 100% overall efficiency, what horsepower will this take from the engine?

8. If the pump in Problem 7 has a displacement of 1.3 in.^{3}/rev, what will the operating speed of the engine have to be?

9. A pump delivers 50 gal/min at 1500 lbf/in^2. What is the theoretical torque if the drive speed is 1200 rpm?

10. A pump that operates at 2000 rpm delivers 25 l/min at a pressure of 15 MPa. What is the theoretical torque at the pump shaft?

11. What is the volumetric efficiency of a pump with a displacement of 2.2 in.^{3}/rev that delivers 15 gal/min at 1725 rpm?

12. Compute the volumetric efficiency of a pump with a displacement of 36 cm^3/rev that delivers 37 l/m of fluid at 1100 rpm.

13. It is known that the volumetric efficiency of a certain piston pump of unknown displacement is 97%. If the pump delivers 15 l/min at 1200 rpm, what is the displacement of the pump?

14. A hydraulic orchard pruner is driven by an engine with an output of 20 bhp. If the pump has a mechanical efficiency of 90%, what portion of the engine power is lost to pump friction?

15. If a pump with a displacement of 16 cm^3/rev has an assumed overall efficiency of 90%, what speed and input brake horsepower would be required to deliver 40 l/min at a pressure of 20 MPa?

16. A positive displacement pump has an overall efficiency of 85% and a volumetric efficiency of 95%. What is the mechanical efficiency of the pump?

17. List the identifying characteristics of four types of nonpositive displacement pumps and four types of positive displacement pumps?

18. Explain what is meant by a hydraulically balanced vane pump. How is this accomplished?

19. What is the theoretical flow rate from a fixed displacement axial piston pump with a nine-bore cylinder group having a 0.035-in. bore and 0.50-in. stroke operating at 2500 rpm?

20. Explain how the size of the pumping chamber is changed in the variable displacement vane pump (NOTE: use a drawing to accompany the explanation).

21. How is the capability of a variable displacement pump changed by the addition of pressure compensation?

22. Compute the size of a reservoir necessary to supply a 2 gal/min pump with adequate fluid to raise a dump cylinder with a 4-in. bore 6 ft.

23. List several factors to be considered when selecting a hydraulic pump. Which one is generally considered most important?

24. What is meant by a pump cavitation? List a number of causes for this problem.

25. How can pump wear be determined?

CHAPTER 8

HYDRAULIC CYLINDERS AND CUSHIONING DEVICES

8-1 INTRODUCTION

Hydraulic cylinders convert fluid flowing under pressure to a linear motion. They are single-acting and returned by either a spring or the force of gravity acting on the load, or they are double-acting and have fluid piped to both ends of the piston. Single-acting cylinders with large bore-to-diameter ratios are called rams. Figure 8-1 illustrates how cylinders and cushioning devices are classified.

The force exerted by a hydraulic cylinder is proportional to the cross section area of the piston and the pressure of the fluid. Double-acting cylinders that have the rod extending through both end caps exert the same force in both directions. Cylinders with single-end rods exert greater force when the cylinder extends. The difference in force between the two is accounted for by the difference in area between both sides of the piston.

Cushioning devices decelerate the load to prevent shocks that could damage cylinders and other equipment. Cylinder cushioning devices use a tapered or stepped rod that enters a sleeve mounted in the end caps of the cylinder as it nears the end of the stroke. A typical hydraulic shock absorber cushions bumping loads with a spring-returned hydraulic cylinder that forces the fluid through an adjustable orifice when the load strikes the end of the rod.

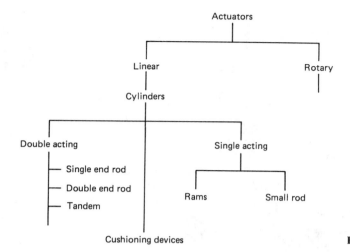

Figure 8-1 Actuator classification.

8-2 CYLINDERS

Cylinder actuators are defined by their ability to exert a linear force and hold it at any specified position indefinitely. Several factors are used to select hydraulic cylinders, including:

- Purpose of the cylinder
- Construction
- Required force
- Temperature
- Load
- Duty cycle
- Mounting
- Acceleration and deceleration forces

The purpose of the cylinder describes its type. The cylinder may move a load resistance in one direction only with a gravity or spring return, or move it in both directions under pressure, or have equal velocities as it travels in both directions. Cylinders, then, may be of the single-acting type, double-acting type or have a double-end rod, "DER," to equalize the areas on both sides of the piston and add support. If the cylinder piston is connected to a single-end rod, it is considered to be of a hydraulically unbalanced design. Double-end rod cylinders are considered to be hydraulically balanced because both sides of the piston have the same area. Tandem balanced cylinders exert twice the force available from single piston cylinders by providing twice the effective piston area in two chambered spaces.

Hydraulic cylinders are constructed of a cylinder barrel, piston and rod, end caps, ports, and seals (Fig. 8-2). The piston provides the effective area against

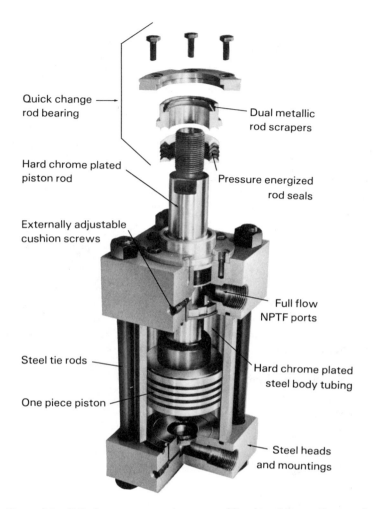

Quick change rod bearing

Dual metallic rod scrapers

Hard chrome plated piston rod

Pressure energized rod seals

Externally adjustable cushion screws

Full flow NPTF ports

Steel tie rods

Hard chrome plated steel body tubing

One piece piston

Steel heads and mountings

Figure 8-2 Cylinder components (*courtesy of Tomkins-Johnson Company*).

which fluid pressure is applied and supports the piston end of the rod. The opposite end of the rod is attached to the load resistance. The cylinder bore, end caps, ports, and seals maintain a fluid tight chamber into which fluid energy is piped. Routing of the fluid determines whether the cylinder extends or retracts. A common feature added to a cylinder is a mechanism called a cushioning device, dash pot, snubber, or decelerator to reduce shock loads that would be caused by bottoming the piston in either extreme stroke position.

Cylinders referred to as *rams* use large cylinder rods approaching the size of the cylinder bore to give maximum support to the load end of the rod. They are used extensively in such single-acting applications as car hoists, dump cylinders, and hydraulic presses.

The output force required from a hydraulic cylinder and the hydraulic pressure available for this purpose determine the area and bore of the cylinder. For a required force and given pressure,

$$\text{Cylinder Area} = \frac{\text{Force}}{\text{Pressure}}$$

$$\frac{\pi d_b^2}{4} = \frac{F}{p}$$

and

$$d_b = \sqrt{\frac{4F}{\pi p}} \qquad (8\text{-}1)$$

where the bore (d_b) is computed in in. (mm), the force (F) is in lbf (N), and the pressure (p) is in lbf/in.² (Pa or bars).

Example 1

Compute the size of the cylinder piston operating from a system pressure of 1,600 lbf/in.² (11 MPa) to move a load resistance of 80,000 lbf (355.8 kN) (Fig. 8-3).

Solution From Eq. 8-1.

$$d_b = \sqrt{\frac{4F}{\pi p}}$$

$$d_b = \sqrt{\frac{(4)(80,000)}{(3.14)(1,600)}} = 8 \text{ in. approximately}$$

In SI metric units

$$d_b = \sqrt{\frac{(4)(355,800 \text{ N})}{(3.14)(11.03 \times 10^6 \text{ N}/m^2)}} = 20 \text{ cm approximately}$$

Similar values are obtained using standard cylinder capacity tables of forces, pressures, cylinder areas, and bore sizes.

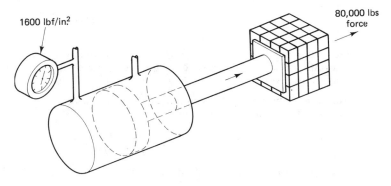

1600 lbf/in²

80,000 lbs force

Figure 8-3 Example 1.

The stroke of the cylinder determines its length. For a given bore size, stroke also determines the flow required for a specified cylinder velocity, and ultimately, the fluid velocity through lines, fittings, and connecting ports. The internal length of the cylinder barrel must be sufficient to extend the anticipated stroke with additional length for clearance. If the cylinder is to decelerate and stop the load, the ends must be equipped with snubbing devices.

The velocity of a fluid flowing through the cylinder line and port is a function of the cross section area and the flow rate. That is,

$$v = \frac{Q(\text{gal/min})(231 \text{ in.}^3/\text{gal})}{A(\text{in.}^2)(12 \text{ in./ft})}$$

and

$$v = \frac{Q}{A} \times 19.25 \qquad \text{ft/min} \qquad (8\text{-}2)$$

where the flow rate (Q) is in gal/min and the area (A) is in in^2. From this relationship, flow rate through ports in ft/min, and piston speed in the cylinder in ft/min can be computed, given the flow rate to the system and a port or cylinder diameter.

Example 2

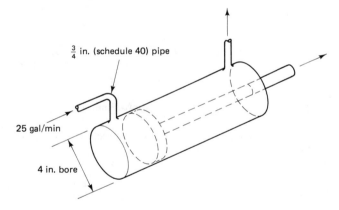

Figure 8-4 Example 2.

Given a flow rate of 25 gal/min through a standard 3/4-in. pipe (schedule 40) cylinder port into a 4-in. cylinder bore, compute the fluid velocity at the cylinder port in ft/sec and the piston velocity in the bore in ft/min (Fig. 8-4).

Solution The fluid velocity at the cylinder port in ft/sec is computed using the area of the 3/4-in. standard pipe and the formula

$$v = \frac{Q \times 19.25}{A \times 60}$$

The area of the 3/4-in. standard pipe is 0.533 in.^{2} (see Table 8-1).

$$v(\text{fluid velocity}) = \frac{25 \times 19.25}{0.533 \times 60} = 15.05 \text{ ft/sec}$$

TABLE 8-1: SCHEDULE OF PIPE SIZES AND PRESSURE RATINGS.

Nominal pipe size (in.)	Outside diameter of pipe (in.)	Number of threads per inch	Length of effective threads (in.)	Schedule 40 (standard)		Schedule 80 (extra heavy)		Schedule 160		Double extra heavy	
				Pipe ID (in.)	Burst press (psi)	Pipe ID (in.)	Burst press (psi)	Pipe ID (in.)	Burst press (psi)	Pipe ID (in.)	Burst press (psi)
$\frac{1}{8}$	0.405	27	0.26	—	—	—	—	—	—	—	—
$\frac{1}{4}$	0.540	18	0.40	0.364	16,000	0.302	22,000	—	—	—	—
$\frac{3}{8}$	0.675	18	0.41	0.493	13,500	0.423	19,000	—	—	—	—
$\frac{1}{2}$	0.840	14	0.53	0.622	13,200	0.546	17,500	0.466	21,000	0.252	35,000
$\frac{3}{4}$	1.050	14	0.55	0.824	11,000	0.742	15,000	0.614	21,000	0.434	30,000
1	1.315	$11\frac{1}{2}$	0.68	1.049	10,000	0.957	13,600	0.815	19,000	0.599	27,000
$1\frac{1}{4}$	1.660	$11\frac{1}{2}$	0.71	1.380	8,400	1.278	11,500	1.160	15,000	0.896	23,000
$1\frac{1}{2}$	1.900	$11\frac{1}{2}$	0.72	1.610	7,600	1.500	10,500	1.338	14,800	1.100	21,000
2	2.375	$11\frac{1}{2}$	0.76	2.067	6,500	1.939	9,100	1.689	14,500	1.503	19,000
$2\frac{1}{2}$	2.875	8	1.14	2.469	7,000	2.323	9,600	2.125	13,000	1.771	18,000
3	3.500	8	1.20	3.068	6,100	2.900	8,500	2.624	12,500	—	—

Schedule 40 (standard)   Schedule 80 (extra heavy)   Schedule 160   Double extra heavy

The velocity of the piston in the bore in ft/min is computed similarly.

$$v(\text{piston velocity}) = \frac{25 \times 19.25}{12.56} = 38.32 \text{ ft/min}$$

The working pressure and temperature of the cylinder fluid affect the selection of material used for seals, packings, and rod wipers. Higher pressures, as a rule, require stiffer materials. As temperature increases, sealing materials tend to soften. Where friction, abrasion, and severe loading also accompany increases in temperature, metallic automotive-type piston rings and asbestos rod packing have received widespread application. Synthetic fluid composition for temperature extremes affects many materials commonly used for piston and rod seals and requires the use of metallic, asbestos, or dissimilar synthetic seal materials.

Cylinder loading results from the force exerted on the piston rod and is affected by both the nature of the load itself and the position of the load with respect to the cylinder. Side thrusts are exerted on the piston rod support if the load is not properly aligned to resist the force exerted by the piston rod on the center line of the cylinder bore. Misalignment has a tendency to cause erratic cylinder operation, seizing and wear at the support bearing.

Piston rod column sizes required to sustain applied loads that are in alignment with the centerline of the cylinder bore are determined from the strength of the rod material, the force applied to the rod column in compression, the mounting situation of the cylinder itself, and the stroke over which the load is to be applied. As rod column length is increased, the tendency to deflect and buckle under loading increases. Common yield strength values for metals used as piston rods are in excess of 100,000 lbf/in.[2] in tension. The force applied to the column in compression is computed in lbf as a function of the load applied externally, or from the pressure applied to the piston area internally. The mounting situation is determined by how the cylinder body is connected to the stationary member (Table 8-2).

The procedure to compute piston rod column sizes and cylinder length under end thrust conditions is as follows:

1. Determine the column strength factor from the mounting arrangement (Table 8-2).
2. Calculate the corrected length of the rod using the formula

   Corrected Length = Actual Stroke × Column Strength Factor

3. Determine the thrust (lbf or N) that will be imposed on the end of the column using the piston bore size and the maximum relief valve pressure using the formula

   Force (F) = Pressure (p) × Area (A)

4. Determine the appropriate piston rod diameter using Table 8-3 by locating the next higher value for corrected rod length opposite the maximum load. Then read the recommended rod size at the top of the chart.

TABLE 8-2 TYPICAL MOUNTING SITUATIONS FOR HYDRAULIC CYLINDERS (*COURTESY OF CARTER CONTROLS, INC.*).

Mounting situation	Examples	Column strength factor
Cylinder rigidly mounted Piston rod guided and supported		Stroke x 0.5 $\dfrac{\text{Stroke}}{2}$
Cylinder rigidly mounted or front trunnion mount Piston rod supported		Stroke x 1.0
Center trunnion mount Piston rod supported		Stroke x 1.5
Rear flange, clevis or rear trunnion mount Piston rod supported		Stroke x 2.0
Rear flange mount Piston rod unsupported		Stroke x 4.0

5. Compute the stop tube length if one is applicable (Fig. 8-5). Stop tubes are used on the cylinder rods between the piston and the end cap to prevent overextension of the rod when the cylinder is fully extended. Misalignment, due to long strokes and unsupported weight, sometimes causes rod jackknifing or buckling if the piston rod is allowed to extend the piston fully to the

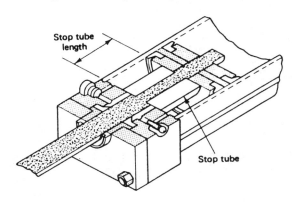

Figure 8-5 Stop tube length.

TABLE 8-3 PISTON ROD DIAMETER FOR MAXIMUM LOADING*

Load (lbf)	Piston rod diameter in./mm												Load (kN)
	⅝/16	1/25	1⅜/35	1¾/44	2/51	2½/64	3/76	3½/89	4/102	4½/114	5/120	5½/140	
100	65/165												0.45
150	50/127												0.67
250	38/97	100/254											1.11
400	29/74	78/198	155/394										1.78
700	21/53	58/147	112/284	190/483									3.11
1,000	17/43	48/122	94/239	160/406	210/533								4.45
1,400	14/36	40/102	78/198	132/335	175/445	275/699							6.23
1,800	10/25	30/76	58/147	98/249	130/330	210/533	300/762						8.00
3,200		25/64	50/127	85/216	110/279	170/432	257/653						14.23
4,000		22/56	44/112	75/191	98/249	158/401	228/579	300/762					17.79
5,000		20/51	39/99	67/170	88/224	140/356	200/508	280/711					22.24
6,000		18/46	35/89	60/152	78/198	127/323	182/462	252/640	300/762				26.68
8,000		15/38	30/76	52/132	67/170	110/279	156/396	220/559	290/737	300/762			35.58
10,000		12/30	26/66	46/117	60/152	96/244	138/351	192/488	260/660	295/749	300/762		44.48
12,000		11/28	23/58	42/107	54/137	92/234	125/318	178/452	236/599	254/645	275/699	300/762	53.38
16,000			18/46	35/89	46/117	76/193	108/274	152/386	200/508	228/579	220/559	275/699	71.17
20,000			14/36	29/74	40/102	68/173	95/241	136/345	180/457	183/465	190/483	240/610	88.96
30,000				19/48	30/76	52/132	77/196	110/279	145/368	159/404	170/432	212/538	133.44
40,000				10/25	21/53	40/102	64/163	94/239	125/318	140/356	155/394	192/488	177.92
50,000					12/31	30/76	54/137	84/213	111/282	130/330	135/343	168/427	222.40
60,000						13/33	44/112	74/188	101/257	110/279	120/305	148/376	266.88
80,000							28/71	57/145	85/216	97/246	110/279	136/345	355.84
100,000								43/109	70/178	84/213	97/246	124/315	444.8
120,000								29/74	58/147	73/185	85/216	112/284	533.76
140,000									44/112	62/157	73/185	124/315	622.72
160,000									10/25	38/97	85/216	112/284	711.68
200,000											64/163	93/236	889.60
250,000												69/175	1112.0
300,000												10/25	1334.0

* Tabled values are stroke in in./cm.

end of the cylinder. Stop tubes space the piston away from the cylinder end to prevent this condition and increase bearing life. Oversized cylinder rods are not considered an acceptable substitute for cylinder rod stops since they are more expensive, add weight to the system, and if misalignment does occur, promote bearing and seal wear because of increased stiffness of the oversized rod. Stop tube length is computed using the formula

$$\text{Stop Tube Length} = \frac{\text{Corrected Length} - 40 \text{ in.}}{10} \qquad (8\text{-}3)$$

Additional bearing surface at the rod support end of the cylinder and/or double piston construction may be necessary for unusually long strokes or heavy loads.

Example 3

Compute the diameter of a cylinder rod to be used in a cylinder with a bore diameter of 3 in., operating at a maximum pressure of 2000 lbf/in.^{2}, center trunnion mounted with a 20-in. stroke (Fig. 8-6). Compute the length of the stop tube if one is applicable.

Solution

1. The column strength factor from Table 8-2 equals 1.5.
2. The corrected length of the rod is then computed from

$$\text{Corrected Length} = 20 \times 1.5 = 30 \text{ in.}$$

3. The thrust on the rod is computed from system maximum pressure and the diameter of the cylinder.

$$F = p \times A$$
$$F = 2000 \times 7.07 = 14{,}143 \text{ lbf}$$

4. The piston rod diameter is determined from Table 8-3 by first locating the next larger value, 16,000, in the left hand column, and moving across to 35, the corrected rod length. Finally, reading upward to the value at the top, a rod diameter of 1¾ in. is indicated.

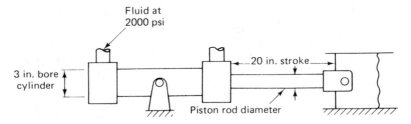

Figure 8-6 Example 3.

5. Finally, stop tube length is computed from Eq. 8-3

$$\text{Stop Tube Length} = \frac{\text{Corrected Length} - 40 \text{ in.}}{10}$$

$$\text{Stop Tube Length} = \frac{30 - 40}{10} = -1$$

The minus value indicates that a stop tube for a corrected cylinder length under 40 in. is not appropriate in this application.

8-3 CUSHIONING DEVICES

Cushioning devices are provided in the ends of hydraulic cylinders or as separate units when loads must be decelerated. Hydraulic shocks to the cylinder and system are reduced by slowing down the piston just before it contacts the cylinder end caps. The sleeve or spear-type cushioning device is often used as the decelerating mechanism. The energy absorbed by decelerating a body in motion is converted to heat and dissipated to the environment through the fluid and cylinder body.

Refer to Fig. 8-7. As the piston completes its extension stroke, the slightly tapered end cushioning sleeve enters the rod end cap blocking the normal flow of fluid. The flow of fluid is then rerouted through the bypass port and needle valve at a controlled rate, decelerating the piston. Cushioning at the head end of the cylinder during the retraction stroke is accomplished in the same manner by the action of the tapered cushioning spear plunger as it enters the end cap. This blocks the normal flow of fluid and causes fluid to be rerouted through the bypass port and metering valve to decelerate the piston. The check valve, shown only on the blank end of the cylinder, is used to direct fluid at system pressure to the full area of the piston during acceleration.

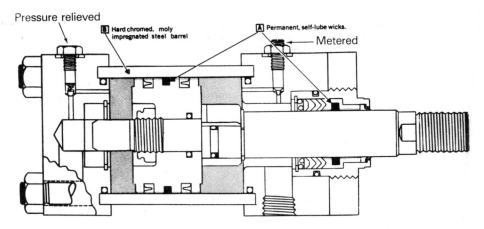

Figure 8-7 Cylinder cushioning devices (*courtesy of Carter Controls, Inc.*).

The force exerted by a body put into motion is derived from Newton's law

$$F = Ma \qquad (8\text{-}4)$$

where the mass (M) is equal to the weight of the body (w) divided by the acceleration (g) due to gravity, and the acceleration (a) is that given to the body. In notation,

$$M = \frac{w}{g} \qquad (8\text{-}5)$$

The force then becomes

$$F = \frac{w \times a}{g} \qquad (8\text{-}6)$$

where the force (F) is in lbf (N), the weight (w) is in lbf (N), the acceleration or deceleration of the body (a) is in ft/sec^2 (m/s^2) and the acceleration due to gravity (g) is in ft/sec^2 (m/s^2).

An acceleration force factor (g_a) can be defined as

$$g_a = \frac{a}{g} \qquad (8\text{-}7)$$

where the acceleration factor (g_a) is without dimension. Values for the acceleration factor (g_a) may be read from graphs (Fig. 8-8) or computed from the acceleration (a) given to the body.

The acceleration (a) of a body, for example, attached to a hydraulic cylinder, is computed through a stroke (S) to a velocity (v). That is

$$a = \frac{v^2}{2 \times S} \qquad (8\text{-}8)$$

where the acceleration (a) is in ft/sec^2, the velocity (v) is in ft/sec and the stroke (S) is in ft. If the velocity is given in ft/min and the stroke is in in., the formula becomes

$$a = \frac{\dfrac{v^2}{3,600}}{\dfrac{2 \times S}{12}}$$

and

$$a = \frac{v^2}{S} \times .00166 \qquad (8\text{-}9)$$

If the acceleration due to gravity (g) is given the value 32.2 ft/sec^2, the acceleration factor (g_a) thus becomes

$$g_a = \frac{v^2 \times .00166}{S \times 32.2}$$

and

$$g_a = \frac{v^2}{S} \times .0000517 \qquad (8\text{-}10)$$

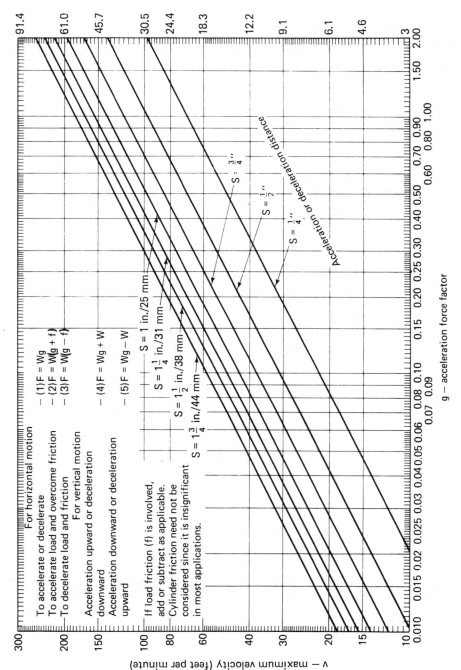

Figure 8-8 Acceleration force factor graph.

The force exerted on or by the system that is attributable only to the weight being accelerated or decelerated equals

$$F = w \times g_a \qquad (8\text{-}11)$$

if tabled values for g_a are used, and

$$F = \frac{w \times v^2}{S} \times .0000517 \qquad (8\text{-}12)$$

for computed values of g_a.

Example 4

Compute the force required to decelerate a free body weighing 8,000 lbf (35,584 N) from 120 ft/min (36.576 m/min) within a distance of 1.5 in. (0.0381 m) (See Fig. 8-9.)

Solution First, compute the acceleration factor (g_a) using Eq. 8-10

$$g_a = \frac{v^2}{S} \times 0.0000517$$

$$g_a = \frac{(120)^2}{1.5} \times 0.0000517$$

and

$$g_a = 0.496$$

Finally, compute the force to accelerate body using Eq. 8-11

$$F = w \times g_a$$

$$F = 8,000 \times 0.496 = 3,968 \text{ lbf}$$

In SI metric units the solution is computed using the same method and Eqs. 8-7 and 8-11.

$$g_a = \frac{\dfrac{v^2}{2 \times S}}{g} = \frac{v^2}{2gS}$$

and

$$g_a = \frac{(36.576 \text{ m/min} \times 1/60 \text{ min/sec})^2}{(2)(9.8 \text{ m/s}^2)(0.038 \text{ m})} = 0.496$$

Finally, solving for the force to decelerate the body using Eq. 8-11

$$F = w \times g_a = (35,584 \text{ N})(0.496) = 17.649 \text{ kN}$$

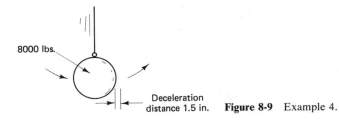

Figure 8-9 Example 4.

When a friction factor (f) acts to impede the motion of a body, it must be added to the total force when the body is accelerated and subtracted from the total force when the body is decelerated. For tabled values of the acceleration factor (g_a), the basic acceleration or deceleration formula becomes

$$F = w \times g_a \pm w \times f$$

and

$$F = w(g_a \pm f) \tag{8-13}$$

For computed values of g_a

$$F = w \left[\left(.0000517 \times \frac{v^2}{S} \right) \pm f \right] \tag{8-14}$$

If the coefficient of friction factor (f) is greater than the acceleration factor (g_a) and negative, then a negative force (F) results from these two formulas.

The additional force being imparted to a vertically moving body being accelerated or decelerated must also be considered and consists of the load resistance of the body itself (w). This situation exists, for example, when a hydraulic elevator must be accelerated or decelerated, and the load must be added or subtracted appropriately. For tabled values of the acceleration factor (g_a), the total force during acceleration or deceleration equals

$$F = w(g_a \pm f) \pm w$$

Specifically, the following conditions may exist:

Accelerate upward	$F = w(g_a + f) + w$	(8-15)
Decelerate downward	$F = w(g_a - f) + w$	(8-16)
Accelerate downward	$F = w(g_a + f) - w$	(8-17)
Decelerate upward	$F = w(g_a - f) - w$	(8-18)

Similar formulas are obtained if computed values are substituted for the acceleration factor (g_a).

For bodies moving horizontally, the additional force (F_a) imparted to the body by an actuator that continues to have fluid power applied to it during deceleration must be added to compute the total force. For a body actuated by a cylinder, for example, the formula to compute the total force during deceleration, including friction, is

$$F = w(g_a - f) + F_a \tag{8-19}$$

for tabled values of the acceleration factor (g_a), and

$$F = w \left[\left(.0000517 \times \frac{v^2}{S} \right) - f \right] + F_a \tag{8-20}$$

for computed values of g_a. This is the case existing when a cushioning device is used on the rod side of the piston to decelerate the load as fluid continues to be supplied to the blind side of the piston. A similar condition exists when a separate cushioning device is used to decelerate a load moving under a continued force supplied by a cylinder actuator.

Example 5

Compute the total force necessary to decelerate a load of 1,500 lbf (6672 N) traveling at 50 ft/min (15.24 m/min) with a coefficient of friction (f) of 0.12 within a distance of 3/4 in. (0.01905 m). The cylinder actuator moving the load resistance has a 3-in. (0.0762 m) bore operating at a maximum relief valve pressure of 750 lbf/in.[2] (5.172 MPa) (Fig. 8-10).

Solution Since the load is moving horizontally and decelerating under an applied force (F_a), the formulas used are Eq. 8-19

$$F = w(g_a - f) + F_a$$

for tabled values of the acceleration factor (g_a), or Eq. 8-20

$$F = w \left[\left(.0000517 \times \frac{v^2}{S} \right) - f \right] + F_a$$

for computed values of g_a.

The maximum additional force (F_a) applied by the cylinder to the load resistance during deceleration is

$$F_a = p \times A$$

and

$$F_a = 750 \times 7.065 = 5,299 \text{ lbf}$$

Using tabled values for the acceleration factor (g_a) (see Fig. 8-8), the velocity of the load resistance, 50 ft/min, is located first in the left margin of the table. Moving horizontally to the right, locate the line indicating the length of the stroke (S) for deceleration. This is 3/4 in. Then, moving vertically down to the lower margin, locate the acceleration factor, which is approximately 0.172.

Substituting in Eq. 8-19

$$F = w(g_a - f) + F_a$$
$$F = 1,500(0.172 - 0.12) + 5299$$

and

$$F = 78 + 5299 = 5377 \text{ lbf}$$

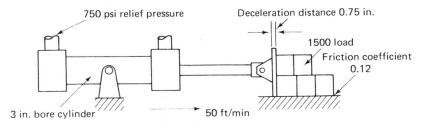

Figure 8-10 Example 5.

Using Eq. 8-20 for computed values for the acceleration factor (g_a), the final force (F) would be computed similarly.

$$F = w \left[\left(.0000517 \times \frac{v^2}{S} \right) - 0.12 \right] + F_a$$

$$F = 1,500 \left[\left(.0000517 \times \frac{(50)^2}{.75} \right) - 0.12 \right] + 5298.75$$

and

$$F = 1,500(.052) + 5299$$

$$F = 78 + 5299 = 5377 \text{ lbf}$$

which is the same value obtained using tabled values for g_a.

In SI metric units, solving for g_a using Eq. 8-7,

$$g_a = \frac{v^2}{2gS} = \frac{(15.24 \text{ m/min.} \times 1/60 \text{ min/sec})^2}{(2)(9.8 \text{ m/s}^2)(0.019 \text{ m})} = 0.172$$

Finally, solving for the deceleration force using Eq. 8-19,

$$F = w(g_a - f) + F_a$$

and

$$F = (6672 \text{ N})(0.172 - 0.12) + (23 \, 576 \text{ N}) = 23 \, 922 \text{ N}$$

The pressure generated on the cushion side of the piston is a function of the force applied and must be taken into account to prevent rupturing the cylinder. An example illustrates this computation.

Example 6

Assuming a total deceleration force of 7,500 lbf on the stationary end of a cylinder with a 3-in. bore and 1½-in. single end cylinder rod, compute the fluid pressure at the rod end of the cylinder (Fig. 8-11).

Solution Fluid pressure at the rod end of the cylinder is computed from the force applied during deceleration and the effective area. The effective area of the cylinder

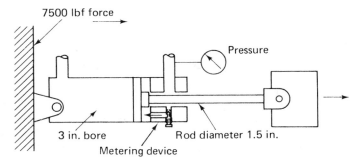

Figure 8-11 Example 6.

Hydraulic Cylinders and Cushioning Devices   Chap. 8

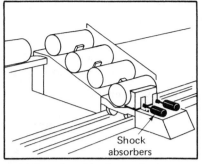

(a) Stopping of loads transferring from one conveyor to another—paper rolls, metal coils, slabs, other large loads.

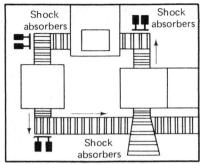

(b) Uniform deceleration of products in automatic assembly operations.

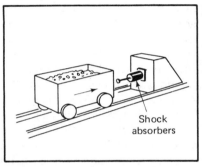

(c) Hydraulic bumpers for railroad cars; endstops for inplant material transfer cars.

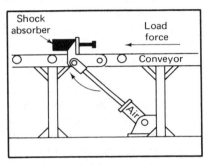

(d) Disappearing stops in material handling transfer lines.

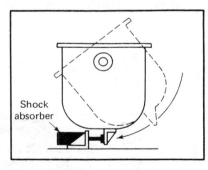

(e) Cushioning for roll-over applications of all types.

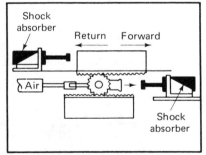

(f) Forward and reverse cushioning of shuttle devices.

Figure 8-12 Typical deceleration application using shock absorbers.

at the rod end is

$$A_1 = A_2 - A_r$$

and

$$A_1 = 7.065 - 1.766 = 5.299 \text{ in.}^2$$

Pressure in the cylinder is then computed from

$$p = \frac{F}{A}$$

and

$$p = \frac{7500}{5.299} = 1414 \text{ lbf/in.}^2$$

This is the pressure the cylinder or cushioning device must withstand. Since it is at the rod end that has the smallest effective area, given the same deceleration force in both directions, it will generate the highest unit pressure (lbf/in.²) that must be contained by the cylinder.

External hydraulic energy absorbing devices also are employed many times to reduce shock loading of machinery that would result in failure. Carriage stops on turntables, doors, transfer machines, turnovers, and small cranes requiring controlled deceleration use these devices (Fig. 8-12). In operation, fluid is metered through adjustable orifices for a wide range of impact loads and velocities. In Fig. 8-13, for example, liquid is metered through the orifice plate valve along the tapered metering pin as the piston is forced into the cylinder. Adjustment is accomplished by removing the protective plug and rotating the adjustment screw with a hexagonal socket key for more or less damping. The hardened and chrome plated tubular piston resists buckling and side loads under impact. The metered liquid is stored in the spring-loaded separator chamber which also repositions the piston.

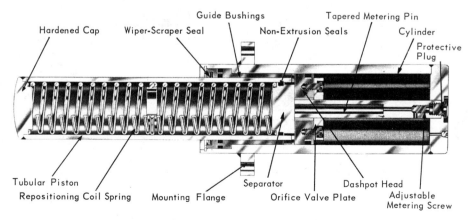

Figure 8-13 Typical hydraulic shock absorber with full-range adjustability (*courtesy of Enidine Corporation*).

8-4 SUMMARY

Hydraulic cylinders convert the high energy level of the fluid to a linear motion by applying the fluid to the movable piston. Movement thereby accomplishes the output objective of the system and lowers the energy level of the fluid, after which it is returned to the reservoir at low pressure.

Cylinders are sized not only to meet the force and distance requirements of the machine, but also to provide the strength needed to prevent buckling. The most severe example is rear flange mounting with the piston rod unsupported, which requires an increase in the stroke by a column strength factor of four to provide the necessary strength. One means to prevent overstroking is to install stop tubes on the cylinder rod to limit the stroke.

Hydraulic cylinders extend and retract to complete a cycle and sometimes include stepped or spear type cushioning devices in the ends to prevent shock loading and noise associated with inertia and overrunning loads. External cushioning devices are another means to prevent shock loading and use a hydraulic shock absorber principle. They are constructed with spring-returned, adjustable pistons and internal metering devices. In both cases, the force to be absorbed by the cylinder or external cushioning device is computed from the magnitude of the load, hydraulic force, acceleration, direction of travel, and the friction factor.

STUDY QUESTIONS AND PROBLEMS

1. Ignoring friction and fluid losses, how large would the blank end of a cylinder operating at 2000 lbf/in.² have to be to raise 25 tons?

2. Compute the linear speed of a 2-in. bore cylinder receiving fluid at 5 gal/min.

3. A cylinder has a bore of 4 in. and a rod diameter of 2 in. Compute the velocity of the cylinder as it extends and retracts if 10 gal/min is supplied to the system.

4. A 10-kN car and 5-kN hoist are supported 1.8 m above the floor by a cylinder with a 300-mm bore. What pressure will this develop in the system?

5. If the car in Problem 4 is lowered at 0.33 ft/sec, what would be the velocity of the fluid through the return line if the cylinder were plumbed with 1-in. Schedule 40 pipe (*clue:* see Table 8-1 for pipe dimensions).

6. A cylinder with a 7-cm bore and 2.5-cm rod receives fluid at 40 l/min. What is the velocity of the cylinder rod extending and returning?

7. If the cylinder in Problem 6 has a stroke of 40 cm, what is the maximum cycle rate that might be achieved?

8. A hydraulic cylinder with a 2.5-in. bore and a 1.25-in. cylinder rod receives fluid at 12 gal/min through 3/4-in. Schedule 40 pipe. What is the velocity of the fluid through the return line when the cylinder is retracting (*clue:* draw an illustration showing the direction of flow).

9. Calculate the corrected length and diameter size for a piston rod in a rear trunnion mounted hydraulic cylinder that must exert a 5-ton force through a 24-in. stroke.

10. A center trunnion mounted cylinder with a 35-mm diameter cylinder rod and 100-cm stroke is extending under a force of 35 kN. How far can the cylinder safely extend without exceeding the recommended stroke?

11. A rear flange mounted cylinder extends 14 in. under a load of 10,000 lbf. What size cylinder rod would be required?

12. For the cylinder in Problem 11, what stop tube length would be required?

13. Compute the force required to decelerate a free body weighing 2000 lbf from 150 ft/min within a distance of 2 in.

14. A 3-in. bore cylinder operating at 2000 lbf/in.^{2} is driving a load of 3000 lbf horizontally at 20 ft/min. Compute the force necessary to decelerate the load within a distance of 2 in. if the coefficient of friction is 0.15.

15. A hydraulic cushioning device has a 2-in. bore and a 0.75-in. cylinder rod. If the fluid pressure on the rod end of the cylinder increases as a result of absorbing a total deceleration force of 6000 lbf, what is the pressure generated?

16. Neglecting friction, what force could be expected from the rod end of a cylinder with a 5-in. bore if the blank end receives fluid at 2500 lbf/in.^{2}?

17. A filled drum weighing 2 kN rolls against a cushing device at 1 m/s. If friction is negligible, what force would be required to decelerate the drum within 2 cm?

18. If in Problem 17 the force is absorbed through a hydraulic cushioning device with a cylinder bore of 350 mm, what pressure could be expected?

19. A 2000 lbf elevator with a 1000 lbf cargo descends at 2 ft/sec. If the friction factor equals 0.08, and the elevator decelerates within a distance of 6 in., what force downward would the cushioning device have to absorb?

20. In Problem 19, what would be the deceleration distance if the elevator were ascending and the cushioning device absorbed the same deceleration force?

CHAPTER 9

HYDRAULIC MOTORS

9-1 INTRODUCTION

Rotary actuators convert the energy in the pressurized fluid to the turning motion of a shaft. They exert both limited rotation, as in the case of torque motors, and continuous rotation as with hydraulic motors (Fig. 9-1). Limited rotation actuators use several novel arrangements of pistons and vanes to achieve rotation, from a few degrees to several rotations. Continuous rotation actuators, commonly called hydraulic motors, use internal geometry similar to gear, vane, and piston pumps to achieve continuous angular motion of the shaft.

In addition to off-the-shelf units, there are a number of specialized hydraulic motors. Radial piston motors, for example, exert a high torque, low speed characteristic because of the size and geometry of the piston arrangement around a common crankshaft. Two-speed operation of these units is achieved by delivering pressurized fluid to the pistons in sequence or in tandem pairs. Another application of the radial piston motor concept is wheel motors, used for agricultural equipment and off-the-road vehicles. Wheel motors are available as stationary bolt-on units and as steerable yoke units.

The machine tool industry and other precision applications make extensive use of electrohydraulic pulse motors to provide precise positioning and velocity control. These units combine a low power digital stepping motor with a high power precision hydraulic piston motor to achieve a position or speed in proportion to the input pulse rate of the control signal.

181

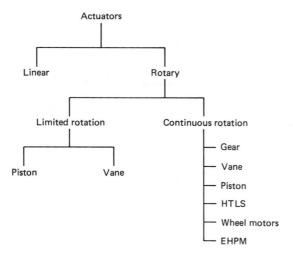

Figure 9-1 Actuator classification.

9-2 TORQUE, SPEED, POWER, AND EFFICIENCY

Rotary actuators convert system pressurized fluid to torque at the output shaft. Theoretical torque at the output shaft is derived by equating the basic fluid horsepower formula to the brake horsepower formula (as was done in Eq. 2-11) and solving for the value of T. That is,

$$\frac{p \times Q}{1,714} = \frac{T \times N}{5,252}$$

$$T = \frac{p \times Q \times 5,252}{N \times 1,714} \qquad \text{in lbf-ft}$$

It is observed in the formula that

$$\frac{Q}{N} = \frac{V_m}{231} \qquad (9\text{-}1)$$

where the volume displacement of the motor (V_m) is in in.^{3} per revolution. If torque in lbf-in. is desired, then

$$T = \frac{p \times V_m \times 5,252 \times 12}{231 \times 1,714}$$

and

$$T = \frac{p \times V_m}{6.28}$$

or

$$T = p \times V_m \times 0.16 \qquad (9\text{-}2)$$

where torque (T) is computed in lbf-in., pressure is in lbf/in.^{2}, and displacement of the motor (V_m) is in in.^{3} per revolution.

Hydraulic Motors   Chap. 9

Example 1

Compute the theoretical torque from a rotating hydraulic motor operating at 2000 lbf/in.^{2} with a displacement of 0.5 in.^{3} per revolution.

Solution   Theoretical torque is computed from Eq. 9-2

$$T = p \times V_m \times 0.16$$

and

$$T = 2000 \times 0.5 \times 0.16 = 160 \text{ lbf-in.}$$

It should be noticed that the torque is not affected by changes in speed or flow rate so long as the pressure remains constant, since from Eq. 9-1,

$$\frac{Q}{N} = \frac{V_m}{231}$$

Also notice that the theoretical speed (N) of the motor is computed by solving for N

$$N = \frac{231 \times Q}{V_m}$$

where the speed (N) of the motor is in rpm. Similarly, theoretical flow through the motor (Q) in gal/min is computed by solving Eq. 9-1 for Q:

$$Q = \frac{V_m \times N}{231}$$

Example 2

Compute the theoretical speed of a fluid power motor with a 2 in.^{3} (0.0328 l/rev) per revolution displacement receiving fluid at the rate of 7.5 gal/min (28.39 l/min).

Solution   The speed can be computed from Eq. 9-1

$$N = \frac{231 \times Q}{V_m}$$

and

$$N = \frac{231 \times 7.5}{2} = 866 \text{ rpm}$$

In SI metric units, where the displacement of the motor is given in liters and the flow rate is in liters/min, Eq. 9-1 simply becomes

$$N = \frac{Q}{V_m} = \frac{28.39 \text{ l/min}}{0.0328 \text{ l/rev}} - 866 \text{ rev/min}$$

Example 3

Compute the effective flow rate through a fluid power motor with a displacement of 0.55 in.^{3} per revolution turning at 3400 rpm.

Solution   Flow rate is computed using Eq. 9-1

$$Q = \frac{V_m \times N}{231}$$

and

$$Q = \frac{0.55 \times 3400}{231} = 8.09 \text{ gal/min}$$

Motor efficiency is measured primarily by how well the fluid horsepower input is converted to useable brake horsepower at the motor output shaft. Overall motor efficiency can be computed by comparing the useable brake horsepower output to the fluid horsepower input. That is,

$$e_o = \frac{\dfrac{T \times N}{5252}}{\dfrac{p \times Q}{1714}} \times 100$$

and

$$e_o = \frac{T \times N}{p \times Q \times 3.06} \times 100 \qquad (9\text{-}3)$$

where the torque (T) is measured in lbf-ft, the speed (N) is in rpm, the pressure (p) is measured across the motor inlet and outlet ports in lbf/in.^{2}, and the flow rate (Q) is in gal/min. Notice that this formula is the reciprocal of the formula for overall pump efficiency.

Example 4

Compute the overall efficiency of a motor operating at a pressure of 2,000 lbf/in.^{2}, a flow rate of 7.5 gal/min, a speed of 1,200 rpm and exerting a torque of 32 lbf-ft.

Solution   Overall efficiency is computed from Eq. 9-3

$$e_o = \frac{T \times N}{p \times Q \times 3.06} \times 100$$

and

$$e_o = \frac{32 \times 1,200}{2,000 \times 7.5 \times 3.06} \times 100 = 84\%$$

The volumetric efficiency of a fluid power motor can be measured by comparing the effective volume flow rate (Q_e) indicated by the product of the displacement of the motor (V_m) times its speed (N), to the actual volume flow rate (Q_a) supplied to the motor. That is,

$$e_v = \frac{Q_e}{Q_a} \times 100$$

$$e_v = \frac{\dfrac{V_m \times N}{231}}{Q_a} \times 100$$

and

$$e_v = \frac{V_m \times N}{Q_a \times 231} \times 100 \qquad \text{as a percent} \qquad (9\text{-}4)$$

where the displacement of the motor (V_m) is in in.^{3} per revolution, the speed (N) is in rpm, and the actual volume flow rate (Q_a) is in gal/min.

A major factor in volumetric efficiency that gives an indication of motor condition can be measured by observing the flow rate of fluid returning from the case drain (q) to the reservoir and comparing this value to the total actual flow rate (Q_a) supplied to the motor. The percent lost to the case drain can then be computed using the formula

$$\text{Percent Case Drain Loss} = \frac{q}{Q_a} \times 100 \qquad (9\text{-}5)$$

where the case drain loss (q) and the actual flow rate (Q_a) are measured in gpm. On small motors with minimum leakage, percent case drain loss may be negligible and difficult to compute.

Example 5

Compute the volumetric efficiency of a motor with a displacement of 0.6 in.^{3} per revolution (0.00983 l/rev) turning at 2,400 rpm that is receiving fluid at the rate of 6.5 gal/min (24.61 l/min). How much fluid might be expected to return to the reservoir from the case drain if 50% of the loss in volumetric efficiency can be attributed to internal leakage?

Solution   Volumetric efficiency for this motor is computed from Eq. 9-4

$$e_v = \frac{V_m \times N}{Q_a \times 231} \times 100$$

$$e_v = \frac{0.6 \times 2{,}400}{6.5 \times 231} \times 100 = 95.9\%$$

In SI metric units, where the displacement is given in liters/rev, and the delivery in liters/min, Eq. 9-4 becomes

$$e_v = \frac{(0.0098 \text{ l/rev})(2400 \text{ rev/min})}{(24.61 \text{ l/min})} \times 100 = 95.9\%$$

The total fluid flow lost from slippage through the high pressure ports and chambers of the pump as well as internal leakage that returns through the case drain is

$$\text{Total Fluid Loss} = 0.041 \times 6.5 = 0.267 \text{ gal/min}$$

If half this loss can be attributed to internal leakage

$$\text{Case Drain Loss} = 0.50 \times 0.267 = 0.134 \text{ gal/min}$$

Mechanical efficiency (e_m) for the motor is then computed from related efficiencies the same as for hydraulic pumps. That is,

$$e_o = e_v \times e_m$$

and

$$e_m = \frac{e_o}{e_v} \times 100 \qquad \text{as a percent}$$

where the overall efficiency (e_o) and volumetric efficiency (e_v) are also computed as percent. This is a restatement of Eqs. 7-14 and 7-15.

Example 6

Compute the mechanical efficiency of a fluid power motor that has an overall efficiency of 88% and a volumetric efficiency of 97%.

Solution   Mechanical efficiency of the motor is computed from

$$e_m = \frac{e_o}{e_v} \times 100$$

and

$$e_m = \frac{88}{97} \times 100 = 90.72\%$$

9-3 LIMITED ROTATION ACTUATORS

Limited rotation actuators, called torque motors, have a wide variety of applications where a limited specified degree of rotation at the output shaft is required. Rotation is usually limited to 720°. They are used extensively in industry for actuating clamping devices, material handling, rotating cams for braking mechanisms, tumbling and dumping, positioning and turning, and many other situations where an economical application of fluid power for limited rotation is desirable (Fig. 9-2). Vane-type limited actuators apply fluid force to the cross section area of single or multiple vanes. Rack and pinion type actuators apply fluid force to the cylindrical chambers which move the rack to drive the pinion gear.

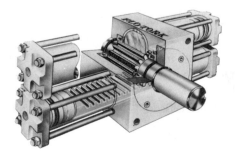

(a)

Figure 9-2   (a) Rack and pinion limited rotation actuator (*courtesy of Flo-Tork, Inc.*); (b) Vane type limited rotation actuator (*courtesy of Bird-Johnson*); (c) Typical applications of limited rotation actuators (*courtesy of Bird-Johnson*).

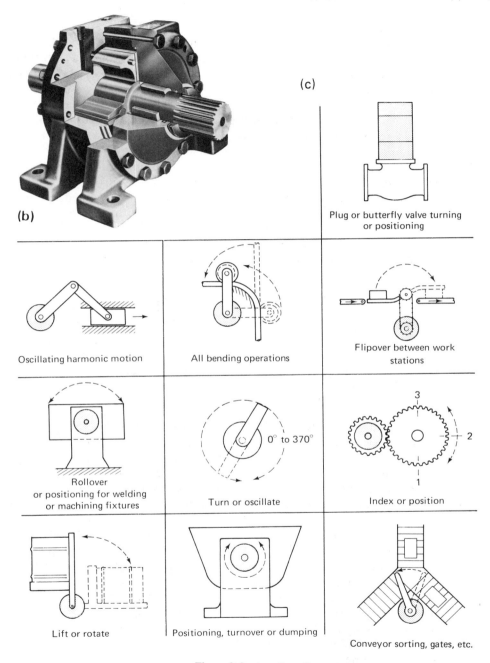

(b)

(c)

Plug or butterfly valve turning
or positioning

Oscillating harmonic motion

All bending operations

Flipover between work
stations

Rollover
or positioning for welding
or machining fixtures

0° to 370°

Turn or oscillate

3

2

1

Index or position

Lift or rotate

Positioning, turnover or dumping

Conveyor sorting, gates, etc.

Figure 9-2 (*continued*)

9-4 CONTINUOUS ROTATION ACTUATORS

Continuous rotation actuators, called hydraulic motors, provide sustained rotation in either direction. Some hydraulic motors are also convertible to serve as hydraulic pumps if a mechanical drive is applied to the output shaft, but this is not usually recommended without special provision because of port timing and other internal part arrangements. Vane motors, for example, have spring loaded vanes, whereas vane pumps usually do not. This is not the case with axial piston motors, however, which are widely advertised as combination pump-motors.

Hydraulic motors differ from pumps in other respects. Because the case is pressurized from an outside source, case drains are provided to protect shaft seals (Fig. 9-3). These may be piped directly to the low pressure reservoir, or through a crossover check valve arrangement to the exhaust port of the motor. External drain lines or crossover check valve arrangements are needed only for series circuits or meter out circuits. This is necessary during reversing, braking, and other operating conditions which would otherwise subject the case drain to system pressure. Maximum pressure at the case drain is usually 100–250 psi. Port timing is an additional factor that may be different between pumps and motors.

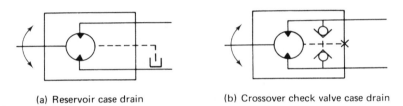

(a) Reservoir case drain   (b) Crossover check valve case drain

Figure 9-3 Hydraulic case drain arrangements.

Hydraulic motors are available as fixed or variable displacement units so that speed variation with rotation in either direction is possible.

External gear positive displacement motors operate in the reverse manner of their pump counterparts (Fig. 9-4). They are available in sizes to 20 in.³ per revolution. Fluid supplied to the inlet port circulates around the outside of the gear teeth driving both gears, although only one gear is connected to the motor output shaft. The gear teeth seal where they mesh and between their ends and the motor housing. Fluid is trapped in cavities formed by the gear teeth and the motor housing and transported around the outside diameter of the gears to the low pressure port side of the motor. Industrial gear motors are the least expensive to manufacture, have overall efficiencies to 90%, and operate in the speed range of 1000 to 2500 rpm. Recent developments in gear tooth materials and technology have extended the speed range of gear motors in excess of 20,000 rmp for motor sizes in the 0.063–0.093 in.³ per revolution displacement range. Their small size, high speed, and high power (4–5.5 hp) make these motors ideal for spindle drives in the machine tool industry.

Vane motors operate similarly to vane pumps with the exception that unlike their pump counterparts, spring loading is used to insure positive contact between

Symbol

Figure 9-4 External gear positive displacement motor (*courtesy of Webster Electric Company, Inc.*).

the vanes and the eccentric cam ring (Fig. 9-5). Seals, high speed bearings, and case drains also are given special attention.

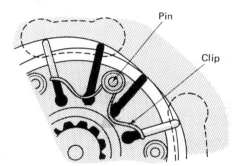

Pin

Clip

Figure 9-5 Vane motor showing spring loaded vanes (*courtesy of Sperry Vickers*).

Fluid entering the motor under high pressure acts against the rectangular surfaces of the vanes while the chamber volume increases, and then is exhausted as it decreases (Fig. 9-6). Balanced vane motors admit fluid from two ports located 180° apart to reduce side thrusts on the supporting shaft and bearings, and discharge the fluid 90° later through two similarly located discharge ports. One-piece motor body and cartridge element construction permit interchange to adapt the unit to a wide variety of flow capacities and ease repair with minimum disassembly and number of spare parts.

Fluid viscosities between 55 and 275 SSU are recommended at normal operating temperatures of 120°F with a maximum of 180°F. Most vane motors are not recommended for use with water-based emulsion fire-resistant fluids.

Gerotor motors like that shown in Fig. 9-7 operate in principle similar to gerotor pumps. Port timing is different, however. Fluid entering the motor moves into the chambers between the inner and outer gear as the pump rotates. After the crossover point, the fluid is discharged at low pressure. Because the case is pressurized, case drains must be provided by one of the means explained previously. The internal seal of the rotating chambers is maintained by contact between the tips of the inner and outer rotor. The displacement of the motor and

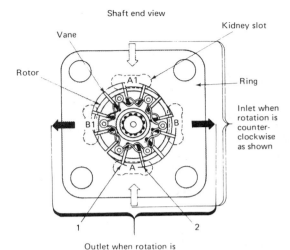

Shaft end view

Vane

Kidney slot

Rotor

A1

Ring

Inlet when rotation is counter-clockwise as shown

B1

B

A

1

2

Outlet when rotation is counterclockwise as shown

Figure 9-6 Balanced vane motor operation (*courtesy of Sperry Vickers*).

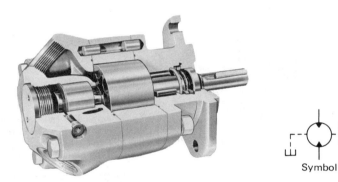

Symbol

Figure 9-7 Gerotor motor (*courtesy of HPI Nichols*).

its output are determined by the space formed by the extra tooth in the outer rotor and the length of the assembly.

Piston motors are the most like their pump counterparts and incorporate only minor changes to effect the conversion. They are available in both axial and radial designs. In-line-type motors are similar to pumps and make use of both the fixed displacement swash plate and variable displacement adjustable yoke. Bent axis motors are available with fixed displacement angles as well as variable displacement angles. Radial piston motors are used widely in low speed, high torque applications. Piston motors also operate at the highest efficiencies, speeds, and pressures available.

Fixed displacement in-line swash plate motors commonly use nine pistons with the swash plate at 15° (Fig. 9-8). They are available in single or double shaft versions. Reversal of the motor is accomplished only by reversing the direction of fluid flowing through the motor. When used as a motor, hydraulic pressure on the piston creates tangential forces on the angled swash plate (Fig. 9-9). This gives

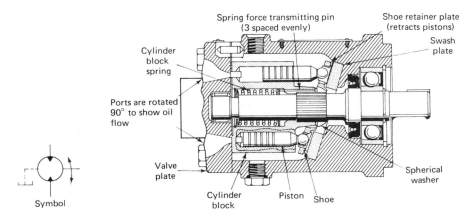

Figure 9-8 Fixed displacement in-line piston motor (*courtesy of Sperry Vickers*).

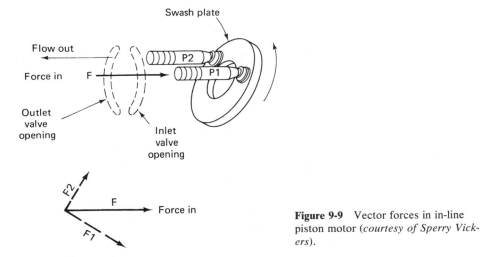

Figure 9-9 Vector forces in in-line piston motor (*courtesy of Sperry Vickers*).

the required turning moment to the shaft. As fluid is directed to the rotor through the inlet and kidney-shaped port in the valve plate, it imparts a force (F) on the pistons in line with their axis. The axial force is resolved into its component angle forces (F_1) and (F_2). While the force (F_1) is directed perpendicular to the surface of the stationary swash plate and is not available to accomplish work, the force (F_2) is in a direction parallel to the surface of the swash plate and so is free to move down the angled surface imparting motion to the piston group, cylinder block, and output shaft. Fluid under pressure is directed to the pistons through the kidney-shaped inlet port while the pistons are descending the swash plate angle, and directed from the pistons to the exhaust port as they ascend the swash plate angle.

Variable displacement in-line piston motors differ from fixed swash plate motors in that the angle of the swash plate is changeable (Fig. 9-10). Angles of

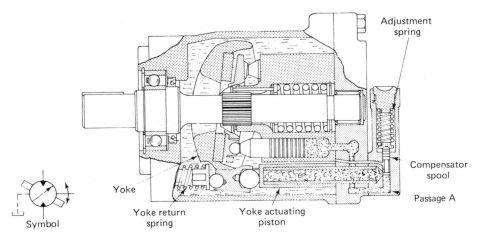

Figure 9-10 Variable displacement in-line piston motor (*courtesy of Sperry Vickers*).

+15° and −15° from the center position are common. Reversal of the motor is accomplished by tilting the yoke over center by the action of a servo yoke actuating piston or manual control. The torque from variable displacement motors varies with the change in volume displacement caused by tilting of the yoke. If the motor receives fluid at constant flow rate from the pump, a decrease in torque is accompanied by an increase in speed. The fluid horsepower of the motor is constant since the product of pressure (p) and flow rate (Q) input are constant. Neglecting losses, the horsepower output is also constant and

$$Bhp = \frac{T \times N}{5,252}$$

is a constant. Increases or decreases in torque resulting from increases or decreases in motor displacement will be accompanied by inverse decreases or increases in motor output speed.

As the angle of the actuating yoke is reduced to approach 0°, the speed of the variable displacement motor increases. Further decrease in the angle of the yoke will cause the motor to stall and cause excessive pressure in the system which must be released to redirect the constant flow provided by the pump. Motors that are internally compensated to provide predetermined torque and speed conditions (see Fig. 9-10) use a spring balanced pressure-actuated piston to appropriately position the angle of the yoke. The spring is loaded to set the yoke at the minimum operating angle. In this position, maximum speed is attained at minimum torque. As the load resistance at the motor output shaft increases and speed is reduced, the pressure rises at the compensator spool and inlet where fluid is supplied at constant volume. The increased pressure actuates the pressure-sensitive compensator spool which admits fluid to the yoke-actuating piston, and in turn, positions the yoke at a greater angle. This increases the displacement of the motor with a corresponding loss in motor speed, but with an inversely corresponding increase in torque.

9-5 HIGH TORQUE–LOW SPEED MOTORS

Radial piston motors such as that in Fig. 9-11 have received widespread use in heavy duty applications. These include crane hoisting and turning apparatus, rolling mills, road rollers, large self-propelled trenchers, railroad transport, and hydraulic production machinery.

Radial motor designs permit a short overall housing and shaft length capable of sustaining high radial and axial loads without damage. The pistons are located around and perpendicular to the axial output shaft and attach to a common crank pin or cam through connecting rods. In some applications, high pressure fluid is applied to the extremes of the cylinders to exert an inward force on the pistons

Figure 9-11 High torque-low speed radial piston motor (*courtesy of Sunstrand*).

through a rotating spool valving arrangement. In others, fluid enters through the stationary pintle to apply a force to the pistons outward radially. The inside of the housing is ground to a cam configuration to impart a tangential force to the connecting roller beams and rods that cause the output shaft to rotate.

In a typical radial piston motor such as that shown in Fig. 9-12, torque is generated by hydraulic thrust on an eccentric. The crankshaft with integral eccentric is carried in large capacity tapered roller bearings in the end housings for increased load. Oil is directed to and from the five radial cylinders through a distributor sleeve in the back of the housing where the feed and return lines are attached. The inlet and outlet passages are separated by an annular groove ring seal. High pressure oil is fed into the crankshaft and distributed to the pistons through machined ports in the pentagon and pistons. Oil flowing to the head of each piston generates a thrust that is transmitted to the crankshaft through the eccentric which orbits as the crankshaft turns. When the motor freewheels, return springs allow the piston to follow the eccentric.

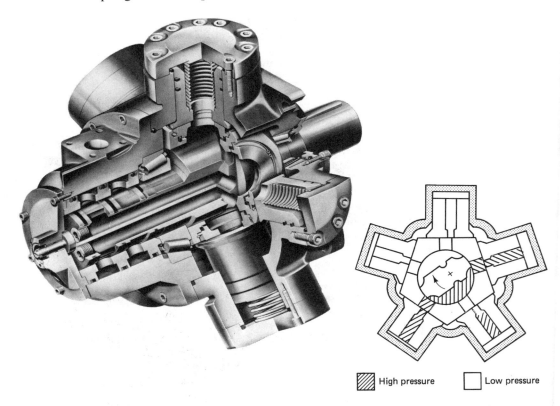

High pressure   Low pressure

Figure 9-12 Radial piston motor operation (*courtesy of Rotary Power, Inc.*)

Two-speed operation is available with the radial piston motor shown in Fig. 9-13. Again, a stationary eccentric which orbits with rotation of the crankshaft holds the pistons at right angles with the flats in the main housing, and rotation is

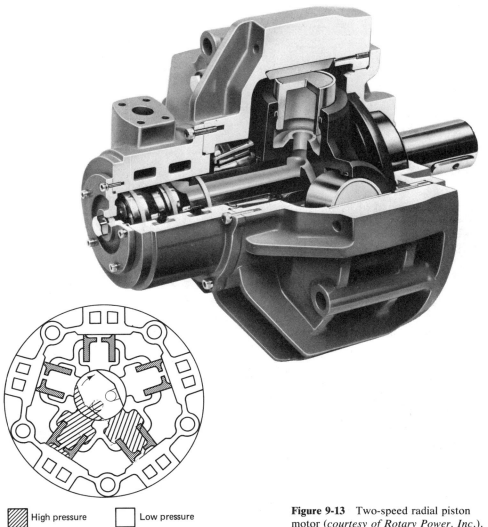

High pressure ▨   Low pressure ☐

Figure 9-13 Two-speed radial piston motor (*courtesy of Rotary Power, Inc.*).

caused by plumbing oil through the axial passages in the crankshaft to and from the distributor plate in the end housing. What is novel about the arrangement is the means used to vary the speed and torque. In the low speed, high torque mode of operation (Fig. 9-14), oil is directed through two ports in the end housing against the combined area of both concentric pistons. In the high speed, low torque mode of operation, the oil is directed through one port in the end housing against the area of the center pistons only. Speed and torque ratios are determined by the relative sizes of the concentric pistons. If the pistons have equal areas, the speed will be doubled and the torque divided by two when the control valve is shifted to the high-speed mode.

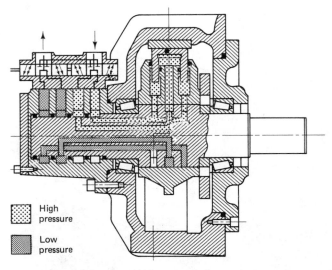

High
pressure

Low
pressure

(a) Low-speed high-torque displacement

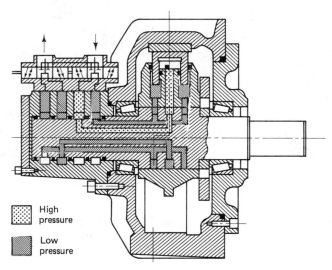

High
pressure

Low
pressure

(b) High-speed low-torque displacement

Figure 9-14 Two Speed Operation of Radial Piston Motor (*courtesy of Rotary Power, Inc.*).

9-6 WHEEL MOTORS

Wheel motors commonly use a radial piston design in a narrow envelope that will fit in a standard size rim. Front wheel tractor drives, combine drives, and utility drives for special purpose vehicles make use of these motors. What is novel about the design is that the motor is constructed to provide power at a location that is remote from the engine, often using the pump that powers the existing hydraulic system. The wheel is mounted where it is needed with hoses and rigid tubing connecting the motor to the hydraulic system through a control valve at the operator's station.

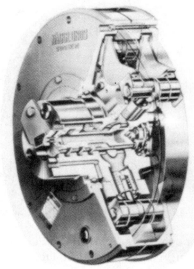

Figure 9-15 Wheel motor (*courtesy of Bird-Johnson Company*).

Figure 9-15 illustrates the internal construction of a radial piston motor suitable for wheel-motor applications. The pistons are contained in a stationary cylinder housing and connected to rollers which bear on the cam ring. The rotary valve distributes oil from the inlet to each cylinder in sequence. As the oil pressure builds against the load resistance, it causes the pistons to be displaced outward, forcing the rollers against the cam ring and the outer casing of the motor to rotate. At the end of each piston stroke, oil escapes through the outlet at low pressure. The flow of fluid through the motor is shown in Fig. 9-16. Here it can be seen that the even number of opposing pistons assures hydraulic balance and prevents loading the main bearing of radial piston forces. Two-speed operation is available by having the oil directed to each piston in sequence in the high-speed mode, and to opposite pairs of pistons in the low-speed mode. This is accomplished with a two-speed valve. In high-speed operation, the torque is halved. Freewheeling can be accomplished by permitting the housing to rotate with internal circulation of hydraulic fluid only.

As with other fluid power motors, care must be exercised to assure that the case drain provided prevents overpressure and damage to seals. Outlet port back pressure must exceed the case drain pressure by 5–10 lbf/in.^{2} to assure that the motor cavity remains full of oil for lubrication and cooling of the internal parts. When the motor overruns and acts as a hydrostatic brake, related circuitry and the motor must be protected by a maximum pressure relief valve circuit. Finally, an adequate supply of oil must be provided to the inlet port during deceleration or

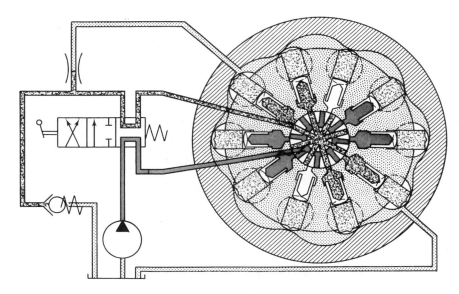

Figure 9-16 Fluid flow through radial piston motor.

overhauling load conditions to prevent the motor, acting as a pump, from cavitating. A pressure of 50 lbf/in.² is considered sufficient for this purpose.

Figure 9-17 illustrates a wheel motor mounted to a steerable yoke. Wheel motors of this type use control valves that direct the fluid to 1/3, 2/3 or all of the pistons to provide three speed ranges of operation. Smaller displacement pumps can be used to reach maximum operating speeds by supplying oil to only some of

Figure 9-17 Steerable wheel motor
(*courtesy of John Deere Company*).

Hydraulic Motors   Chap. 9

the pistons, or full-size pumps can be used with a three-speed valve to shift displacements on the go without stopping. Twelve pistons and 15 cam lobes provide 180 power strokes per revolution. The timing is such that a smooth flow of power is transmitted continuously, and radial forces are balanced at maximum displacement to cancel radial forces. Hydraulic fluid is fed to the motor through the pivot pins of the steering yoke and this feature allows the pressure lines to be connected at the fixed mounting, thereby eliminating line flexing during turning. Freewheeling is accomplished by an integral pump that destrokes the pistons when the motor is disengaged. This allows all the engine power to be directed to the main drive system on a priority basis, or the vehicle to be towed at high speed in the freewheeling mode without an external power source.

Figure 9-18 illustrates the assembly view of the motor to show how the cam lobe drive mechanism and valving work. The piston carrier has one drilled passage for each piston. The manifold and piston carrier interface becomes a rotary valve directing oil to and from the pistons as the wheel turns. Once pressurized, each piston is forced radially outward, and the piston follower generates torque. As the motor turns in response to the combined torque produced by the pistons, rotary power is developed. The valving is held together by hydrostatic force that is varied with system pressure and power requirements. The cycloidal cam profile of the casing allows the pistons to generate a smooth torque output, and follower motion gives the pistons a zero acceleration at the top and bottom dead center positions. It should be noticed that the cam and manifold are fixed with respect to the motor mounting, while the piston carrier and axle are free to rotate.

Figure 9-18 Wheel motor assembly view (*courtesy of John Deere Company*).

Gear reduction units also can be used to increase the effective displacement and torque of off-the-shelf hydraulic motors. Speed ranges from 100 to 1000 rpm can be selected by matching the motor and reduction unit ratios. This arrangement allows the unit to be used for forward drive, reversing, braking, and as part of a hydrostatic drive system.

The unit in Fig. 9-19 is a combination external gear motor of a balanced design coupled to a planetary gear set with a reduction of 5.41 : 1. This effectively

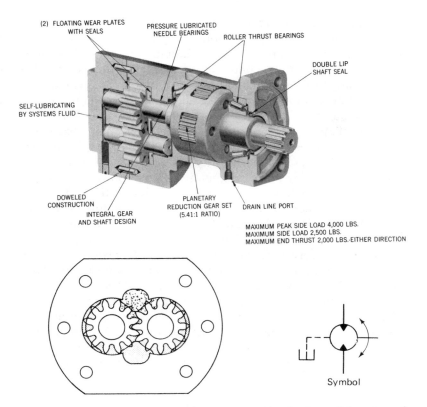

Figure 9-19 High-torque low-speed gear motor and reduction unit (*courtesy of Borg Warner*).

multiplies the displacement, flow rate, and torque of the motor by the same amount. At the same time, output shaft speed is reduced to 1/5.41 of motor speed.

9-7 ELECTROHYDRAULIC PULSE MOTORS (EHPM)

Electrohydraulic pulse motors offer the designer, builder, and user of machines an unusual combination of power and precision for position and velocity control. EHPMs provide precision positioning, a high transverse rate, and very simple control and installation. They are essentially *bolt on and use* units.

An EHPM combines a low power digital electric stepping motor with a high power precision hydraulic motor (Fig. 9-20). Conventional square wave digital pulses are used as input commands to the electric stepping motor. This, in turn, controls the direction, speed and position of the high-torque hydraulic motor shaft. Precision control of the shaft is accomplished without requiring the use of external feedback transducers. Since closed-loop feedback is not used, the EHPM system is defined as open loop. The output shaft speed of the hydraulic

Figure 9-20 Electrohydraulic pulse motor (*courtesy of Sperry Vickers*).

motor is in direct proportion to the input pulse rate to the electric stepping motor. Output shaft position is directly related to the number of input pulses. The rotational direction of the output shaft is determined by the sequence in which the input pulses are applied to the various phases (or coils) of the electric stepping motor.

In operation, low power digital electrical pulses, from a standard digital electronic or NC control, are fed to the electric stepping motor that rotates depending on the sequence, direction, pulse rate, and number of pulses.

The stepping motor opens the four-way spool valve in either direction by rotating the spool through a gear set, in or out of the ball nut (Fig. 9-21). Hydrau-

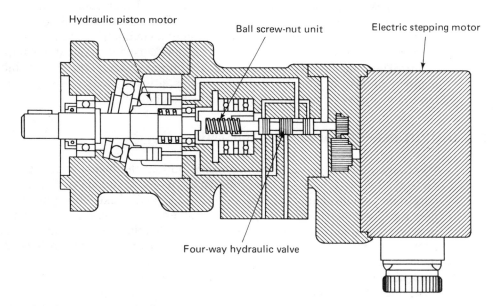

Hydraulic piston motor   Ball screw-nut unit   Electric stepping motor

Four-way hydraulic valve

Figure 9-21 Electrohydraulic pulse motor operation (*courtesy of Sperry Vickers*).

lic fluid is allowed to pass to the hydraulic motor through the spool at a rate determined by the spool opening. The ball nut is coupled to the hydraulic motor shaft, and as the hydraulic motor rotates, the ball nut turns with the spool, tending to close it. When electrical impulses stop, the spool stops rotating. The hydraulic motor continues to rotate until the attached ball nut completely closes the spool. The hydraulic motor remains positioned under load because of the high stiffness of the drive.

Positive logic level input pulses are fed to the 3-2 phase electronic control (Fig. 9-22). The electric stepping motor is a five-phase motor and requires that 3 and 2 phases be alternately excited. The logic level command signal is properly sequenced into 3-2 phase excitation and amplified to power the electric stepping motor. This controls the four-way hydraulic valve that, in turn, controls the hydraulic motor output.

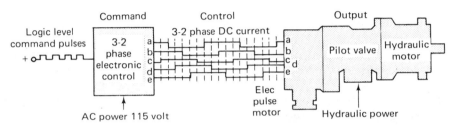

Figure 9-22 Electrohydraulic pulse motor control system (*courtesy of Sperry Vickers*).

EHPMs are controlled by digital signals and therefore are well suited for industrial applications where digital controls such as numerical control, process controllers, mini-computers, and other digital logic controls are found. The convenience and availability of digital control can be used in applications requiring 1-20 hp. These include NC position and contouring systems, turret indexing, cut to length systems, synchronized drive systems, grinder feeds, conveyer drive systems, tension systems, winders, and process control.

9-8 OUTPUT PERFORMANCE AND TESTING

Hydraulic motors are performance tested both at the bench and as part of complete systems to determine operation characteristics. Bench testing of the motor determines such characteristics as torque, speed, brake horsepower, effective flow rate, volumetric efficiency, and overall efficiency; whereas testing the motor in a system determines its compatibility with other components. Hydrostatic drives, for example, that couple variable displacement pumps to fixed displacement motors, add braking loads and other conditions to the motor during the application of overhauling loads which test the appropriateness of the reservoir size, line sizes, oil coolers, and so on, for particular applications.

Motor performance data are collected in tabular form and then plotted on graphs. Independent variables are specified (abscissa) and dependent variables (ordinate) are read directly or computed. Independent variables usually include displacement, pressure, and speed. Dependent variables usually include torque, fluid horsepower, flow rate, and volumetric efficiency. Horsepower and speed are oftentimes interchanged and specified as independent or dependent variables depending on the purpose and application of the graph. Other variables may also be interchanged to accommodate the purposes of specific manufacturers.

Graphs are commonly drawn to represent single independent versus dependent variables. Figures 9-23, 9-24, and 9-25 are typical for a high speed gear motor and illustrate this example. In Figure 9-23, output horsepower is plotted against speed for four pressures. The pressure range helps the fluid power system designer select the appropriate motor for a given horsepower requirement, since for a given pressure requirement and motor speed, the output horsepower can be selected directly from the graph. Figure 9-24 illustrates the plot of input flow rate in gpm against motor speed for the same four pressures. As pressure increases, the flow rate increases slightly for any given speed, indicating increased slippage

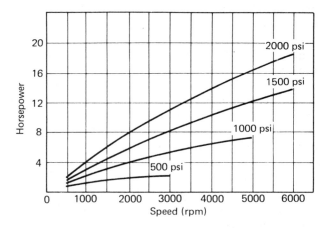

Figure 9-23 Fluid horsepower vs. motor speed.

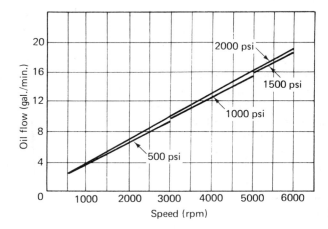

Figure 9-24 Oil flow vs. motor speed.

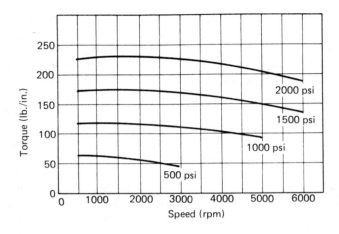

Figure 9-25 Motor torque vs. motor speed.

through the motor with increased load. Figure 9-25 shows the accompanying droop in torque that occurs because of fluid slippage and increased mechanical friction.

The tendency for characteristic curves to droop is an indication of the volumetric and mechanical efficiency of the unit, and is largely influenced by the design of the unit. Gear motors exhibit the greatest droop and least efficiency, but are less expensive to manufacture and purchase than equivalent vane and piston motor counterparts. Given the approximate speed range and pressure requirements, gear motors appropriate for a specific task can be selected. Where more exacting requirements must be met over a wide range of speed and loads, for example in airborne applications, close-fitting piston motors are more appropriate, even though they have a higher initial cost.

Another variable affecting motor performance is the viscosity of the fluid. As fluid temperature increases, it has a tendency to thin and increase slippage. Oil that becomes too thin reduces volumetric efficiency below an acceptable minimum. Conversely, selecting a more viscous oil will increase the volumetric efficiency but may also increase mechanical friction and heat associated with excessive fluid shear. Care should be exercised to match the required viscosity of the fluid to the system at the expected operating temperature.

While torque is generally considered to be constant with respect to speed for a given pressure in fixed displacement motors, the droop in the torque curves seen in Fig. 9-25 indicates that this is not the case. To produce a flat torque curve the speed range requires that pressure must be increased from the source to make up the fluid lost through slippage. In matching the pump to the motor in the system, this is taken into account by oversizing the pump so that the additional volume can be made up with increased pressure at higher motor speeds. This is also the case if constant speed is to be maintained at increased torque and pressure. As the load is increased on the motor, pressure and slippage increase, with some loss in speed. To maintain constant speed, pressure and fluid volume flow rate must be increased.

9-9 SUMMARY

Hydraulic motors convert the high energy level of the fluid to a more useable form by applying the fluid to the movable internal mechanism of the actuator. Movement thereby accomplishes the output objective of the system and lowers the energy level of the fluid, after which it is returned to the reservoir at low pressure.

Rotary actuators include continuous as well as limited rotation devices. Limited rotation devices, sometimes called torque motors, rotate from a few degrees to several turns in both directions, and receive widespread application in the production and material handling industries. Another continuous rotation hydraulic actuator is the high torque, low speed motor, used extensively where rotation speed is low, but where substantial torque must be generated. Gear reduction units are commonly connected to the actuator in a common case to keep the size of the unit small. Applications include windlasses, crane hoists, wheel motors, mixers, and others.

Rotary actuators are classified by type as gear, vane, and piston motors. Their displacement may be either fixed, as is the case with most gear motors, or variable, as can be the case with vane and piston motors. Cost and efficiency are typically lower with gear motors and higher for piston motors that represent the highest available standard in the industry with respect to precision engineering and machining. Efficiencies are also higher, although the mechanism of the motor is usually less tolerable of dirt, adverse conditions, and system degradation. Where a high level of system and fluid maintenance is provided, and costs are secondary to system quality, piston motors are preferred.

Variable displacement piston motors receive widespread use in industry. Displacement may be varied either manually, or the motor may be sensitive to changes in pressure brought about by changes in the load at the output shaft. If the motor receives fluid at constant volume from the pump, a decrease in displacement is accompanied by an increase in the speed of the output shaft. The torque from the variable displacement motor varies with the change in volume displacement and the horsepower, neglecting losses, is constant since the product of the pressure and volume flow rate remains relatively the same. Variable displacement motors, then, provide the capability for a constant horsepower output.

Electrohydraulic pulse motors are used on machinery where an unusual combination of power and precision for position and velocity control are required. Recent development in this technology has made them available in bolt-on versions suitable to a variety of applications. The output shaft speed of an EHPM is proportional to the impulse rate to the stepping motor, and the position of the output shaft is directly related to the number of input pulses. Rotation direction of the output shaft is determined by the sequence in which the input pulses are applied to the various phases of the electric stepping motor. Since EHPMs do not use a feedback loop, they are defined as open loop systems.

Hydraulic motors are performance tested both at the bench and as part of complete hydraulic systems to determine characteristics such as speed, torque, brake horsepower, and efficiency. Other variables, such as fluid temperature rise and pressure drop are considered when particular applications are treated. Motor performance data are typically collected in table form and then transposed to graphs using suitable independent and dependent variables.

STUDY QUESTIONS AND PROBLEMS

1. Compute the speed of a hydraulic motor with a fixed displacement of 2.2 in.^{3}/rev receiving fluid at 5 gal/min.
2. What flow rate at 4000 lbf/in.^{2} would it take for a hydraulic motor to deliver 20 FHP?
3. What is the displacement of a hydraulic motor receiving 18 gal/min at 1500 rev/min?
4. What flow rate would be required to drive a hydraulic motor with a displacement of 20 cm^3/rev turning at 2000 rev/min?
5. What FHP would the motor in Problem 4 deliver at a pressure of 25 MPa?
6. What theoretical torque could a 5 hp motor be expected to deliver at a rated speed of 1000 rev/min?
7. If in Problem 6 the pressure remains constant at 1750 lbf/in.^{2}, what effect would doubling the speed have on torque? How about halving the speed? What about at stall? Construct a table listing *flow rate, speed, torque,* and *power* to explain what happens.
8. What is the SI metric equivalent of Eq. 9-1?
9. Compute the theoretical torque from a hydraulic motor with a displacement of 2.5 in.^{3}/rev operating at 1500 lbf/in^2.
10. What theoretical torque could be expected from a hydraulic motor with a displacement of 75 cm^3/rev operating at 15 MPa?
11. Assuming no volumetric losses, what is the operating speed of a hydraulic motor with a 28 cm^3/rev displacement receiving fluid at 45 liters/minute?
12. At what pressure must a hydraulic motor with a displacement of 3.6 in.^{3}/rev operate to deliver a torque of 50 lbf/ft?
13. A hydraulic motor with a displacement of 1.1 in.^{3}/rev operates at 1500 rev/min. Assuming no volumetric losses, what must be the delivery to the motor?
14. Compute the overall efficiency of a hydraulic motor that operates at a pressure of 2500 lbf/in.^{2}, a flow rate of 5 gal/min, a speed of 1500 rev/min, and a torque of 20 lbf-ft.
15. What is the overall efficiency of a hydraulic motor with a displacement of 4.2 in.^{3}/rev that delivers 78 lbf-ft of torque at 1000 rev/min and 1750 lbf/in.^{2}?
16. In a bench test of a gear motor, it was found that a maximum overall efficiency of 84% was achieved at a pressure drop of 3400 lbf/in.^{2} across the ports when the ratio of the flow rate to speed was 0.01. Compute the torque and displacement of the motor.
17. Compute the volumetric efficiency of a hydraulic motor with a displacement of 0.55 in.^{3}/rev turning at 1800 rev/min and receiving fluid at 5 gal/min. How much fluid can be expected to return through the case drain if 75% of the loss in volumetric efficiency can be attributed to internal leakage?
18. If the expected wear over the life of a hydraulic motor with a displacement of 4 in.^{3}/rev results in a case drain loss of 3%, how much fluid will be returned through the case drain line at 1200 rev/min?
19. Compute the mechanical efficiency of a motor with an overall efficiency of 85% and a volumetric efficiency of 95%.
20. Describe the major difference between piston motors and pumps.

CHAPTER **10**
_____

VALVES

10-1 INTRODUCTION

Valves control the pressure, rate of flow, and direction of fluids in accordance with basic principles of flow. Several valves classified by function are listed in Table 10-1.

Pressure control valves limit and reduce pressure, sequence hydraulic operations by stepping system pressure, and counterbalance such external loads as vertical presses so they can't fall by maintaining back-pressure on the underside of the press cylinders. They are also used to unload slack cycle circuits at low pressure to reduce power consumption and provide switch signals at specified pressures to interface hydraulic systems with adjacent electrical controls.

Flow control valves vary the fluid flow rate using restrictions in fluid passages which may be fixed, variable, or flow and pressure compensated.

Directional control valves are used to check, divert, shuttle, proportion, and by other means manage the flow of fluid in one, two, three, four, or more ways. Pressure and flow compensation are commonly included in directional control valves.

Valves may be direct acting because of the arrangement of the internal mechanism design, or pilot operated from an adjacent or remote location. The actuating force to operate the valve can be supplied manually by the human operator, directly by the fluid under pressure or from a pilot circuit, or by electrical devices such as solenoids or servo electric drives. The internal mechanisms which physically accomplish the valving action use a variety of elements including poppets, diaphragms, flat slides, balls, round shear-action plates, and rotating or sliding spools (Table 10-2).

Other considerations related to valves in hydraulic transmission control include appropriate valve sizing, control panels, remote circuitry servo valve controls, and valve testing.

TABLE 10-1 VALVE FUNCTIONS CLASSIFICATION

Pressure control	Flow control	Directional control
Pressure relief	Fixed	Two way
Hydraulic fuse	Variable	Check
Pressure reducing	Compensated	Shuttle
Sequencing	Deceleration	Three way
Unloading	Flow divider	Four way
Counterbalance	Electrohydraulic	Limit switches
Pressure switches	Servo	Proportional electrohydraulic Servo

TABLE 10-2 VALVE PHYSICAL CHARACTERISTICS CLASSIFICATION

MODE OF OPERATION	Pressure control	Flow control	Direction control
Direct acting	√	√	√
Pilot operated	√	√	√
METHOD OF ACTUATION			
Manual		√	√
Fluid	√	√	√
Electrical		√	√
ACTUATING ELEMENT			
Poppet	√		√
Diaphragm			√
Plug	√		√
Ball	√		√
Rotating spool		√	√
Sliding spool	√	√	√

10-2 PRESSURE CONTROL VALVES

Relief valves are the most common of the pressure control valves. They are located near the pump outlet to protect the pump and provide the system plumbing and components with protection against pressure overloads. They also limit the output force exerted by cylinders and rotary motors. Circuits using positive fixed displacement pumps must have pressure relief valves.

Figure 10-1 illustrates the operation of a direct-acting spring-type ball relief valve. Spring pressure acts to keep the ball element against the seat and the valve in the closed position. System pressure at the port marked "inlet" acts against

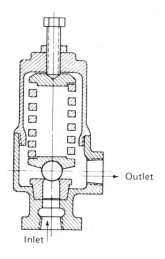

Outlet

Inlet

Symbol

Figure 10-1 Direct-acting spring-type ball relief valve (*courtesy of Sperry Vickers*).

the exposed area of the ball. When the force of the fluid (pressure × area) becomes greater than the opposing resistance offered by the spring, the ball is forced from its seat, the valve opens, and fluid is directed to the reservoir at low pressure through the port marked ''outlet.'' The pressure at which the valve opens is called the *cracking pressure*. The pressure at which the rated flow passes through the valve is termed *full flow pressure*. The pressure at which the valve ceases to pass fluid after being opened is called the *closing pressure*. Adjustment within the pressure range of the valve is made with the adjustment screw which acts to compress the check valve spring.

Compound pilot drained pressure relief valves increase pressure sensitivity by reducing the pressure override usually encountered with valves using only the direct-acting force of system pressure against the valve element. The valve in Fig. 10-2 illustrates the principle of operation. In schematic (a), fluid pressure acts on both sides of piston 1, which is held closed on its seat by a relatively light bias spring 2. In schematic (b), when the pressure increases sufficiently to move the pilot poppet valve 4 from its seat, fluid behind the piston goes to drain, and the resulting pressure imbalance on the valve piston causes it to move in the direction of lower pressure. This action compresses the piston spring 2 and opens the discharge port, preventing a further rise in pressure. The pressure setting is adjusted with the adjustment screw that increases or decreases the pressure required to unseat the pilot drain poppet valve.

Pressure reducing valves are used to supply branch circuits with fluid at a pressure lower than system pressure. Essentially, they step the pressure down to the requirements of the branch circuit by restricting fluid flow when branch circuit pressure reaches the pre-set limit determined by the rating and adjustment of the spool positioning spring. A pressure reducing valve is seen in Fig. 10-3. Fluid passes unobstructed from *C* to *D* in illustration (a). The pressure reducing spool valve is held open by spring 2, and leakage around the free floating spool passes at low pressure to drain. As the pressure of the system at the outlet of the valve

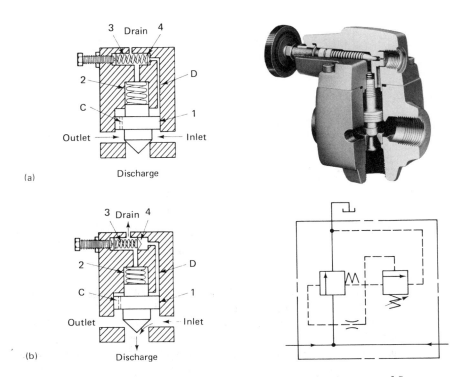

Figure 10-2 Compound pilot drained pressure relief valve (*courtesy of Sperry Vickers*).

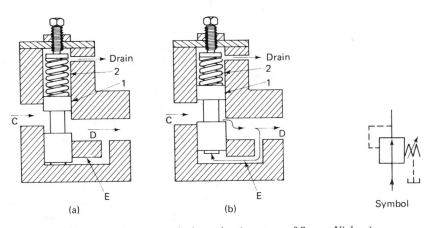

Figure 10-3 Pressure reducing valve (*courtesy of Sperry Vickers*).

increases, it acts through passage *E* against one end of the spool. An imbalance exists because the opposing end of the spool is open to drain. Further increase in pressure at the outlet causes the spool valve to compress the positioning spring and restrict the flow of fluid at the outlet until the pressure drops to the specified level. In this position, pressure at the outlet will remain constant regardless of inlet pressure. Reverse flow across the spool valve from D to C at the limiting pressure will cause the valve to close. If unobstructed reverse flow is necessary in the circuit, additional provisions, such as a directional control check valve, must be incorporated.

Sequence valves control the order of operation of two parallel branch circuits. Common examples of applications of this circuit include the operation of two work cylinders in sequence such that the second one activates after the complete extension of the first. Or a cylinder clamping device to secure a workpiece is closed, followed by the activation of a rotary actuator to complete a drilling operation.

The operating principle of the sequence valve is seen in Fig. 10-4. In the closed position A, fluid passes through the valve from C to D at low pressure. When the first step in the sequence has been completed and the clamping cylinder extends and stalls against the workpiece, system pressure increases to act against the indicated area of the piston. Continued increase in pressure causes the piston to compress the spring and unseat the valve, thereby directing the flow of fluid at high pressure through port E. Fluid pressure is maintained in both branches of the circuit at high pressure so long as the sequence valve is open. Adjustment of the sequence valve is accomplished by compressing or extending the piston spring with the capscrew. If a return flow of fluid from port E to C is necessary to retract the second cylinder, a return check valve must be incorporated in the sequence valve.

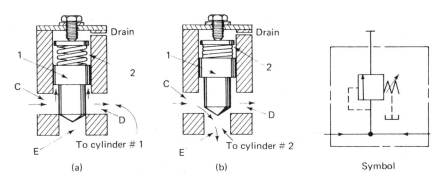

Figure 10-4 Sequence valve operating principle (*courtesy of Sperry Vickers*).

Unloading valves are used to discharge the fluid of fixed displacement pumps to the reservoir when they are not in use in the circuit. Their function reduces power and heat losses associated with discharging fluid at high pressure across direct acting pressure relief valves. A typical application incorporates an unloading valve in a two-pump circuit with the smaller high-pressure low-volume pump unloading the high-volume low-pressure pump at a predetermined pressure. The operation of the unloading valve is illustrated in Fig. 10-5. Pilot pressure acts against the differential area of the piston which is held in the normally closed position by the action of the spring. As pilot pressure increases, the piston

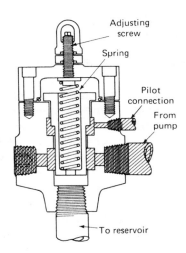

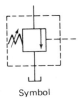

Symbol

Figure 10-5 Unloading valve (*courtesy of Mobil Oil Corporation*).

moves upward, discharging the pump to the reservoir. The power savings gained by unloading the large volume pump rather than discharging it over a direct acting relief valve equals

$$Fhp = \frac{p \times Q}{1,714}$$

where the pressure (p) equals the discharge pressure across the direct acting relief valve and the flow (Q) equals the rate output of the pump. For a traverse feed pump operating at 300 lbf/in.^{2} at 20 gal/min on the low pressure side, the savings would be greater than

$$Fhp = \frac{300 \times 20}{1,714} = 3.5 \text{ hp}$$

assuming the high-volume pump were isolated from the system during the high-pressure portion of the cycle.

Counterbalance valves are used to maintain back pressure on cylinders that are generally mounted vertically, such as presses that position and sustain vertical loads. They are not used, typically, to support varying loads that would require frequent variance of pilot pressure. For this purpose, a pilot operated check valve should be substituted for the counterbalance valve.

The counterbalance valve in Fig. 10-6 might be operated either by direct or remote pilot. If direct pilot operation is used, pressure on the rod end of the ram must exceed the pressure setting of the valve. This is usually 10% greater than the pressure required to sustain the vertical load. In Fig. 10-6, the pressure at the rod end of the cylinder resulting from the load resistance alone is

$$p = \frac{F}{A} = \frac{20,000}{20} = 1000 \text{ lbf/in.}^2$$

To operate the counterbalance valve using a direct pilot pressure 10% greater than normal system pressure, the valve would have to be set at 1,100 lbf/in^2.

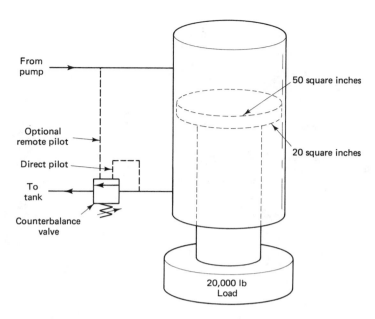

Figure 10-6 Application of a counterbalance valve.

If the counterbalance valve were to be operated by a remote rather than a direct pilot, the pilot pressure setting could be much lower. In Fig. 10-6, for example, pressure at the blank end of the cylinder must be raised only by the amount necessary to increase the pressure at the rod end of the cylinder by 100 lbf/in.², 10% over existing pressure, to operate the valve. At the blank end of the cylinder, the increase in force to raise the pressure on the rod end 100 lbf/in.² is

$$F = p \times A = 100 \times 20 = 2{,}000 \text{ lbf}$$

and the pressure at the blank end to provide this force is

$$p = \frac{F}{A} = \frac{2{,}000}{50} = 40 \text{ lbf/in.}^2$$

This indicates that 40 lbf/in.² from the pump portion of the circuit might be used as a pressure setting for the counterbalance valve rather than 1,100 lbf/in.² if the pilot is operated directly from pressure at the rod end of the cylinder.

The load holding valve shown in Fig. 10-7 is an example of a direct pilot operated check valve used to prevent dropping of a loaded hydraulic cylinder, either for safety purposes or because of seepage across the control valve spool. Where the loaded cylinder is raised, fluid enters through port 1 or 3 to lift check valve 4, which is lightly sprung, and then flows out port 2 to raise the loaded cylinder. When the directional valve is moved to the neutral position, check valve 4 closes, blocking the return of oil through the valve. To lower the load, pilot oil is directed back through port 2 to unseat check valve 5. Notice that check valve 5 is held closed by a heavy spring 7 through a piston with a larger face area 6. When

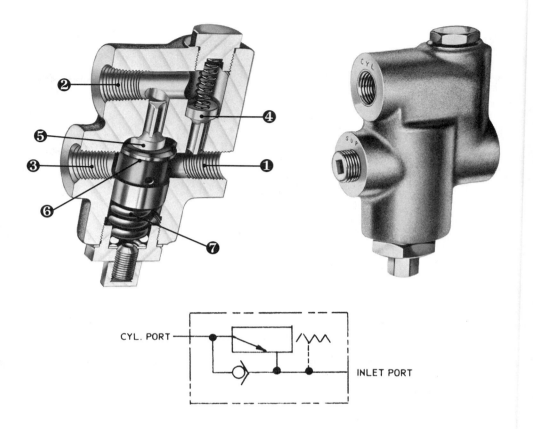

CYL. PORT

INLET PORT

Figure 10-7 Direct pilot operated check valve (*courtesy of Webster Electric Company, Inc.*).

pilot pressure unseats valve 5, the pressure acts on the larger area of piston 6. This has the effect of reducing the pressure required to keep valve 5 unseated during the lowering stroke.

A double-acting cylinder pilot operated check valve such as that shown in Fig. 10-8 is used to position varying loads on both the extension and retraction strokes of the cylinder. It is used on elevators, forklifts, and boom and crane equipment and effectively locks the cylinder any time the direction control valve is shifted to the neutral position. The schematic shows how the valve would be installed in the circuit. To move the cylinder rod out, the oil flow is directed to the blank end of the cylinder through check valve B, and also moves a pilot piston to unseat check valve A to allow oil from the rod end of the cylinder to return to the reservoir. To move the rod in, the process is reversed. Oil flow unseats check valve A to move the rod in and at the same time pilot operates check valve B to allow the oil from the blank end of the cylinder to return to the reservoir. In the neutral position, both check valves A and B are closed, locking the cylinder from drifting in either direction.

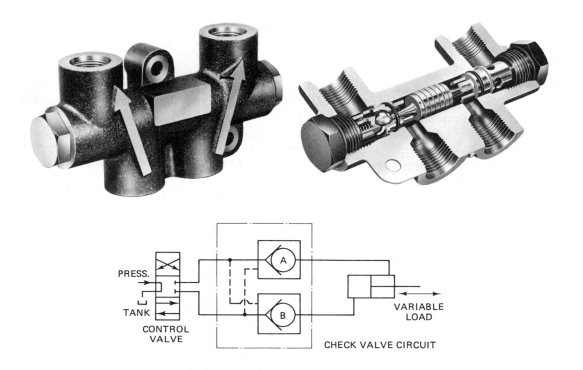

Figure 10-8 Double-acting pilot operated check valve (*courtesy of Webster Electric Company, Inc.*).

Pressure control valves from some manufacturers serve different functions, depending on the arrangement and connection of ports. Direct-acting or balanced spool valves, for example, can be used as pressure relief, pressure reducing, pressure sequencing, unloading, and counterbalance valves with only minor changes in port, pilot, and drain connections. Typically, the solution of the problem of valve selection is determined by the application, performance characteristics in a given cycle of operation, and cost factors.

10-3 FLOW CONTROL VALVES

Flow control valves accurately limit the fluid volume flow rate from fixed displacement pumps to or from branch actuator circuits. They provide velocity control of cylinders, or speed control of hydraulic motors. Typical applications include regulating cutting tool speeds, spindle speeds, surface grinder speeds, and the travel rate of vertically supported loads moved upward and downward by fork lifts and dump lifts. Flow control valves also are used to allow one fixed displacement pump to supply two or more branch circuits fluid at different flow rates.

Flow control valves present a fixed or variable restriction to the flow of oil, may be compensated for pressure and/or temperature changes, may vary the flow rate with respect to time as in the case of deceleration valves, or may actually throttle the flow rate by compensating for changes in system demand during the operation cycle.

Typically, fixed displacement pumps are sized to supply maximum system volume flow rate demands, and where industrial applications feed two or more branch circuits from one pressurized manifold source, there will be an oversupply of fluid in any circuit operated by itself. Mobile applications that supply branch circuits, such as the power steering and front end loader from one pump, pose a similar situation. If left unrestricted, branch circuits receiving an oversupply of fluid would operate at greater than specified velocity, increasing the likelihood of damage to work and personnel.

Speed control of circuits receiving an oversupply of oil is accomplished in three ways:

1. By metering the fluid supplied *to the* actuator.
2. By metering the fluid returned *from the* actuator.
3. By bleeding excess fluid back *to the* reservoir.

A flow control valve of the adjustable but noncompensated type is seen in Fig. 10-9. Regulated flow passes through the stepped needle valve in one direction, and free reverse flow passes through the poppet check valve in the opposite direction. The first three turns of the adjustable two-step needle valve from the fully closed position provide fine adjustments for low flow settings. The last three settings provide conventional throttling control to the fully open position.

Some branches of the valve industry describe the flow characteristics of control valves in terms of a *flow* coefficient C_v. This is not to be confused with the *velocity* coefficient described in Chapter 5. Here the flow coefficient is defined as the flow in gallons of water per minute at 60°F, at a pressure drop of 1 lbf/in.^{2} across the valve. There is no similar coefficient in SI units.

The volume flow rate for a hydraulic fluid flowing through a control valve using the flow coefficient is determined from

$$Q = C_v \sqrt{\frac{\Delta p}{Sg}} \qquad (10\text{-}1)$$

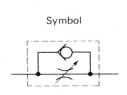

Symbol

Figure 10-9 Adjustable noncomponsated flow control valve with reverse bypass (*courtesy of Parker Hannifin, Manitrol Division*).

where the flow rate (Q) is in gal/min, the flow coefficient (C_v) is determined by each manufacturer for specific types and sizes of valves, the pressure drop across the valve (Δp) is in lbf/in.^{2} and the specific gravity of the fluid (Sg) is at a specified test temperature and viscosity. Flow coefficients in the range of 0.20–7.50 are typical for a test fluid viscosity of 150 SSU (32 cSt) at 140°F.

Example 1

Compute the flow coefficient for a flow control valve similar to that in Fig. 10-9 having a pressure drop of 300 lbf/in.^{2} while passing 30 gal/min of fluid with a Sg of 0.85 at 140°F. If the return flow rate is the same and flow coefficient increases to 5.41, what will be the pressure drop across the valve during return flow?

Solution Rearranging Eq. 10-1 to solve for the flow coefficient (C_v)

$$Q = C_v \sqrt{\frac{\Delta p}{Sg}}$$

and

$$C_v = \frac{Q}{\sqrt{\frac{\Delta p}{Sg}}}$$

Substituting appropriate values in the formula, the flow coefficient for the direction of controlled flow is

$$C_v = \frac{30}{\sqrt{\frac{300}{0.85}}}$$

and

$$C_v = \frac{30}{18.79} = 1.597$$

Rearranging Eq. 10-1 to solve for the return flow pressure drop (Δp) across the valve

$$Q = C_v \sqrt{\frac{\Delta p}{Sg}}$$

$$\sqrt{\Delta p} = \frac{Q\sqrt{Sg}}{C_v}$$

and

$$\Delta p = \frac{Q^2 \times Sg}{C_v^2}$$

Substituting appropriate values in the formula, the pressure drop across the valve on return flow is

$$\Delta p = \frac{(30)^2 \times 0.85}{(5.41)^2}$$

and

$$\Delta p = \frac{765}{29.27} = 26 \text{ lbf/in.}^2$$

Pressure compensated flow control valves allow for the passage of the same volume flow rate at the valve outlet regardless of pressure fluctuations at the valve inlet or outlet. Figure 10-10 illustrates the action of a controlled orifice valve. Fluid passes through the valve from the inlet to the outlet, first through a fixed and then through a controlled variable orifice. Pressure drops slightly across both orifices at flow rates less than rated flow. Increase in fluid flow causes an increase in the pressure drop across the fixed orifice at the inlet, unbalancing the sliding valve that compresses the valve spring as it moves to close the controlled variable orifice at the valve outlet. Reducing the size of the controlled orifice acts to balance the pressure across the fixed orifice, and spring action tends to return the valve to its fully open position. The regulating of the valve spool that fluctuates to maintain a constant pressure drop across the fixed orifice keeps the volume flow rate at the output nearly constant. Adjustable pressure compensated flow control valves use both controlled and variable orifices to compensate for pressure changes at the valve inlet or outlet, but change the initial flow rate through the valve by varying the size of the controlled orifice.

Temperature compensation is added to flow control valves to make allowance for changes in fluid viscosity that result from temperature change. Since increases in temperature decrease fluid viscosity, flow rates through passages tend to increase as the temperature of the working fluid rises. Compensation must either provide an additional restriction controlled by a temperature sensitive element, such as a tapered metal rod that can move in and out of a fixed orifice as temperature changes, or remove the effects of viscosity by providing that fluid pass an adjustable fixed orifice with a knife sharp edge. This eliminates the effects of the wall attachment properties of the oil and subsequent changes in flow that usually occur with temperature changes. Flow variations of less than 5% for

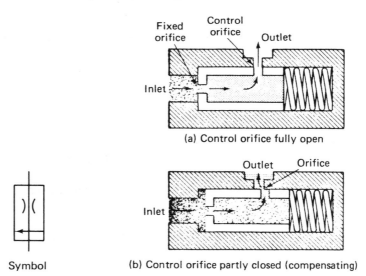

Symbol

(a) Control orifice fully open

(b) Control orifice partly closed (compensating)

Figure 10-10 Controlled orifice flow control (*courtesy of John Deere Service Publications*).

fluids within the temperature range of 68° to 140°F are reported using sharp-edged orifices.

Deceleration valves are used to reduce or increase fluid flow rate at some intermediate position of the actuator. They operate independently from built-in cylinder and other actuator cushioning devices. For example, an operator may want to stop or slow down a hydraulic cylinder in some intermediate position from a higher velocity to a lower velocity to begin a machining or other low speed operation. This would be accomplished by placing a normally open deceleration valve in the circuit. Similarly, deceleration valves may also be used with flow control valves to speed up operations by allowing fluid flow to increase at some intermediate position by placing a normally closed valve in the circuit.

Most deceleration valves are manually operated tapered spool element valves that are cam operated by a follower on the actuator or workpiece. Figure 10-11 illustrates a simplified circuit operating in the forward and reverse positions in which a drill head is decelerated before contact is made with the workpiece. On extension (Fig. 10-11a), the flow of oil is directed to the blank end of the cylinder, and fluid exiting the rod end flows back to the reservoir through the normally open deceleration valve. At an intermediate position (Fig. 10-11b), before the drill head contacts the workpiece, the cam on the cylinder rod depresses the deceleration valve spool, gradually cutting off the flow of oil and forcing the return to be rerouted through the flow control valve. When the actuator reverses its motion on the return stroke (Fig. 10-11c), the flow of oil to the rod end of the cylinder passes through the internal free flow check valve. The spool is held open by the return spring until the tapered spool element is actuated by the tapered cam. The amount of restriction imposed by the deceleration valve is dependent on spool travel, and may be varied from imposing a partial restriction to fully closing the valve. Some valves incorporate an adjustable needle valve to control fine metered flow when the valve is in the closed position. Return flow through the valve unseats the integral poppet check valve to permit unrestricted flow regardless of the position of the spool valve.

Flow divider valves direct the volume flow rate of fluid to two or more circuits, usually on a priority or proportional basis. While they are flow control valves, a part of their function is also directional control.

Mobile hydraulic systems make extensive use of flow divider valves. For example, the flow may be divided between the power steering circuit and the front end loader circuit of a wheel tractor, or between the left and right tracks of a power steering circuit of a crawler tractor.

On circuit demand, a simple priority flow divider valve directs the flow of oil from one supply or inlet port to two outlet ports. The valve in Fig. 10-12 directs oil first to the priority outlet, and when the demand from this circuit is satisfied, directs oil to the secondary outlet. For example, a power steering circuit may require 3 gal/min and the front end loader 10 gal/min on a wheel tractor. At full speed the pump will supply 15 gal/min. The power steering circuit will receive the first 3 gal/min available from the pump. When the pump speed is increased and more than 3 gal/min is available, the secondary circuit will receive the remainder of the oil. In operation, the valve spool is held to the right by the action of the

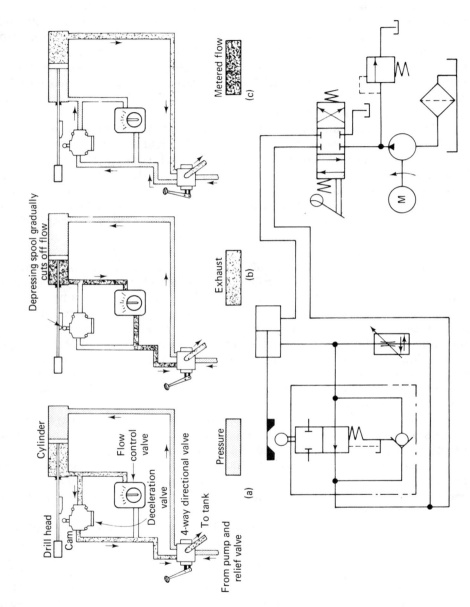

Figure 10-11 Deceleration valve circuit and graphic symbol schematic (*courtesy of Sperry Vickers*).

Cylinder

Drill head

Cam

Flow control valve

Deceleration valve

4-way directional valve

To tank

From pump and relief valve

Pressure

(a)

Depressing spool gradually cuts off flow

Exhaust

(b)

Metered flow

(c)

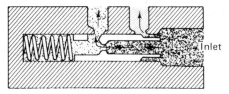

Priority Secondary
outlet outlet

Figure 10-12 Priority flow divider valve (*courtesy of John Deere Service Publications*).

return spring and low pressure oil, and directs flow to the primary outlet. Oil at higher inlet pressure acts against the right side of the spool. As the output of the pump increases to a predetermined amount, for example 3 gal/min, the pressure difference across the orifice through the center of the valve spool increases until inlet pressure overbalances both the pressure on the priority outlet side of the spool and the force of the spring, thereby moving the valve spool to the right. This allows the remaining oil to be directed to the secondary port. So long as the flow is in excess of 3 gal/min, the front end loader or other equivalent circuit will get the remainder of the oil. When the flow drops to 3 gal/min or below, the priority valve will move to the right to direct all oil to the power steering circuit.

10-4 DIRECTIONAL CONTROL VALVES

Directional control valves manage the flow path of the fluid in the system. They function to stop, start, check, divert, shuttle, divide proportionally, and by other means direct the flow of oil in one, two, three, four, or more flow paths or ways. Pressure and flow compensation, in addition to the major control function, are frequently built into these valves. Directional control valves are also common as an integral part of such components as mobile hydraulic pumps to add flexibility through internal directional control. Other modern applications include electro-hydraulic servo valves that control the flow of fluid in response to programmed and feedback signals from electronic and other logic master control systems.

Directional control valves may be direct acting or pilot operated from an adjacent or remote location. They may be manually actuated physically by an operator, hydraulically by pilot pressure, pneumatically by air pressure, and electrically by solenoids or servo drives receiving control signals from control panels, servo systems, and programmed signals. Valving action is commonly accomplished by the action of balls, round shear plates, and rotating and sliding spool elements. The majority of directional control valves use sliding spool elements.

Three terms receive widespread use in the description of directional control valves—*position, way,* and *port.* Standard symbols are used to represent their meaning.

Position refers to the number of positions within the valve body that the valve shifting mechanism or element, such as the sliding spool in a spool valve, can assume in directing the flow of oil through the valve. The symbol used to describe the number of valve positions is called the valve envelope. It is divided into sections to indicate the number of finite positions that the valve element can assume. A finite number of positions is indicated by solid lines that divide the envelope. An infinite number of positions is indicated by dotted lines between each segment and the addition of horizontal bars to the envelope (Fig. 10-13).

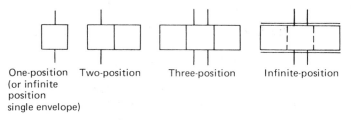

One-position   Two-position   Three-position   Infinite-position
(or infinite
position
single envelope)

Figure 10-13   Control valve envelope with ports.

The term *way* is used to mean flow path through the valve, including reverse flow. One-way, two-way, three-way, and four-way valves are common. One-way valves will allow fluid to flow in only one direction through the valve. These are simple check valves. Two-way valves allow fluid to flow in both directions in one fluid passage through the valve. These are simple shut-off valves. Four-way valves are reversing valves allowing fluid to flow in both directions in two passages to operate a cylinder, motor, or other device in the forward and reverse direction. The flow path through a valve in each position is indicated by lines within each position segment of the control valve envelope showing the flow path and direction of flow with an arrow.

A *port* is a plumbing connection to a valve passage. For example, one-way check valves must have two ports, an inlet and an outlet. Similarly, two-way valves also have two ports. Four-way valves must have four ports to allow reverse flow to an actuator, but may have six or more ports, some of which are through or bypass ports.

A complete control valve designation typically includes the number of ways, positions, and ports in that order, with other descriptions of flow conditions within the valve added to clarify the operational characteristics of the valve more fully. The arrangement of the ways in the center position, for example, is necessary to clearly describe flow conditions in a four-way valve. This will be included in the discussion of four-way valves. Several control valve symbols and their designations are illustrated in Fig. 10-14.

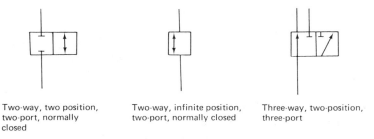

Two-way, two position,   Two-way, infinite position,   Three-way, two-position,
two-port, normally   two-port, normally closed   three-port
closed

Figure 10-14   Control valve symbols.

Three-way, three-position, three-port, all ports blocked in center position, normally centered

Four-way, three-position, four-port, all ports blocked in center position, normally centered

Four-way, infinite-position, four-port, all ports blocked when valve is in normally centered position

Figure 10-14 (*continued*)

Manual turn handle on-off valves may be classified as two-way directional control valves even though partial opening also constitutes flow control. Fluid can flow two ways through the valve. Globe valves, turn cocks, gate valves, and needle valves may be categorized under this classification (Fig. 10-15). They are typically fitted with a turn handle and are actuated manually. Because of restriction to the flow of fluid and leakage around the stem, they are usually kept in the fully open or closed position. Needle valves, often used for uncompensated flow control adjustment, are an exception to this statement. Because these valves are not hydraulically balanced, they may require substantial operating force when the surface area of the valve element is exposed to system pressure.

A high capacity two-position, two-way solenoid-operated valve with spring return is illustrated in Fig. 10-16. Valves of this type have capacities to 300 gal/min at pressure to 3000 lbf/in^2. Pilot-operated versions of this valve are also available. The hydraulically balanced spool requires pilot pressures of only 100 lbf/in.^{2} for shifting, making other mediums such as air appropriate to supply pilot pressure. High shocks usually associated with two-way valves are reduced by

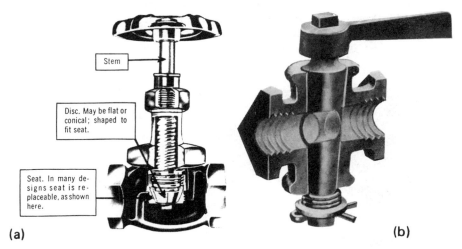

Stem

Disc. May be flat or conical; shaped to fit seat.

Seat. In many designs seat is replaceable, as shown here.

(a)

(b)

Figure 10-15 Manual turn handle on-off valves; (a) Globe valve; (b) Turncock; (c) Gate valve (*courtesy of Sun Oil Company*).

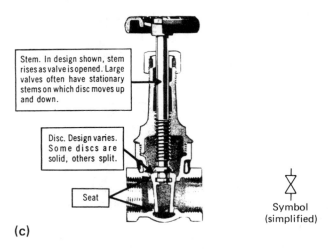

Stem. In design shown, stem rises as valve is opened. Large valves often have stationary stems on which disc moves up and down.

Disc. Design varies. Some discs are solid, others split.

Seat

(c)

Symbol
(simplified)

Figure 10-15 (*continued*)

metering fluid out of the top or bottom of the valve pilot chambers, thus controlling the travel rate of the valve spool.

Direct acting check valves are perhaps the simplest of the flow control valves. They are one-way valves and function to stop flow in one direction and offer nearly unrestricted flow in the opposite direction. They have wide application in circuitry that holds the load resistance in position when the flow of fluid from the pump stops, and in others that abruptly reroute fluid direction. Check valves are available in both straight through and right angle designs and use a variety of internal valve elements to accomplish the valving action, including flappers, ball checks, and poppets. Flapper valves are available in both small and large sizes, since they offer minimum obstruction to return flow. They may be held closed by spring pressure, gravity, or a reverse flow pressure differential. Ball check valves unseat against spring pressure by the action of pressurized fluid against the exposed cross section of the ball element. Once the ball element is

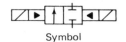

Symbol

Figure 10-16 High capacity solenoid controlled pilot actuated two-way valve (*courtesy of Parker Hannifin*).

unseated, fluid flows around the ball in the case of in-line check valves, or past the unseated ball in the case of right angle check valves. Poppet valves operate similarly to ball check valves. The valve is kept in the normally closed position by the action of the return spring against a guided piston or plug (Fig. 10-17). This assures proper alignment and seating of the accurately fitted poppet valve against the beveled seat. Opening of the valve is caused by fluid pressure acting against the exposed area of the poppet.

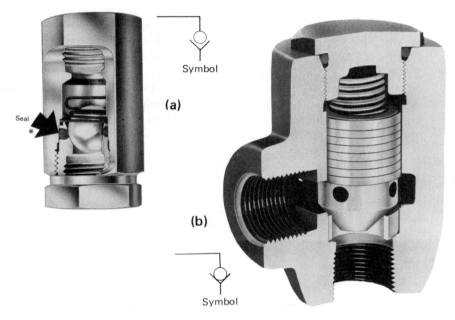

Figure 10-17 (a) In-line check valve (*courtesy of Kepner Products Company*); (b) Poppet type check valve (*courtesy of Miller Fluid Power Division Flick-Reedy Corporation*).

Pilot-operated check valves permit free flow in one direction and are closed in the opposite direction. These valves are opened or closed hydraulically by piping fluid, usually at system pressure, to the control spool (Fig. 10-18). Pilot

Symbol

Figure 10-18 Pilot operated check valve (*courtesy of Miller Fluid Power Division Flick-Reedy Corporation*).

fluid may be isolated from system fluid, however, permitting other materials than system fluids and pressures to be used to pilot-operate the check valve. Pilot-opened check valves function to permit reverse flow in the normally closed direction when pilot pressure opens the valve. This makes them particularly applicable in circuits where a preload is desirable, or in circuits where a constant load must be supported.

Shuttle valves are direct-acting double check valves with a cross bleed. That is, they permit reverse flow between two of the three connecting legs of the circuit. Shuttle valves may be biased—that is, spring loaded in one direction to give priority to flow; or unbiased, permitting priority to be determined by circuit conditions. Shuttle valves are essentially OR logic elements. An input at either *A* OR *B* will give an output at *C*. This characteristic of shuttle valves permits many interesting hydraulic circuit variations. An application of a shuttle valve is illustrated in the braking circuit shown in Fig. 10-19 which offers two control stations, either *A* or *B*, from which to operate the hydraulic brakes of a mobile piece of equipment. The shuttle valve allows an instant change of control from one control station master cylinder to the other.

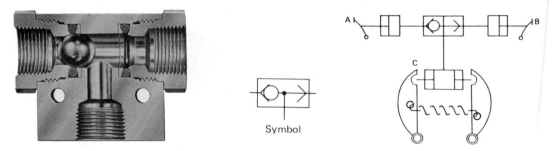

Figure 10-19 Shuttle valve application in hydraulic brake circuit (*courtesy of Kepner Products Company*).

Three-way valves provide three flow paths for oil. They are usually two-position valves, but may have a center neutral position. The three ports of three-way valves may be connected in circuits to accomplish a variety of functions, although their primary function is to pressurize and exhaust one actuator, such as a cylinder or fluid motor. The arrangement possibilities of the three ports are illustrated in Fig. 10-20.

Usually, single-acting cylinder actuator systems use three-way valves to alternately connect the pressure port to the working cylinder, and then block the pressure port to return the fluid from the cylinder to the reservoir. This allows the system to return to its original position by gravity or through the use of helper springs (Fig. 10-21). Another application of three-way valves with single-acting systems alternately directs the flow of pressurized fluid first to one port and then to the other. In this mode, the valve might alternately actuate two rotary actuators, or direct the fluid source alternately to two circuits, such as a front end loader or a back hoe circuit on a mobile piece of equipment. A third application distributes pressurized fluid from two pressurized ports to the third port. The

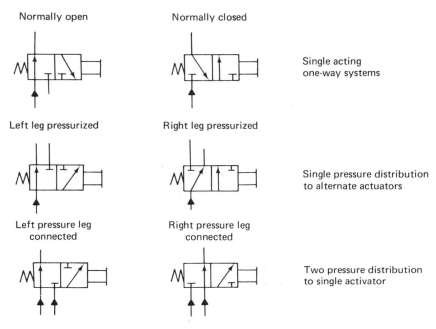

Normally open   Normally closed

Single acting
one-way systems

Left leg pressurized   Right leg pressurized

Single pressure distribution
to alternate actuators

Left pressure leg   Right pressure leg
connected   connected

Two pressure distribution
to single activator

Figure 10-20   Three-way valve applications.

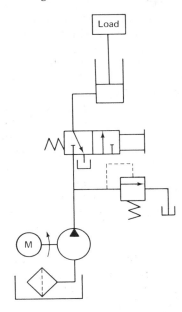

Figure 10-21 Three-way valve used
with single-acting system.

third port may be connected to a rotary actuator, the speed of which is controlled
by alternately connecting two fixed displacement pumps of different sizes.

    Two three-way valves may be paired to act as a four-way valve. Several
operational features are inherent to such a circuit (Fig. 10-22). In the circuit

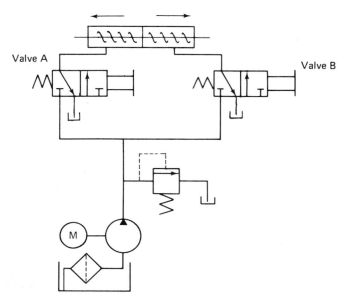

Figure 10-22 Two three-way valves in parallel.

illustrated, both pressure ports from the pump are blocked and both working or actuator ports are connected to the reservoir. The cylinder actuator is held at rest in the center position by the return springs. If valve *A* is operated, the piston inside the double-acting cylinder will move to the right. If valve *A* is released, the valve is spring returned to the closed position, the cylinder port is opened to the reservoir and the cylinder piston spring returns the piston to the center position. Alternately operating and releasing valve *B* causes the piston in the cylinder actuator to react similarly, but in the opposite direction. Actuation of valve *A* followed by actuation of valve *B* will cause the piston to move right and then to stop, since equal pressure is applied to both sides of the piston. This operational characteristic allows a third position for the cylinder piston at any place between full stroke in either direction. Other benefits are also cited from the use of two three-way valves in place of one four-way valve.[1] First, close coupling of three-way valves reduces pressure drop with the result that cylinder velocities can be increased. Secondly, in pneumatic systems, air consumption may be reduced by simpler plumbing usually associated with valves located in the same proximity as the cylinder.

Four-way valves provide four main flow paths for fluid through the valve body. They are usually two-position or three-position valves with four ports, although valves with more than three positions and four ports are common, particularly in mobile hydraulics applications that often use infinite-position valves for fine control and additional bypass, or through ports for other circuits.

[1] Staff Report, "Directional Valves for Profit-Making Designs," *Hydraulics and Pneumatics Magazine,* March 1973, HP-2.

Valves   Chap. 10

Four-way valves function primarily to pressurize and exhaust the two working ports of the valve to operate the actuator in both directions. The four ports of a four-way valve are typically identified as pump (P), reservoir (T), and the two working or actuator ports, (A and B) (Fig. 10-23). Sliding spools and rotating shear plates are the most common internal valve elements used to accomplish shifting. Flat sliding shear plates, rotating plugs, and other valve elements are less common.

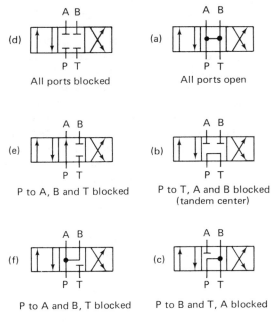

(d) All ports blocked

(a) All ports open

(e) P to A, B and T blocked

(b) P to T, A and B blocked (tandem center)

(f) P to A and B, T blocked

(c) P to B and T, A blocked

Closed center

Open center

Figure 10-23 Open and closed center valve positions.

The center position of four-way spool valves is important in determining which valve is appropriate for a particular application. The center position of a four-way three-position valve also influences the operation of the rest of the hydraulic system when the valve is in the neutral position. Two basic center arrangements are available: open center and closed center. Several options in both open and closed center valves are available.

Open center valves connect the pump to reservoir ports in the center position. System pressure drops and the pump, which is usually of the fixed displacement type, is unloaded. This reduces power consumption during the time that the system is idling and provides circulating fluid for cooling purposes. The load resistance usually coasts to a stop, and before movement can be initiated, the system must come to pressure. Several advantages accrue from open center systems in applications given the following conditions:

1. System idling time consumes a high percentage of operating time.

2. A fixed rather than variable displacement pump is to be used.

3. A single actuator is to be independently powered, since when the valve is in the center position, system pressure is reduced by unloading the fluid to the reservoir.

4. Manual or electric operators are to be used to actuate the control valve, since system pressure is usually too low without additional metering provisions to supply adequate pilot pressure.

5. Machine response time is not critical, since the system must be brought to pressure after the control valve is shifted from the center position.

In closed center valves, the pump and tank ports are not connected. When the control valve is shifted to the center position, system pressure remains at operating pressure or above if fluid is discharged over the relief valve. If a fixed displacement pump is used, some pressure surge can be expected to occur when the valve is shifted to the neutral position unless flow compensating devices are incorporated in the control valve. Because system pressure is maintained, pump delivery is immediately available for use in other parts of the system. If leakage is minimum across the valve spool, the valve can be expected to hold cylinders in a blocked position, support loads, and otherwise control actuator movement with a minimum of creep. Accumulators connected in parallel with the pump will also remain charged. Several advantages are inherent in these systems given the following conditions:

1. System idling time consumes a small percentage of operating time.

2. A variable displacement pump can be used, thus limiting wasted power during idling.

3. The system must operate more than one actuator using more than one independent control valve.

4. Pilot oil pressure to operate control valves must come from the system.

5. Machine response time is critical.

Several open and closed center valve positions and representative circuit applications are worthy of mention. These are illustrated in Fig. 10-24. In Fig. 10-24a, the open center valve has all ports connected to the reservoir when the valve is in the center position. The circuit is appropriate for use with a fixed displacement pump when it is not working. The motor in the circuit will coast when the valve is spring centered and will start smoothly when the valve is shifted to either extreme position. Figure 10-24b illustrates a circuit in which two open tandem center valves are used to control two cylinders. In the center position, fluid flows from the pump through each valve in series and unloads to the reservoir. All cylinder ports are blocked, holding their respective positions. The tandem circuit is used extensively in mobile hydraulic applications. The series connection in Fig. 10-24b allows only one valve to control the direction of fluid flow to one actuator when it is shifted to either extreme position, although feathering of the valve hand controls by the operator allows for two or more actuators to be simultaneously actuated to the extent that fluid is available from the pump.

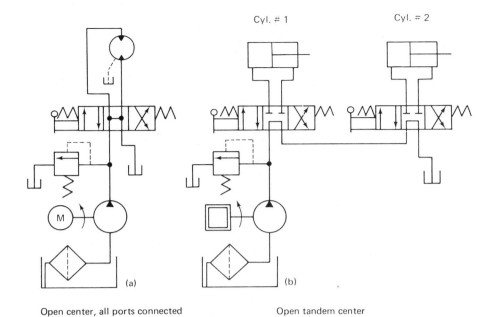

Cyl. # 1   Cyl. # 2

(a)

Open center, all ports connected

(b)

Open tandem center

(c)

Open center, pump and one cylinder
port to drain, other cylinder port
blocked

Optional
P.R valve

(d)

Closed center, all ports blocked

Figure 10-24 Open and closed center valve circuits.

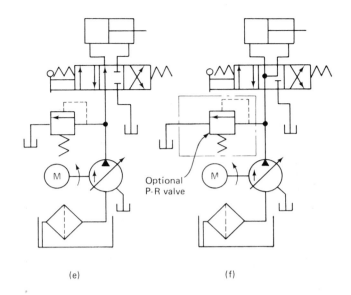

(e)

(f)

Closed center, pump to one cylinder
port, other cylinder port and reservoir
port blocked

Closed center, pump to both
cylinder ports, reservoir port
blocked

Figure 10-24 (*continued*)

Valves are usually stacked in banks with convenient hand controls positioned to give the operator maximum flexibility of manipulation. Four-way, three-position, six-port open tandem center valves eliminate this difficulty by providing a bypass passage to supply fluid independently to each control valve in a parallel circuit arrangement. The open center circuit in Fig. 10-24c connects the pump and one cylinder to the reservoir, and blocks the other port when the control valve is in the center position. This unloads the pump and one side of the cylinder. The blocked cylinder port arrangement gives some capability in cylinder positioning and holding of the load in one direction. Shifting the control valve to supply fluid to the rod end of the cylinder moves the load left. Centering the valve blocks the blank end of the cylinder which might be desirable in supporting stationary or overhauling loads. The cylinder would not sustain overhauling loads in the opposite direction, however, since the rod end of the cylinder is open to the reservoir. The closed center circuit illustrated in Fig. 10-24d blocks all ports when the valve is shifted to the center position. The circuit locks the cylinder in position and will support loads in both directions. The pressure-compensated variable displacement pump maintains pressure in the system at above working pressure, depending on the setting of the pump. Pressure surging can be expected when the control valve is shifted to the center position unless compensation such as tapered spools or other flow restriction in the valve is provided. Figure 10-24e illustrates a closed center circuit that is also appropriate for positioning and holding loads. The pump is connected to one end of the cylinder, and the other cylinder and reservoir port are blocked. Fluid loss past the piston of the cylinder sustaining compressive load on the cylinder rod is made up by the pump. Fluid loss past the control valve spool at the blocked port from the return end of the cylinder, however, will result in cylinder creep to the right. A pressure-compensated variable displacement pump controls system pressure. Another interesting closed center circuit is illustrated in Fig. 10-24f. In this circuit, both ends of the cylinder are pressurized by the pump and the reservoir port is blocked in the center position. When the control valve is shifted to the center position, system pressure is applied to both ends of the single end rod cylinder. This has a regenerative effect causing the cylinder to move rapidly to the right because of the difference in effective areas between opposite sides of the piston. Fluid from the rod end of the cylinder joins the flow from the pump in extending the cylinder causing rapid extension. This circuit may also be used with fluid power motors to prevent cavitation.

Single and multiple section four-way valves may be hand operated or pilot valve operated. Pilot valves themselves may be operated manually or by solenoids receiving electric signals from manually operated or electrically programmed control panels.

A manually operated single body, four-way, three-position, spring centered control valve is seen in Fig. 10-25. The spool is hydraulically balanced to reduce shifting effort and side thrusts. The valve spool is centered from either position by return springs when the control lever is released or held in any shifted position by a detent. Notice the hand operator and centering springs on the symbol.

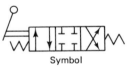
Symbol

Figure 10-25 Manually operated four-way, three-position, spring centered control valve (*courtesy of Sperry Vickers*).

Four-way multiple section valves are assembled by banking or stacking single-valve sections together. They are used to control such applications as earth moving, mining, material handling, agricultural, and ground support equipment. Sectional-type construction allows for working sections to be assembled in any order without affecting performance. Figure 10-26 illustrates a typical road grader circuit that uses a six-valve stack assembly.

Large spool valves for high flow rates require substantial force to operate the spool valve element. Shifting of the spool valve is accomplished by manual- or solenoid-operated pilot control valves, which direct pilot pressure to the ends of

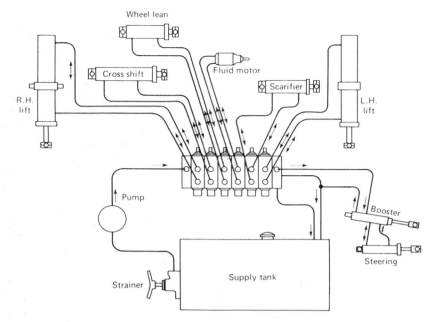

Figure 10-26 Typical road grader circuit.

the main valve spool providing the necessary force to shift the position of the spool. The cutaway solenoid-controlled, pilot-operated control valve in Fig. 10-27 has a flow capacity of 30 gal/min at pressures to 3000 lbf/in². Because the spool is hydraulically balanced, shifting pressures as low as 100 lbf/in.² will actuate the main spool valve element.

10-5 PROPORTIONAL CONTROL VALVES

A proportional valve is one that can be controlled continuously throughout its full operating range. Proportional electrohydraulic control valves are electrically operated control valves that direct and meter the flow of fluid to system actuators in proportion to the position of hand-operated control station levers. In modern hydraulic systems, they are replacing conventional control valves and are particularly suited to industrial control and mobile hydraulics applications including steel mill control, earth moving vehicles, material handling units, articulated arms, and other construction and service equipment.

Essentially, proportional electrohydraulic control valves consist of a control valve driven by a torque motor that receives an electrical signal from a remote controller in proportion to the manual positioning of the control station lever. A feedback signal from the valve spool balances the action of the torque motor to change and/or hold the valve spool position to meter flow in proportion to the control lever position. The operator manipulates the electrical manual control station providing the electrical signal to the control valve proportional to the lever

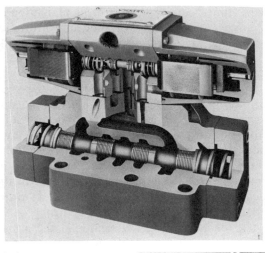

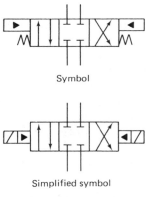

Symbol

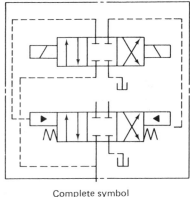

Simplified symbol

Complete symbol

Figure 10-27 Pilot operated industrial directional control valve (*courtesy of Sperry Vickers*).

position. Control systems can be mounted to control panels, or held as portable units off the vehicle at the end of an umbilical control cable. They also can be completely portable by using radio-controlled units. These options offer convenience and flexibility not possible with stationary mounted control valves.

Several operational characteristics are inherent to such a system. Operator fatigue is drastically lowered because of reduced stress and physical exertion requirements. Hydraulic plumbing is simplified because the control valve can be mounted close to the actuator. This also eliminates fluid leakage at the control station. Flexibility of controls allows single or multiple mounted control stations, either on or off the vehicle or both, including two axis joystick operation or other human engineered controls for operator convenience. Radio controls add to this flexibility of operation. Because of the low mass and friction of the control station levers, precise control and feathering are possible with no loss in actuator speed.

Typical delay times between operator manipulation of the control lever and proportional movement of the valve spool are a fraction of a second. Valves are available as single units or can be assembled in banks like conventional control valves.

The proportional electrohydraulic valve shown in Fig. 10-28 consists of a torque motor pilot valve assembly and a four-way power stage valve assembly. The torque motor portion (Fig. 10-28b) consists of a double air gap having two coils and two magnets. The armature is suspended by a frictionless torsion pivot and extends from one air gap to the other. It can move with a limited rocking action. An electrical signal current to the coils alters the permanent magnetic field causing the armature to pivot one way or the other. The polarity and strength of the current determine the direction and magnitude of the armature force about the pivot. Because the armature has a low mass, the torque motor has a high dynamic response bandwidth and bi-directional control.

The flapper attaches to the center of the armature and extends down between the feedback springs of the follow-up piston. A nozzle is located on each side of the flapper such that rocking of the armature varies the size of the two nozzle openings. Pressurized fluid from the pilot supply is carried through the valve stack and enters each pilot stage through a 75-micrometer filter screen. The fluid then passes through two fixed orifices, one upstream of each nozzle. Discharge fluid from the nozzle passes out through a drilling in the power valve spool to a tank cavity in the power valve body. The pressures between each nozzle and fixed orifice are ported to opposite ends of the spool centerline. The level of these pressures holds the follow-up piston in continuous contact with the end of the power valve spool. The differential between the pressures, as varied by the flap-

Figure 10-28 (a) Proportional electrohydraulic control valve (*courtesy of Moog, Inc.*); (b) Torque motor portion.

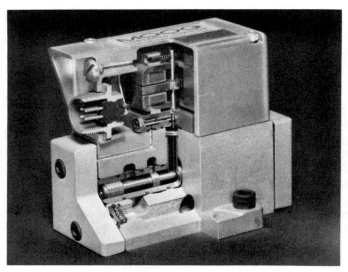

(a)

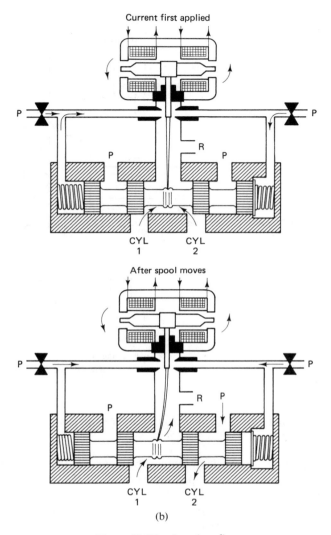

Figure 10-28b (*continued*)

per motion between the nozzles, creates the driving force on the power valve spool and the follow-up piston.

Essentially, the double nozzle flapper valve is a hydraulic resistance bridge circuit with two parallel connecting flow paths. Figure 10-29 shows how the bridge circuit works. Pressurized fluid passes through both paths from a common supply port to a common return port. Each flow path has an upstream orifice and a downstream nozzle, with an output port located between the orifice and the nozzle. The flapper is located equidistant between the two nozzles. Flapper to nozzle spacing determines the resistance of the nozzle. Steady flow, called quiescent flow, which passes through the circuit, is a function of the orifice and nozzle

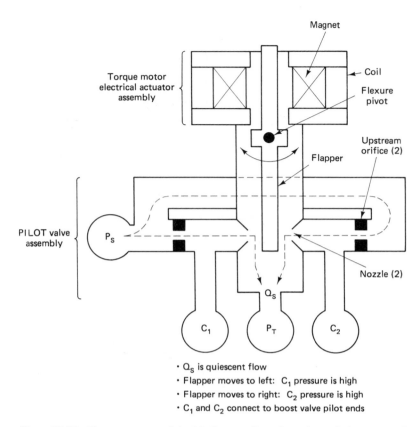

Figure 10-29   Torque motor and double flapper pilot valve schematic (*courtesy of Sundstrand Mobile Controls*).

size. When the flapper is in the neutral position, the bridge circuit is balanced and the pressures at the output ports are equal. Slight movement of the flapper causes the nozzles to change sizes. This is how the flapper imbalances the bridge circuit and causes a differential pressure and flow from the output ports which shift the hydraulic valve main spool, or in the case of two stage valves, to shift the boost valve spool. Flapper displacements as low as 0.004 in. will create a differential pressure greater than 80% of the supply pressure with maximum output flow rates of 50% of the quiescent flow rate.

While the double nozzle valve is found in most high performance proportional valves, sliding spool pilot stage valves are found in low to modest performance mobile and industrial applications (Fig. 10-30). The hydraulic bridge circuit used is very similar except that quiescent flow is across two meter-in control edges and two meter-out control edges of the sliding spool pilot valve. Flow from the meter-out control spool edges drains to a common return port. The output ports to the boost stage of the valve are between each of the meter-in and meter-out edges of the control valve spool. In the neutral position, quiescent flow across meter-in and meter-out edges of the spool balance the pressure between the

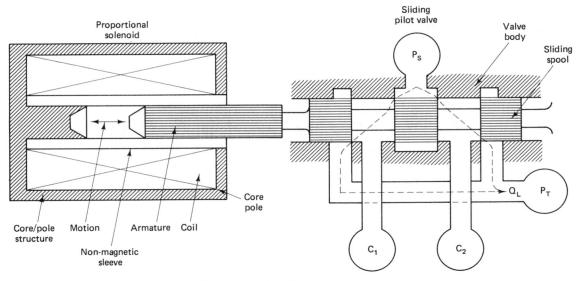

- Q_L is neutral leakage flow
- Spool moves left: C_2 pressure is high
- Spool moves right: C_1 pressure is high
- C_1 and C_2 connect to boost valve pilot ends

Figure 10-30 Sliding spool pilot stage for proportional valve (*courtesy of Sunstrand Mobile Hydraulics*).

boost ports. When an electrical signal is sent to the solenoid actuator, the spool is displaced, imbalancing the bridge circuit by opening one set of meter-in/meter-out control edges and closing the other meter-in edge. This produces a pressure flow situation at one output port or the other to shift the main spool or boost stage of the control valve. Because of the design, shifting the sliding spool pilot stage requires relatively large spool displacements (up to ± 0.030 in.) and requires high actuating forces from the solenoid actuator (with a dither signal to minimize friction and hysteresis effects). The control spool and solenoid mass also limit the dynamic response bandwidths to modest values unless high force output solenoids are used, but this requires high electrical inputs which generate unwanted heat.

The flow capacity and flow control characteristics of proportional electrohydraulic valves are determined by the power stage valve spool configuration and flow characteristics which are altered by flow metering slots. Spools may have a number of configurations. Tandem center spools are used with constant flow pump supplies (fixed displacement pumps) (Fig. 10-31). These spools have a bypass from pressure to return when the spool is centered in order to unload the hydraulic supply. As the spool moves to either side of center, the bypass closes off and pump pressure builds throughout the hydraulic system upstream of the actuator.

Spool configurations for proportional valves operating on constant pressure systems may have either open or closed control ports when the spool is in the

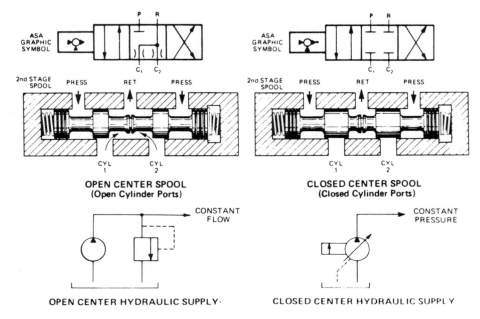

Figure 10-31 Typical valve spool flow characteristics (*courtesy of Moog, Inc.*).

center position. Remember that an open center spool has both control ports open to return when the spool is centered. The tandem center bypass, which allows stacking and independent supply to other loads, also diverts the constant volume pump to tank when the valve spool is in the center position. The pressure lands usually are overlapped so that 10% to 30% of spool travel is necessary to close the return and open the cylinder port to pressure. This spool configuration always is used to operate a man lift, as the open center assures operation of load holding valves at the lift cylinders which are required by OSHA and ANSI. Closed center spools are used with constant pressure supplies (variable displacement pumps). The control is quite accurate and linear, but the requirement of a variable displacement pump increases the cost of the system considerably.

An alternative to the constant pressure system (variable volume pump) is to use a load demand pressure regulator (pilot operated unloading valve) on a constant volume fixed displacement pump. The result is that all valves will have the same supply pressure, and that the system pressure will change to accommodate maximum load. Standby pressure is essentially the same as the losses in the tandem center valve system (200–300 lbf/in.²), and pump pressure builds as the load requires. The advantage of this system over conventional tandem center constant volume systems is that the performance of all control valves is immediate and non-interactive as with true constant pressure systems. In both cases, however, the value of the pressure is dependent upon the valve controlling the maximum load, and when two or more valves are opened simultaneously, the valve controlling the least load will receive priority flow. A fixed displacement pump circuit with a cylinder demand pump unloading valve is shown in Fig. 10-32.

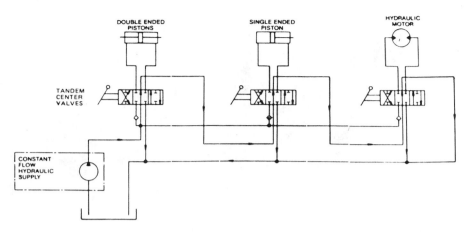

Figure 10-32   Constant flow hydraulic circuit.

Notice that the open center valves require load holding valves at the cylinder ports.

Figure 10-33 illustrates several controllers that are used with proportional control valves. Single-level controllers can be moved back and forth to control the direction of one double-acting cylinder or the forward and reverse rotation of a motor. The center position is usually detented and removes electrical power. Two-axis single-lever controllers are used to direct two actuators hooked to a single piece of equipment such as a loader fork on slides that lift as well as slide from side to side. The single lever is moved back and forth or from side to side. Each direction has its own detent, feel-centering spring, potentiometer and on-off switch. Two units of this style can be mounted side by side to provide convenient two-hand four-axis control. External mounting of the controller requires a sealed weather tight unit with a shroud to prevent accidental switching. Multiple control lever controllers such as that in Fig. 10-33b give the operator control over 2 to 6 lever switches which can drive a single proportional valve or a hydrostatic pump stroker. The center position is neutral, and a "deadman" switch is located on the side to energize the control levers. Custom designed control stations such as that shown in Fig. 10-33c are available for special use applications. And self-contained radio controllers provide up to six proportional and 16 on-off functions. Individual logic codes prevent erroneous control from electrical interference or from other radio controllers. The integral radio transmitter operates from rechargeable batteries.

Electrohydraulic proportional control valves can be used to control hydrostatic transmissions by replacing the usual manual control lever of the variable displacement pump with an electric controller driven by a battery powered potentiometer. The electrical remote control for the hydrostatic pump is provided by a proportional valve mechanism like that shown in Fig. 10-34. In this application the position of the pump stroking mechanism is proportional to the electrical input, so a lever and spring connect from the stroking mechanism to the torque motor. When an electrical signal is applied from the hand controller to the torque

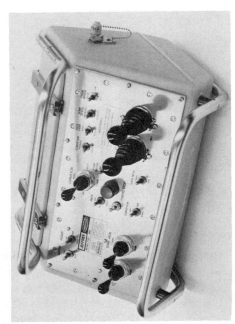

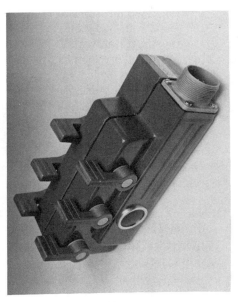

Figure 10-33 Proportional valve controllers (*courtesy of Moog, Inc.*).

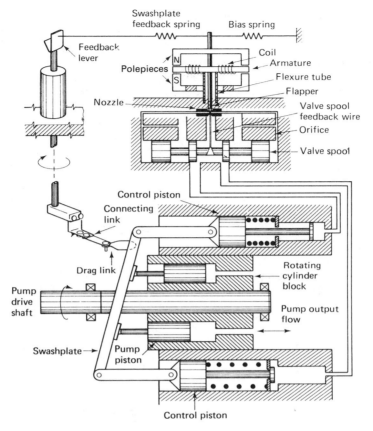

Figure 10-34 Proportional control for hydrostatic pump (*courtesy of Moog, Inc.*).

motor coils, the armature rocks either clockwise or counterclockwise. This torque displaces the flapper between the two nozzles. The differential nozzle flow moves the spool to either the right or left. The spool continues to move until the feedback torque applied to the flapper from the spool valve counteracts the electromagnetic torque. At this point the armature and flapper are returned to center, the spool stops and remains displaced until the electrical input signal changes to a new level. Valve spool position is proportional to the electrical signal. The actual flow from the valve to the stroking mechanism of the pump depends on the load pressure, and the stroking mechanism will move until it reaches a displacement where the feedback spring just counteracts the electrical input torque. At this point the valve has moved back near its centered position. The control loop in this instance is open by its very nature, and it can be closed only by the actions of the operator who must compensate for changes in the speed of the engine which drives the hydrostatic pump or changes in the load.

One of the problems inherent to proportional control valves is that operator effort to shift the controller is not proportional to the output of the hydraulic valve. This can cause the actuator to start too quickly or exert too much force when the load resistance is encountered. For example, quick-shifting of the valve on the tram motors on a tracked vehicle or on a fork lift drive could spill the load and break the machine, particularly at start-up or if the machine were stalled. Here the result could be personal injury as well as equipment damage.

The output force and velocity can be regulated by installing pressure or flow control valves at the actuators, but these have adjustment limitations and often consume power needlessly. A more recently developed solution incorporates an inexpensive electrical programmable velocity control in the valve itself. Figure 10-35 illustrates the characteristics of such a circuit as it would be applied to a hydraulic cylinder or motor. The upper half of the figure illustrates the input voltage from a variable controller which, in this case, is switched abruptly forward (positive) and then reversed (negative). The lower half of the figure illustrates adjustable (ramped) outputs at ports A and B, which are programmed with respect to acceleration, velocity, and deceleration. No matter how fast the controller is shifted, the output of the actuator is programmed to ramp the start-up acceleration, maximum velocity, and stopping acceleration in both directions within the safe limits of the machine and operating personnel.

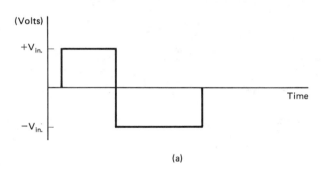

(a)

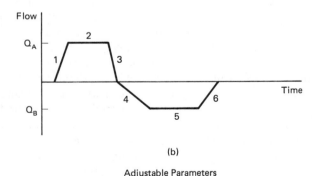

(b)

Adjustable Parameters

Port A Adjustments	Port B Adjustments
1 Acceleration	4 Acceleration
2 Max. velocity	5 Max. velocity
3 Deceleration	6 Deceleration

Figure 10-35 Ramp control for proportional valve circuit (*courtesy of Moog, Inc.*).

10-6 SERVO VALVES

A servo valve produces a continuously controlled output as a function of an electrical input signal. It does this by incorporating a feedback loop that compares the velocity, position, or force of the output actuator with the command signal sent to the control valve, and then makes corrections at the output for differences in the two. The most common type of servo valve is a four-way valve, but three-way and two-way porting configurations are also common.

In a proportional electrohydraulic valve like that shown in Fig. 10-34, the position of the swash plate stroking mechanism on the pump is controlled by the position of the hand-operated proportional controller. The control loop in this instance is open by its very nature because the operator must compensate for changes in engine speed when the load builds or reduces such that the engine driving the pump is made to slow down or speed up. Closed-loop operation (Fig. 10-36) is achieved by attaching a tachometer to measure the speed of the hydraulic

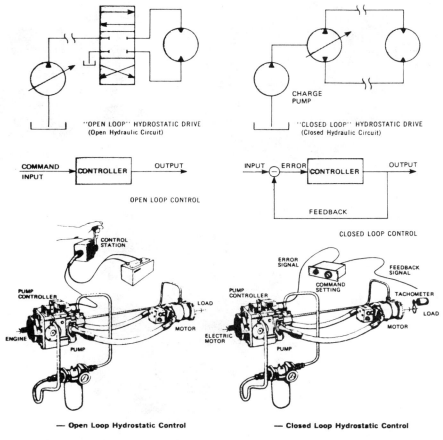

Figure 10-36 Open and closed loop hydrostatic transmission control (*courtesy of Moog, Inc.*).

motor and comparing this value with the command setting of the potentiometer at the control station, and then making adjustments to the input to compensate for differences (the error) between the two.

Servo-valves come in single, two-stage and three-stage designs, depending upon the flow rate and pressure ratings (Fig. 10-37). With single-stage valves, pilot pressure is used to shift the main spool. Staging has the effect of increasing the flow capacity by providing the higher spool pressures and flow required to shift the main valve spool. Single-stage valves commonly provide output flow rates to 15 gal/min at 1000 lbf/in.^{2} drop and supply pressures to 5000 lbf/in.^{2}, although 5 to 10 gal/min is often the realistic limit for single-stage valves. Two-stage valves increase the flow capacity to 40 gal/min or more, and three-stage

Figure 10-37 Single and multiple state servo-valves (*courtesy of Moog, Inc.*).

valves are used with flow rates between 40 and 400 gal/min, although some three-stage valves have capacities to 1000 gal/min.

Both proportional valves and servo valves position the valve spool in proportion to an electrical signal from the controller. The proportional valve electrical signal is regulated by a hand-operated controller with the magnitude determined by the position. Servo valves, on the other hand, receive the initial electrical signal from the command input, and any deviation from that is sensed at the output and fed back into the controller as an error signal. Thus it would seem that proportional valves could be used to replace servo valves which are more expensive, simply by exchanging the command station and output sensing device for the hand-operated controller. In most instances, this cannot be done because the difference lies not in the control of the valve, but in the spool configuration of the valve.

Proportional valves are designed so that at full opening the pressure drop across the valve is only 5% to 10% of system supply pressure. In this respect, they are power delivery valves. They also have a flow deadband of 20% to 30% of spool stroke in the center position. A flow deadband of this magnitude creates a positive null which is desirable for open-loop manual control systems. Proportional valves can be used in closed-loop servo systems, but this requires the use of nonlinear electronic controllers to compensate for the instability associated with a wide null bandwidth.

Servo valves, on the other hand, are precision metering valves. Pressure drop across these valves at full spool opening is commonly 33% of input pressure. This is done to ensure precision flow metering characteristics, and because there is very little spool overlap, the flow deadband is negligible. It must be remembered, however, that the amount of power transmitted through the valve is limited to 67% of the input power.

Servo valves are used to control the velocity, position, force, and acceleration-deceleration rate of the actuator. Most servo valves control the flow rate to the actuator and are called flow control servo valves. Servo valves that control and differentiate pressure to the load are called pressure control servo valves. Both functions can be incorporated in one servo valve with the primary function being flow rate output (velocity control), with load pressure controlled as a function of the load resistance to motion. When flow and pressure are controlled in one servo valve, the valve provides a load flow rate as a function of the load resistance to motion.

Which function the valve serves is determined by how the output is sensed, the configuration of the valve, and how the controller receives the feedback as a combined error signal. Thus far we have discussed how a tachometer sensing the output speed of a hydraulic motor can be used to stroke the hydraulic pump. This is a flow control device. Here it was not important what caused the speed to change at the output of the hydrostatic transmission output. It could have been caused by a change in the output load, for example as the vehicle traveled up or down a grade, or by a change in load on other parts of the system to cause the drive speed to the pump to change. Either one would affect the speed of the output and if the command signal from the controller specified a certain speed, then the feedback from the tachometer signal would cause an error signal to be sent to the controller.

One means used to reduce the number of sensors, wiring harness, and connectors on the output actuator and simplify the integration of the various feedback signals into an error signal is to incorporate a programmed electronic microprocessor module in the controller, usually called a system enhancer, to compute the velocity, acceleration and load parameters. Figure 10-38 illustrates the schematic for such an example. The riveting machine shown requires control of the compression rate, force levels, time of compression (squeeze), and rate of decompression. The sensing device used to activate the program in the microprocessor is the load cell. The circuit is designed to have the ram advance upwards, contact the rivet, compress and then hold for a specified time, and then retract at a rapid traverse rate.

The ram starts from the bottom position and rapid advances under open-loop servo control until it contacts the rivet. At the instant of rivet contact, the force begins to build up. This force is detected by the load cell and sent to the microprocessor, in the controller, which in turn commands the servo valve to instantaneously reverse direction and reduce the force level to zero. When the moment of zero force level is achieved, the system automatically switches to closed-loop force control. The force is then increased at a controlled rate to a predetermined level. Once the force level is reached, it is maintained for a given time interval, and then the system decompresses at a controlled rate to zero force level. At the zero force level, the microprocessor commands the ram to return to the full down position at a rapid rate and the system stands by for the next cycle. The micropro-

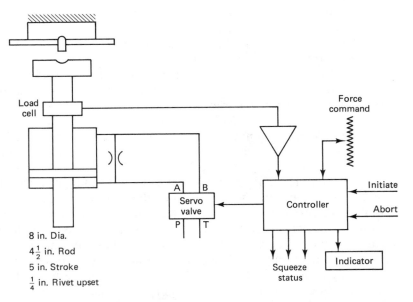

Figure 10-38 Microprocessor application in servo-valve circuit control (*courtesy of Sperry-Vickers*).

cessor contains all the system logic, timers, ramps, and comparators, as well as most of the control circuitry. The only inputs that are required are "initiate" to start the cycle and "abort" to stop the cycle.

Another means to simplify the control of servo valve systems, if the actuator is not subjected to a hostile environment, is to mount the servo valve directly on the actuator (Fig. 10-39). The advantage of this arrangement is to eliminate the plumbing between the valve and actuator, thereby increasing the drive stiffness. This reduces oscillation and increases operating speed and accuracy. Servo actuators of this type use LVDT (linear variable differential transformer) or DCDT (LVDT with solid state exciter and phase sensitive demodulator) transducers to control velocity (flow) or load (pressure).

Figure 10-39 Direct mounted servo-valves on cylinder and rotary actuators (*courtesy of Moog, Inc.*).

10-7 SUMMARY

Hydraulic valves are used to control the pressure, flow rate, and direction of fluid in the system. They may be operated manually, by system pressure or flow rate, or remotely by fluid or electrically controlled pilots from adjacent or remote locations. Pressure control valves protect the hydraulic system and work from damage resulting from peak pressures. They also maintain the safety of personnel. Flow control valves control the fluid output rate of the system, that is, the rate at which work is accomplished. Essentially, this is the fluid horsepower output of the actuator to which fluid is directed. Flow control valves may also divide the flow of fluid on a priority or proportional basis to maximize the use of available flow from the system pump to one, two, or more of the several circuits working alternately in the system. In this application they are also acting as flow sensitive directional control valves. Directional control valves manage the direction of the fluid, including stopping and starting. In mobile applications, control valves are typically actuated by a physical force provided by an operator. Industrial and other applications make extensive use of solenoid valves and pilot-operated valves controlled by solenoids, pilot valves, and other air and oil fluid pilot sources. Proportional and servo valves are used extensively to control fluid flow rates and, more recently, for direction control in modern hydraulic systems. Valve response occurs as the result of current applied to the valve operator, which is usually a torque motor. The infinite control capability of these valves is dependent on input current from a programmed or feedback source. Mobile hydraulic applications that have typically relied on direct or pilot-directional valve operation provided by a manual operator in a fixed location, such as the machine cab, are now using proportional electrohydraulic valves to increase versatility and accuracy control. Systems may be operated by control stations located remotely using umbilical cords, or by radio control stations carried by the operator. Leakage usually associated with pilot-operated control stations located in the vehicle cab and operator fatigue are also eliminated or reduced significantly.

STUDY QUESTIONS AND PROBLEMS

1. List the three functions of valves.
2. What is the difference in application between a pressure relief valve and a hydraulic fuse?
3. Operationally, what is the difference between a pressure relief valve and a pressure reducing valve?
4. What is the purpose of a sequence valve and how is this typically accomplished?
5. What is the difference between the operation of a sequence valve and that of an unloading valve?
6. Compute the fhp loss across a direct-acting pressure relief valve returning 10 gal/min to the reservoir at a pressure setting of 2000 lbf/in².
7. What is the function of deceleration valves, and how is this accomplished?
8. What three types of circuits apply flow control valves, and what is the purpose of each?
9. What is the purpose of flow dividers, and how are they used?
10. What flow rate could be expected through a control valve with a flow coefficient of 1.8 if at the operating temperature of the machine the fluid has a Sg of 0.907 and the pressure drop is 400 lbf/in.²?
11. A manufacturer lists the flow coefficient for a certain control valve as 3.5 at a flow rate of 40 gal/min and a fluid Sg of 0.92 when the machine is at operating temperature. What would be the pressure drop across the valve?
12. A flow control valve shows a pressure drop of 1000 lbf/in.² at 64 gal/min. If the fluid has a Sg of 0.977 at a temperature of 155°F, what is the flow coefficient for the valve?
13. The graph in Fig. 10-40 shows the flow vs. pressure drop curve for a flow control valve when the fluid has a viscosity of 150 SSU and Sg. of 0.89. What is the flow coefficient of the valve at a flow rate of 30 gal/min?

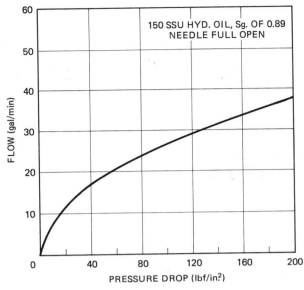

Flow Control Valve Performance Characteristics

Figure 10-40   Figure for Problem 13.

14. What purposes do directional valves serve?
15. What do the three terms *position, way,* and *port* mean when describing directional control valves?
16. How are shuttle valves used?
17. List two applications for three-way valves.
18. What is the difference between an open-center valve and a closed-center valve?
19. Why are directional control valves pilot operated?
20. What is a proportional valve?
21. Why do OSHA and ANSI require open-center valves on man lifts?
22. How does the double flapper in proportional and servo valves develop pressure in the pilot stage?
23. Give three reasons why proportional valves are preferred to manually operated control valves?
24. What means is used to prevent quick starts and excessive force in actuating proportional control valves?
25. Describe two ways that a servo valve differs from a proportional valve.
26. Why are proportional valves not used to replace more costly servo valves?
27. How can the number of sensors used to control the pilot stage of servo valves be reduced?
28. What advantage does mounting the servo valve directly on the actuator have? Is there any disadvantage to this arrangement?

CHAPTER *11*

SEALS
AND PACKINGS

11-1 INTRODUCTION

Fluid power seals prevent internal and external leakage in the system. They are used to seal between the surfaces of static parts such as pump housings, valve bodies, reservoir covers, and fittings where no relative movement between parts occurs, as well as between dynamic parts such as pump and motor shafts and housings, cylinder pistons and walls, and between cylinder rods and seals where a high degree of movement at varying speeds is present. About 80% of the applications involving seals are static.

Seals are most often classified by the material or compound used for their fabrication, by application or use, and by configuration. Common materials include fabric, rubber, leather, metal, elastomers, and plastics. Applications refer to such factors encountered in the system as the properties of the fluid, system pressure, temperature, and the severity of operating conditions such as speed, deflection of moving parts, and abrasion from dirt and metal surfaces. The configuration of seals and packings is defined from their cross section profile. Rectangular, circular, cup, flange, and collar are common types of shapes used for specific applications. Mating grooves and support surfaces which hold these seals are also described by cross section profile and may either match the shape of the seal, as in the case of rectangular metal piston ring grooves, or not match the shape of the seal, as in the case of O-rings, so as to deform their cross section by design to provide an effective seal between adjacent moving surfaces.

Static seals and gaskets are used between mating surfaces where no relative movement occurs. The seal is usually compressed between the two adjacent parts by bolts securing the two stationary parts together, as in the case of a pump housing or reservoir end plate, or joining and compressing several parts held in a stack in applications such as built-up valve assemblies.

Dynamic seals are used between the surfaces of parts where movement occurs and control both leakage and lubrication. In many cases, provision is

252

made to keep dirt and other foreign matter from entering the system by the use of covering boots or rod wipers. The motion encountered by dynamic seals is reciprocating or rotating, or a combination of both, oscillating, and requires contact under pressure between the relatively smooth surface of the hydraulic member and the flexible or adjustable contact surface of the seal. Clearance between the movable part and its supporting member, such as between a pump shaft and the support bearing, or between the piston rod and the rod bushing, must be compensated for by the seal if excess leakage is to be prevented. Out of round running, wobble, vibration, and chatter commonly occur in the proximity of the seal where excessive clearances and other adverse circumstances prevail.

Internal seals such as those used on double-acting pistons prevent leakage between areas of differential pressure. If leakage occurs between the contact surface of the seal and the moving part, loss in volumetric efficiency, system power, control, and the generation of excessive heat will result. Lack of lubrication is usually not a problem since fluid is present on both sides of the sealing ring.

External seals, on the other hand, provide for both a seal against differential pressures, that is, between the internal system pressure and external atmospheric pressure, as well as for controlled lubrication of the seal itself and adjacent bearing support. This lubrication is essentially controlled leakage of fluid from the system and extends the life of the surfaces of the parts moving adjacent to the seal as well as the seal contact edge itself.

The failure of an external seal results in loss of system fluid which usually overrides any accompanying effects caused by loss of volumetric efficiency, system power, heat generation, or control. Usually, seal failure is relative and progressive. As system pressure, fluid temperature, and speed are increased, leakage also tends to increase. As a seal begins to fail, for example from abrasion or aging, it will become progressively worse in its performance. Bursting seals, with sudden loss of fluid, are not the most common failures to this part of the system. Successive failure is more the common rule. Practices which prevent successive failure to external seals list periodic maintenance as the highest priority to insure against substandard seal performance, which ultimately results in loss and local accumulation of fluid from system leaks.

Seal materials are selected to be compatible with system hydraulic fluids and operating conditions. Operating conditions include system pressure, fluid and adjacent part operating temperatures, surface cycling speeds, and the severity of operating conditions such as the presence of dirt, abrasion, and vibration. Generally speaking, if the seal is an organic compound, the content of the materials used to fabricate the seal should be dissimilar to the composition of the fluid and inert to its effects. Similarities between chemical composition of fluids and seals tends to cause softening, swelling, and loss of necessary sealing qualities. Some fluids, synthetics for example, are known to dissolve many conventional sealing compounds.

Many materials have been used for seals. Early sealing techniques used only natural leather and rubber cups on piston heads for internal seals, and stuffing boxes for external seals filled with such materials as sawdust, leather, cotton rope, and fibrous materials impregnated with graphite which also possessed lubricating qualities. The fluid used with these seals was a petroleum-base fluid or water. Compatibility between the material used to fabricate the seal and the chemical composition of the fluid was no great problem because they were dissimilar and inert to the effects of each other.

More recent development of seals composed of hydrocarbon compounds, fluorocarbon, silicone, metals and improved rubber compounds, and both conventional and fire-resistant fluids containing many special purpose additives now present the user with a variety of combinations from which to choose. Improved versions of those sealing materials originally available, such as impregnated leathers and improved rubber base compounds, also are available.

11-2 STATIC SEALS

Static seals are designed to fill the space between two attached parts. The method of attachment is not in itself of primary importance to the seal as long as it provides the compressive force necessary to effect a reliable seal between the surfaces of the two attaching parts and the two faces of the gasket, and does not damage the gasket during installation and maintenance.

Because these two conditions are rather easy to meet, many types of attachment as well as many diverse applications make use of static seals. End covers for hydraulic pumps, motors, and valves are common uses (Fig. 11-1). Less common, but equally important applications include pressure end caps, plug seals, through wall line seals, rivet head seals, bolt head seals, and tube fitting seals (Figs. 11-2 and 11-3). Some special applications attach several gaskets to a

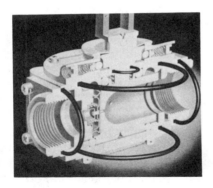

Figure 11-1 O-Ring static seal application (*courtesy of Parker Seal Company*).

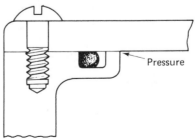

Figure 11-2 End cap O-ring static seal.

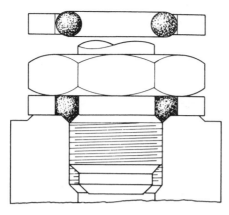

Figure 11-3 Through-wall static seal (*courtesy of Parker Seal Company*).

retainer plate to achieve an effective seal for multiple parts, some of which have irregular shapes.

The compressive force necessary to effect a reliable seal between two attached parts is usually expressed in pounds per linear inch or Newtons per linear cm of seal. Factors which affect the compressive force include the gasket cross section, material, seal hardness, and the percent of gasket deformation desired.

Static seals and gaskets are designed for installation in both nonconfined and confined spaces. In the case of nonconfined gaskets, such as some flange seals, head gaskets, and others, the compressive force exerted by the attaching bolts must be sufficient to effect the seal between the surfaces of the two attached parts and the two faces of the gasket (Fig. 11-4). Impregnated rubber, asbestos, and other similar materials are used. The major disadvantage associated with nonconfined seals is the tendency to deform the two attached parts as well as to damage the gasket during assembly by the compressive force necessary to effect a reliable seal. In addition, the surface finish of the two attaching parts must be parallel and smooth to mate with the gasket surfaces.

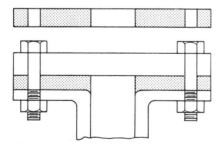

Figure 11-4 Non-confined flange gasket installation.

Gaskets for confined applications, such as O-rings and lathe-cut gaskets, use the groove and ring principle to effect a seal (Fig. 11-5). The groove is machined into one or both of the attaching parts so as to be slightly thinner but wider than the O-ring cross section. Initial sealing between the O-ring and the two surfaces is caused by a slight compression and deformation of the O-ring, usually between

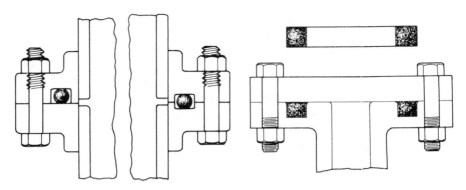

Figure 11-5 Confined gaskets.

15% and 30%, to form the necessary zero clearance to seal. As the pressure increases, the O-ring is forced in a direction opposite the pressure and wedges itself at the parting line of attachment. As the pressure is further increased, deformation of the seal and further wedging increases proportionally. The limit of the seal to withstand increased pressure is dependent not only on the gasket material, but on the fit at the parting line of the two attached pieces, such as at the flange of two pipes being joined, or at the junction of a pump cover and housing, that prevent the seal from extruding. Effective seals for pressures in excess of 10,000 lbf/in.^{2} have been maintained using this technique.

While both O-rings and square cut gaskets are generally interchangeable and may be used in similar applications, they are different in several ways. First, O-rings will seal in any direction against two opposite surfaces—radially, axially or diagonally. Square cut gaskets, on the other hand, are designed to seal between two opposite surfaces only, in directions parallel to the flat surfaces—radially or axially. Secondly, the initial force necessary to effect slight deformation and subsequent zero clearance seal between two surfaces and the O-ring is small. Square cut gaskets, however, require a much higher initial sealing force that may deform attaching parts where light weight flanges and covers are used. Additionally, O-rings are more tolerant of surface irregularities at the groove surface as well as at the mating surfaces than are square cut gaskets in effecting the initial seal. Square cut gaskets, however, do have one distinct advantage over O-rings. They are less expensive, primarily because in quantity they are cut from molded cylinders rather than molded individually as are O-rings.

Other less common applications using the O-ring to solve problems have been made to tube fittings that attach transmission systems, to bolts and cap screws also used with these systems, and to integral gasket seals containing two or more gasket materials (Fig. 11-6). In addition, the use of O-rings to seal rivets used in the construction of reservoirs and pressurized vessels is a recent innovation applied to fluid power technology.

The controlled confinement and clearance necessary for fastener O-ring applications uses an outer retainer ring and an inner elastomer O-ring. The outer retainer ring prevents extrusion of the O-ring and the controlled size of the annular

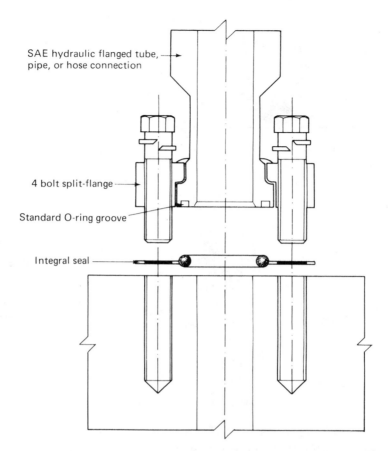

SAE hydraulic flanged tube, pipe, or hose connection

4 bolt split-flange

Standard O-ring groove

Integral seal

Figure 11-6 Integral seal application (*courtesy of Parker Seal Company*).

space into which the seal is compressed to the desired dimension. Seals of these types are available for applications in which liquids and gases are contained at pressures to 5000 lbf/in.^{2} (34.5 MPa) at temperatures from $-80°$ to $400°F$ ($-62°$ to $204°C$).

While most O-ring face seal applications, such as flange end caps or pump covers, make use of an annular groove of rectangular cross-section to contain the O-ring, some applications require a positive method to hold the seal in place (Fig. 11-7). In these applications, mean groove diameter is the same size as the mean O-ring diameter. The two dimensions of the pairs of groove radii are also critical in that they must be large enough to prevent damage to the O-ring seal during installation, but small enough to prevent extrusion during operation.

Calculating the proper size for a given O-ring application is an important part of this sealing technique (Fig. 11-8). Three typical applications for static O-ring seals are used: (1) end caps, (2) male glands, and (3) female glands. To calculate proper groove dimensions for these installations, it must be remembered that the

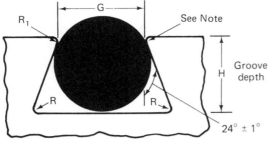

Note: First cut groove to sharp edge at R_1.
Then round off edge to G dimension.

Figure 11-7 Dovetail O-ring retaining gland (*courtesy of Precision Rubber Products Corporation*).

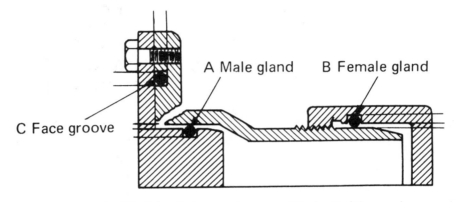

Figure 11-8 Calculating O-ring sizes (*courtesy of Parker Seal Company*).

O-ring is to be compressed 15% to 30% in one direction to accomplish the initial seal when the two stationary parts are assembled and expanded in a direction perpendicular to this because of the cross section deformation. That is, the cross section dimension is reduced in one direction due to compression and expanded a similar amount in a direction perpendicular to the direction of squeeze. The designer's problem, then, is to specify the groove such that an O-ring of given material can be compressed in one direction and expanded in another. In most cases, groove specifications are available from O-ring manufacturers' standard tables. These should be consulted whenever possible.

When gland dimensions from standard tables are not available, the following procedure is recommended for designing glands for pressures to 1500 lbf/in.^{2} (10.3 MPa) (no back-up rings necessary):

1. Select the appropriate O-ring free cross section diameter for the particular application (1/16th-1/4 in. or 1.6–6.4 mm).

2. Select the appropriate O-ring material considering pressure, temperature, and fluid compatibility.

3. Record the percent the selected O-ring cross section will be compressed during installation.

4. Determine the diameter, depth, and width of the groove that will properly accommodate the O-ring.

  a. For male gland applications, the diameter of the groove is calculated by determining the diameter of the cylinder bore and subtracting twice the compressed cross section diameter of the O-ring from this dimension. The compressed cross section of the O-ring will be 70%–85% of the free cross section diameter. The actual depth of the O-ring groove equals the compressed cross section diameter of the O-ring (70%–85% depending on the O-ring selected) minus one-half the clearance between the male plug and the cylinder bore diameter. The width of O-ring groove equals the free cross section diameter of the O-ring plus twice the amount that the O-ring was compressed (15%–30%). Finally, select an O-ring with an outside diameter approximately the same size as the bore diameter.

  b. For female gland applications, the diameter of the groove is calculated by determining the diameter of the tube and adding twice the compressed cross section diameter of the O-ring to this dimension. The compressed cross section diameter will be 70–85% of the free cross section diameter of the O-ring. The actual depth of the O-ring groove equals the compressed cross section diameter of the O-ring (70–85%, depending on the O-ring selected) minus one-half the clearance between the tube outside diameter and the throat inside diameter. The width of the O-ring groove equals the free cross section diameter of the O-ring plus twice the amount the O-ring is compressed (15–30%). Finally, select an O-ring with an inside diameter approximately equal to the tube outside diameter.

  c. For face groove installations, the mean diameter of the groove should equal the mean diameter of the O-ring (inside diameter of the O-ring plus the free cross section diameter), with tolerances added or subtracted to place the O-ring in the groove opposite the side of the applied pressure. The depth of the groove equals 70%–85% of the free cross section diameter of the O-ring. The width of the groove equals the free cross section diameter of the O-ring plus twice the dimension it is compressed (15%–30% depending on the O-ring selected).

Example 1

Compute the groove diameter, the groove width, and the O-ring outside diameter for an application where a male end plug is to make a static seal with a 4.000-in. (10.16 cm) bore diameter cylinder. Use an O-ring with a free cross section diameter of .140 in. (3.56 mm) which is to be given a 20% squeeze on assembly.

Solution In the male gland application, the groove diameter equals the bore diameter size minus twice the compressed cross section diameter of the O-ring. For a 20% squeeze, this equals:

$$\text{Groove diameter} = 4.000 - 2(.80 \times 0.140)$$
$$= 4.000 - 2(.112)$$
$$= 4.000 - 0.224$$
$$= 3.776 \text{ in. (9.59 cm)}$$

The groove width equals the free O-ring cross section diameter (0.140) plus twice the dimension it was compressed (20%).

$$\text{Groove width} = 0.140 = 2(0.20 \times 0.140)$$

$$= 0.140 = 2(0.028)$$

$$= 0.140 = 0.056$$

$$= 0.196 \text{ in. (5 mm)}$$

Finally, the outside diameter of the O-ring should be approximately 4.000 in. (10.16 mm), the same as the bore diameter of the cylinder.

11-3 DYNAMIC SEALS

Dynamic seals prevent leakage between two surfaces moving relative to each other under conditions of varying pressure. Pump and motor shafts, cylinder rod seals, piston ring seals, and valve spool seals are included as part of this category.

To be effective, a dynamic seal must stop the passage of fluid under conditions when no pressure or relative movement is present, as well as when the system is brought to operating pressure and temperature and moving at full speed.

Materials used for dynamic seals must combine the qualities of hardness and initial sealing force with minimum wear, abrasion resistance, and fluid compatibility to insure the longest possible life of the seal under the many and varied conditions which it is expected to operate. Several materials and seal designs are in use to attain this goal. Common materials include steel or cast iron, elastomers, asbestos, rubber, and leather. Design configurations include the square cut ring, O-ring, cup, and V-packing.

Seal designs have as their objective to prevent leakage under varying conditions of pressure and speed as other operating conditions change, and at the same time to extend the life of the seal as long as possible. To accomplish this end, an initial seal must be maintained when the ring is installed by having the design exert a radial force on the ring. As system pressure is increased, additional radial as well as longitudinal force is applied to the ring to increase its sealing capability. When relative movement occurs, friction between the sealing ring and the moving surface is at first very high during breakaway, and then decreases as the speed increases. If initial and operational sealing force is too low, leakage will occur. If initial and operational sealing force is too high, high friction will result in excessive wear and premature failure of the seal.

System pressure affects sealing in two ways. If the radial force applied to the seal is held constant, increase in pressure will cause a thicker oil film to pass between the seal and the moving part which will result in increased system leakage. If pressure is also directed against the seal to increase the radial force proportionally as pressure increases, then the oil film will remain approximately the same thickness resulting in no change in seal lubrication and leakage. Seal wear and life, however, will be shortened.

Two other factors that interact to affect seal design are the coefficient of friction and the operating speed. The coefficient of friction is largely dependent on the surface finishes of the seal and the metal surface against which it rests. The coefficient of friction and operating speed interact to produce lubricating conditions like those in Fig. 11-9. It is seen that as speed increases from zero, the coefficient of friction decreases substantially at first, and then increases gradually. As speed increases, the thickness of the oil film increases, indicating that a given seal design operates at peak efficiency only within a certain range of speeds.

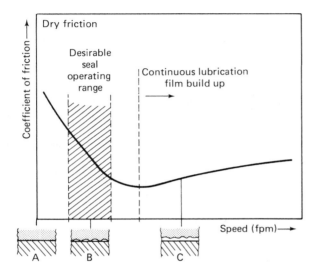

Figure 11-9 Coefficient of friction with operating speed (*courtesy of Disogrin Industries*).

Leakage and friction affect service life, and this is the prime consideration of the seal user. When leakage becomes excessive, the seal must be replaced. High seal friction may also affect system operation, as in the case of gravity or spring return single-acting cylinders, as well as actually shorten the service life of the seal.

The leakage from a double-acting cylinder is approximately equal to

$$V = \pi \times D \times S(l_2 - l_1) \tag{11-1}$$

where V is in in.^{3} per cycle, the diameter of the sealing surface (D) is in in., the length of the stroke (S) is in in., and the thickness of the oil film on the forward and return strokes ($l_2 - l_1$) is in in., respectively. From the equation it is seen that as either the seal diameter (D) or the stroke (S) are increased, leakage also increases. Film height ($l_2 - l_1$) is affected by several factors related to the oil, including viscosity, operation of the system, rod velocity, and system pressure. The effects of velocity, fluid viscosity, and system pressure are shown in Fig. 11-10. It is seen that leakage increases roughly as the square of the speed and proportionally with oil viscosity. Leakage with system pressure increases initially at an appreciable rate and then at a lesser rate with further increase in pressure,

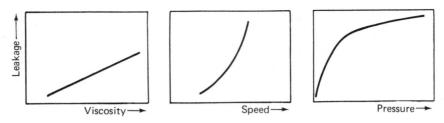

Figure 11-10 Effects of speed, viscosity, and pressure on leakage (*courtesy of Disogrin Industries*).

indicating that system pressure is also directed against the seal to increase the radial force between the seal and the metal surface with which it is in contact.

Example 2

Compute the average oil thickness between a seal and a 2-in. (5.08 cm) double-acting cylinder rod with a stroke of 10 in. (25.4 cm) that loses 1 gal (3.79 l) of hydraulic fluid each 10,000 strokes.

Solution   Rearranging Eq. 11-1 and dividing by 10,000 to solve for oil thickness per stroke

$$V = \pi \times D \times S \times (l_2 - l_1)$$

$$l_2 - l_1 = \frac{V}{\pi \times D \times S \times 10,000}$$

$$l_2 - l_1 = \frac{231}{3.14 \times 2 \times 10 \times 10,000}$$

and

$$l_2 - l_1 = 0.000\ 3678 \text{ in. } (0.009\ 342 \text{ mm})$$

11-4 MATERIALS AND COMPOUNDS

Matching Fluids and Seal Compatibility: Materials and compounds selected for seals must be compatible with the fluid, system operating conditions, and temperatures during both periods of operation and idleness. Metallic seals are inert to the effects of most fluids and offer a suitable solution to sealing for high temperature and pressure installations. The seal is not fluid tight in the case of dynamic seals, however, and is fairly demanding of accurate tolerances. O-rings and other elastomeric seals are suitable for medium pressures, offer nearly zero leak capability and are tolerant of such phenomena as cylinder "breathing," but are vulnerable to the deteriorating effects of severe operating conditions. Elevated temperatures and pressures also have an adverse effect on many of these compounds. Asbestos used in static as well as fabricated dynamic seals has desirable characteristics for withstanding the effects of the fluid and elevated temperatures, but offers less desirable mechanical characteristics in effecting the seal under higher compressions. For static installations such as flange covers, for example, very high compressive forces are necessary to effect a proper seal. Selecting the right material for the seal, then, is a compromise and balance between the seal material, the make-up of the fluid, and the system environment which is mostly determined by safety considerations and desired performance. Several common sealing materials, their hardness characteristics, temperature range, and recommended service are listed in Table 11-1.

Selected Physical Properties: Selected physical properties describing the behavior of elastomers are hardness and friction, volume change, compression set, tensile strength, elongation modulus, tear strength, thermal effects, squeeze, stretch, coefficient of thermal expansion, and permeability.

Hardness is the physical property that describes the resistance of an elastomer and other materials to indentation. It is measured with a durometer for sealing materials on the "Shore A" scale. This scale measures hardness on a scale between 0 and 100. Higher durometer readings indicate a greater resistance to denting and a harder material. Lower durometer readings indicate less resistance to denting and a softer material. It is general practice to use softer materials for lower pressures and harder materials for higher pressures. A durometer hardness of 70 on the Short A hardness scale is the most widely used for sealing compounds. A Shore A hardness of 80 is most often specified for rotary motion to eliminate the tendency toward side motion and bunching in the groove. Softer 50 and 60 Shore A O-rings and square cut gaskets are used with static seals on rough surfaces, whereas harder 80 and 90 Shore A materials have less wiping action at the surface and a greater tendency to break during installation. An advantage with the use of harder sealing materials is, however, the reduction of breakaway friction, since softer materials have a greater tendency to deform and flow into surface irregularities at the place of contact between the seal and the moving part.

Indentation hardness and creep can be related to the tendency for elastomer compounds to extrude into the clearance space between the gland and the adjacent surface. When system pressure is applied, the seal is forced to the low pressure side of the gland. And if the pressure is high enough to exceed the capacity of the sealing material to bridge the gap, part of the seal will be extruded into the clearance space. Movement between the two surfaces when extrusion occurs causes nibbling and eventual failure of the seal. Reducing the clearance space will decrease the tendency for material to extrude by reducing the gap that must be bridged between the gland and adjacent surface, but this has the limitation that a minimum running clearance must be maintained between moving parts to prevent seizing. The second means used to reduce nibbling is to increase the hardness of the material. Thus, the final selection of an elastomer sealing compound is a compromise between a material that is soft enough to seal and compensate for wear, but hard enough to resist extrusion.

The relationship between the Shore A hardness of sealing compounds and the ability to resist extrusion is illustrated in Fig. 11-11, where four curves for compounds with Shore A hardnesses of 60, 70, 80, and 90 are plotted against clearance gap and system pressure. No extrusion occurs below the curve where

TABLE 11-1 COMMON SEALING MATERIALS AND APPLICATIONS

Fluid and seal compatibility

Seal material	Durometer hardness	Temperature range	Service
Metallic	Not applicable	Low to 500°F (260°C)	Petroleum-base and synthetic fluids, phosphate esters; suitable for high temperatures, pressures and severe operating conditions for extended periods of time.
Leather (treated or impregnated)	60-90	−65°F to 225°F (−54°C to 107°C)	Petroleum-base fluids, some synthetics, phosphate esters; medium and high pressure installations.
Asbestos	Varies between types and applications		
Neoprene rubber	70	−65°F to 300°F (−54°C to 149°C)	General purpose industrial use, Freon 12, weather and salt water resistant.
Nitrile rubber (Buna N)	60-90	−65°F to 225°F (−54°C to 107°C)	Petroleum-base fluids and mineral oils; used for some rotating seals, low temperature resistance, high extrusion resistance.
Silicone rubber	60-70	−80°F to 450°F (−62°C to 232°C)	Water and petroleum-base fluids, phosphate esters; low tensile strength and tear resistance, high thermal resistance, recommended for static seals only.
Fluoro-Elastomers (Viton and Fluorel)	75-90	−20°F to 400°F continuous; 600°F for short durations (−29°C to 204°C)	Petroleum-base, synthetic, di-ester, silicate ester, and halogenated hydrocarbon fluids; high temperature fluids applications, ultra low compression set.
Polyurethane	70-90	−60°F to 200°F (−59°C to 93°C)	Petroleum-base fluids, high resistance to ozone, sunlight, and weathering; good tensile strength abrasion, and tear properties; low water, chlorinated hydrocarbon resistance as well as compression set.
Ethylene propylene	70-80	−65°F to 300°F (−54°C to 149°C)	Phosphate ester fluids (Skydrol), silicone oils, ketones, and alcohols (not for use with petroleum- or di-ester-base fluids); high resistance to tearing and extrusion.

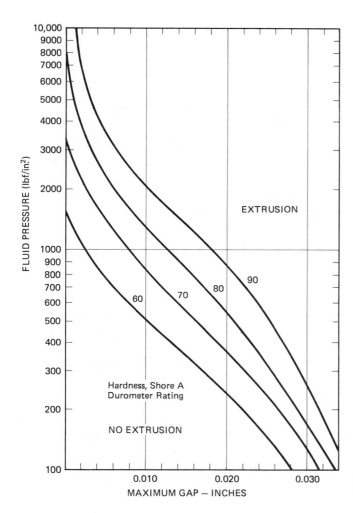

Figure 11-11 Resistance to extrusion of Shore A hardness compounds.

the combination of maximum clearance gap and system pressure do not exceed the limits of the compound. Above the curve the combination of system pressure and clearance gap are such that extrusion and nibbling will occur. The object is to select a compound that is hard enough to resist wear and extrusion for the clearance gap and pressures that will be encountered, but still soft enough to maintain a seal with continued use. For example, an application where the clearance will be 0.005 in. could use a compound with a Shore A hardness of 80 for pressures up to 2000 lbf/in^2. However, if the clearance were to increase to 0.020 in., the practical limit for a compound with a Shore A hardness of 80 would be 550 lbf/in.^{2} before extrusion would occur.

Indentation hardness as well as creep are measured using ASTM Test Procedure D 2240 and a durometer similar to that shown in Fig. 11-12. Creep is defined as the time-dependent part of strain resulting from stress. Similarly, drift or strain relaxation is the deformation in either vulcanized or unvulcanized rubber that

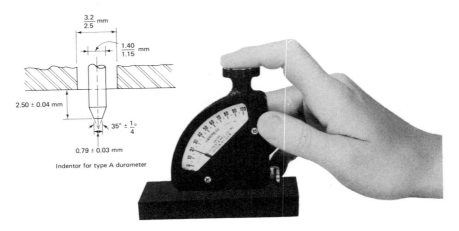

Figure 11-12 Durometer (*courtesy of Shore Instrument and Manufacturing Co., Inc.*).

occurs with a lapse in time after immediate deformation by the indentor. The cross section of the indentor is shown in the insert. When applied to the test specimen, both hands of the durometer move together. After indicating the hardness of the specimen, the driving hand recedes proportionate to the extent of creep in the specimen. The maximum hand remains at the highest reading. Both the maximum reading and the reading of the driving hand after the stated time interval are recorded for the specimen.

Volume change describes the increase or decrease in size of an elastomer occurring as the result of contact with the hydraulic fluid. While a certain amount of swell is desirable to compensate for wear and improve seal effectiveness, excess increase in size or swell is undesirable, particularly in dynamic seals where friction and the tendency to abrade are increased by the addition of volume and accompanying softness. Twenty percent is considered the maximum allowable swell for dynamic O-ring seals, while 40%–50% is allowable for fixed or confined seals providing the fluid does not extract plasticizers which would cause the seal to dry out and leak.

Compression set describes the tendency for an elastomer to lose its resilience. It is calculated as the ratio of the loss in thickness of an elastomer after it has been deformed to the original thickness expressed as a percent. That is

$$\text{Compression Set} = \frac{\text{Loss in Thickness}}{\text{Original Thickness}} \text{ as a percent} \qquad (11\text{-}2)$$

Compression set is determined by compressing an O-ring between two heated plates for a given time and then measuring the thickness to determine recovery. Temperature and the size of the O-ring affect compression set. Increase in temperature tends to increase compression set, whereas increase in the O-ring size tends to decrease compression set.

If the seal is subjected to the effects of the fluid, and these cause swell, the final squeeze on the seal will be

$$\text{Final Squeeze} = \text{Initial Squeeze} + \text{Swell} - \text{Compression Set} \qquad (11\text{-}3)$$

Example 3

An elastomer O-ring is to be used in a static end cover installation. The O-ring has a diameter of 0.275 in. and is given an initial squeeze at assembly of 15%. Fluid swell accounts for a 20% increase in squeeze (actually about 45% by volume), and compression set accounts for a reduction of 8%. What is the final available squeeze on the O-ring in in. and as a percent?

Solution

$$\text{Final Squeeze} = \text{Initial Squeeze} + \text{Swell} - \text{Compression Set}$$

$$\text{Final Squeeze} = (0.275 \times .15) + (0.275 \times .20) - (0.275 \times 0.08)$$

and

$$\text{Final Squeeze} = 0.0413 + 0.0550 - 0.0220 = 0.0743 \text{ in.}$$

As a percent

$$\text{Final Squeeze} = \frac{0.0743}{0.2750} = 27.02\%$$

As a result of the swelling, final squeeze is greater than initial squeeze and the effect of compression set is more than offset by swell.

Tensile strength, elongation modulus, and tear strength are properties related to the mechanical strength of the elastomer. Each of these properties affects the operation of the seal, particularly in dynamic applications, since physical contact and relative movement have a natural tendency to stretch, abrade, tear, and wear the seal. These properties also have extensive application in quality control testing. Ultimate elongation and modulus are computed as a percent.

Tensile strength is a measure of the mechanical sturdiness of the material under test conditions which elongate a sample in tension until it separates. It is computed as the force per square inch (per square cm) area of cross section of the material and expressed in lbf/in.^{2} (Pa). That is,

$$\text{Tensile Strength} = \frac{\text{Separation Force}}{\text{Cross-Section Area}} \text{ in lbf/in.}^2 \text{ (Pa)} \qquad (11\text{-}4)$$

Tensile strength values exceeding 1000 lbf/in.^{2} (6.89 MPa) are not uncommon for modern polymer compounds.

Ultimate elongation is the length an elastomeric seal will stretch before failure and separation. It is computed as a percent of the free length of the seal. That is,

$$\text{Ultimate Elongation} = \frac{\text{Separation Length}}{\text{Free Length}} \text{ as a percent} \qquad (11\text{-}5)$$

Ultimate elongation has extensive application in the installation of seals that must be stretched over pistons. It is also one of the measures used in testing for quality control purposes. Elongation values necessary to install O-rings in glands over small pistons without breakage often must exceed 150%.

11-5 SEAL CONFIGURATION

Typical configurations for seals include metallic and other rectangular rings, the O-ring, U-packings, V-packings, cup-seals, flange seals, other nonconventional seal configuration combinations, and rod wipers.

Seal configuration describes the cross section area of the seal. Circumferentially, seals and packings are a continuation of the cross-section area, which plays a less significant part in seal design, since in normal operation the seal is assumed to retain approximately the same cross section without excessive twisting throughout its entire continuous length. Metallic rings with an assembly and expansion joint are an exception to this rule. Usually they are machined eccentrically to equalize the pressure exerted radially by the ring against the cylinder wall.

The many seal and packing configurations available are the result of much independent research focused on developing the most efficient seal possible with the materials available for a variety of pressure-temperature applications that are compatible with existing fluids and operating conditions. Many seal configurations, such as O-rings, are standardized and produced by many companies with variations primarily in compound. Most seal configurations, however, have proprietary value to the developer and are produced under patent protection. In some cases, both the configuration and the compound have been developed to perform in concert and are protected jointly by patent.

Seal configurations are determined, in part, by function. Some seal configurations, however, such as the O-ring, are made to perform in many circumstances, both in static as well as in dynamic applications. In all cases, the primary purpose is to effect a long lasting seal between the two parts of the system during periods when no movement occurs and to retain this seal when relative movement between the two parts does occur. This seal is to be retained with little or no leakage as both pressure and temperature increase and fluctuate, speed and cycle rate are varied, clearances change, and dirt and other foreign material act to wear, abrade, and contaminate the system.

Automotive-type piston rings account for a large percentage of the rectangular metallic seals used in hydraulic systems. They are used for medium and high pressures where conditions are moderate to severe, for elevated temperatures, such as near furnaces, and for use where controlled leakage is permissible. Oscillating and rotating piston movements that tend to twist and exert circumferential forces on sealing rings have little or no effect on metallic rectangular piston rings. These rings are also applicable where high differential speeds exist between stationary and moving parts, such as between the piston and cylinder wall, as well as where frequent cycle rates exist. Metallic rectangular piston rings are particularly appropriate in applications where the sealing ring must pass open ports.

Nonmetallic piston rings that deform under higher pressures fail in these applications because they are forced into port openings and damaged by the sharp edges forming the opening of the port.

Metallic piston rings are usually made of cast iron, alloyed for stability and special wear characteristics. However, bronze, high strength nodular (ductile) cast iron, and other materials are also available from ring manufacturers. They are usually plated or given an outer coating of materials such as zinc phosphate or manganese phosphate to prevent rusting and corrosion. The outer portion of the ring is finished in a fine spiral groove to provide initial lubrication and rapid ring seating during the break-in period. Cylinder wall surface finishes are not critical for applications using metallic piston rings. The range of surface finish for satisfactory results is between 4 and 80 rms[1], with a range of 30–40 rms considered excellent. Metallic piston rings also adapt well to a wide range of tolerances between the cylinder and bore resulting from manufacturing inefficiencies, as well as from normal wear and cylinder expansion, called "breathing," during operation.

Metallic piston rings are usually impervious to the effects of fluid. Acids formed in the system have some corrosive effect, but the greatest threat exists from system contaminants and water during the time when the hydraulic machinery is left idle and piston rings can corrode, rust, and stick in the piston grooves.

Leakage past metallic piston rings, for the most part, occurs between the piston ring and the piston lands that form the sides of the groove. This is caused by improper seating between the two surfaces. Dirt and foreign particles, misalignment of the piston from machining, and poor surface finish at the piston groove can account for this improper fit and seal.

O-ring gaskets are used for both static and dynamic sealing applications (Fig. 11-13). They play their major role as static seals but are used extensively as reciprocating and rotating seals. O-rings are also used in combination with metallic piston-type rings as back-up seals to exert radial pressure against the primary metallic sealing ring, as well as to reduce to an absolute minimum the leakage inherent to metallic sealing ring installations. In all cases, the initial seal is affected by deformation of the cross section area and increases in pressure further deform the O-ring cross section in a direction with the pressure to increase sealing effect. The O-ring is considered to be a balanced seal design. *A balanced seal design is one in which an increase in pressure causes an equal force to be exerted on the seal through its cross section toward both surfaces to be sealed.*

Q-ring cross section seals are used for both static and dynamic applications (Fig. 11-14). The design is an adaption of the O-ring with material removed from the two opposite nonsealing sides of the seal and an equal amount added to the two surfaces that come in contact and are deformed during installation. The cross section area of the Q-ring is the same as equivalent size O-rings and is standardized to interchange with them. The Q-ring has several advantages over the conventional O-ring cross section. First, because of its cross section profile, it re-

---

[1] RMS is an abbreviation for root mean square which is the square root of the arithmetic sum of all the squares of the average deviations from the mean surface.

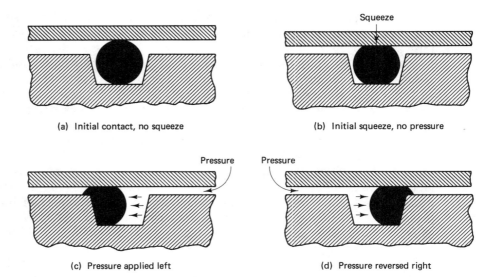

(a) Initial contact, no squeeze

(b) Initial squeeze, no pressure

(c) Pressure applied left

(d) Pressure reversed right

Figure 11-13   How O-rings seal.

quires a lower unit force and deformation of the cross section area to affect the initial seal. Second, operating pressures deform the seal up to 300% less than conventional O-ring seals operating under the same load. Third, abrading damage caused by burrs, sharp edge glands, and other parts of the system have a tendency to damage the Q-ring seals less than conventional O-ring seals during installation. Fourth, the flat surfaces facing the pressure and opposite pressure sides of the seal have a larger initial contact area to maintain a low pressure seal more effectively.

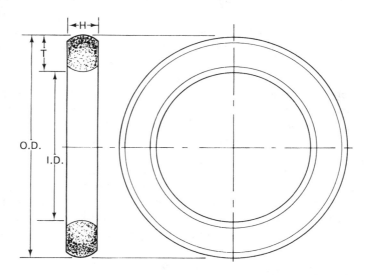

Figure 11-14   Q-ring geometry (*courtesy of Parker Seal Company*).

Harder Shore A hardness materials with better wear and lower friction characteristics may be used with the tendency to extrude significantly reduced. Finally, the tendency for the Q-ring to twist and roll is much less than that of conventional O-rings because the flat faces on the cross section support the Q-ring in the desired position to resist rolling. Both O-ring and Q-ring seals are considered to be balanced designs.

U-seals (Fig. 11-15) and U-packings have extensive uses as dynamic seals, both for reciprocating as well as rotating motion. The initial force necessary to effect a seal between the lips of the seal and the metal cylinder, rod, or housing is caused by a slight deformation of the seal and reduction in volume during installation. System pressure increases the sealing force on the lips proportionally to effect a seal directly related to need. The U-cup design is a balanced seal. Collapse of the seal at low system pressures is prevented by installing a filler or pedestal ring in the U-cup to keep the lips of the seal in contact with the wall surface. Care must be exercised in gland installations to be sure the gland does not force the seal against the end of the packing box or else leakage and failure will occur. Impregnated and treated leather, rubber, and many other elastomeric compounds are used for U-cap seals.

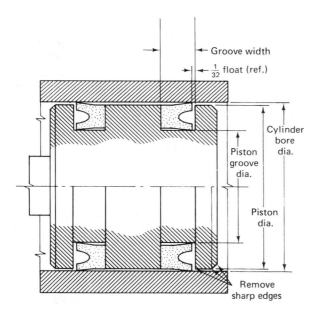

Figure 11-15 U-seal (*courtesy of Parker Seal Company*).

V-packings (Fig. 11-16) are widely used in dynamic applications for high pressures and severe operating conditions. When installed in packing glands to seal reciprocating piston rods or rotating shafts, provisions are usually made to compensate for wear by tightening the gland nut or supporting flange. In some installations, overtightening of the gland nut, which would result in excessive force, lack of necessary lubrication, friction, and wear at the seal, is avoided by using a compensating spring of desired strength supported by the gland nut bot-

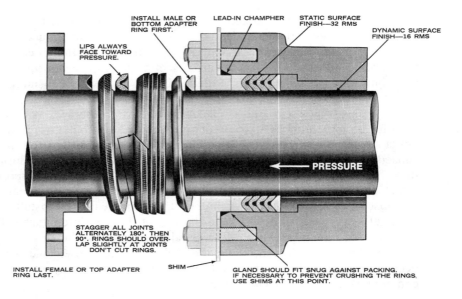

INSTALL MALE OR BOTTOM ADAPTER RING FIRST.

LEAD-IN CHAMPHER

STATIC SURFACE FINISH—32 RMS

DYNAMIC SURFACE FINISH—16 RMS

LIPS ALWAYS FACE TOWARD PRESSURE.

← PRESSURE

STAGGER ALL JOINTS ALTERNATELY 180°, THEN 90°. RINGS SHOULD OVER-LAP SLIGHTLY AT JOINTS DON'T CUT RINGS.

INSTALL FEMALE OR TOP ADAPTER RING LAST.

SHIM

GLAND SHOULD FIT SNUG AGAINST PACKING. IF NECESSARY TO PREVENT CRUSHING THE RINGS, USE SHIMS AT THIS POINT.

Figure 11-16 V-packing Installation (*courtesy of Crane Packing Company*).

tomed or run flush with the mounting surface (Fig. 11-17). The V-packing stack is composed of a minimum of three seals, with the maximum number determined by the pressure, temperature, and operating conditions. As many as seven V-seals may be stacked. Common materials used for V-packings include impregnated and treated leather, rubber, and asbestos. A major advantage of the V-packing is that dissimilar but compatible seals of different materials may be used in combination to offer the system the best available pressure, wear, and friction service characteristics associated with respective seals in the stack.

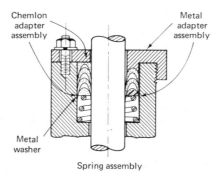

Chemlon adapter assembly

Metal adapter assembly

Metal washer

Spring assembly

Figure 11-17 Spring loaded packing assembly (*courtesy of Crane Packing Company*).

Cup seals have extensive application as piston seals. When pressure is to be applied in both directions, two cup seals are installed on the piston facing in opposite directions with the lips of the cups facing against the direction of applied pressure (Fig. 11-18). Initial sealing force is applied to the seal by a slight reduction in the cross section volume during installation. Follower plates insure the

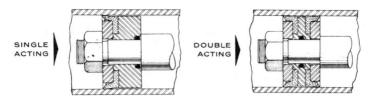

Figure 11-18 Cup seal applications (*courtesy of Crane Packing Company*).

integrity of the seal at low pressures by preventing the seal lips from collapsing. The design of the piston plate, the follower, and clearances are important to the effective support and operation of the seal. Materials commonly used for cup seals include impregnated and treated leather, rubber, and other elastomeric compounds. The cup seal is considered to be an unbalanced design.

Flange seals function essentially in the same manner as cup seals but are used in glands as rod seals rather than on pistons as seals. The seal is also an unbalanced design. Many of the same materials used for cup seals are also used for their fabrication. Gland design and clearances are important to prevent excessive unsupported forces on the seal which would result in extrusion, leakage, and premature failure.

Rod wipers and scrapers are the first line of defense for the hydraulic system, and have as their purpose to exclude dirt and other foreign matter from entry into the system (Fig. 11-19). In some instances, the rod wiper also serves double duty by providing a low-pressure seal against escaping fluid that has bypassed primary seals in the system. The general configuration of rod wipers is a beveled edge facing outward and slanting toward the piston rod. Materials used as rod wipers include brass and other metals and elastomers known for toughness and resilience. The material used for the wiper is strongly influenced by the operating environment, the temperature, and the severity of operating conditions. Typically, mobile hydraulic applications, such as backhoes, and other equipment encountering mud, dirt, coal dust, water, brush, and outside weather place the most severe operating conditions on the wipers.

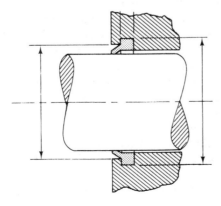

Figure 11-19 Rod wiper geometry (*courtesy of Parker Seal Company*).

Sec. 11-5 Seal Configuration

Back-up rings prevent extrusion and abrading of such seals as the O-ring (Fig. 11-20). They themselves are not seals. Extrusion and abrasion of the seal at pressures above 1500 lbf/in.² (10.3 MPa) are prevented by providing the additional support necessary to keep the seal from excessive flexing and creeping into the clearance space. Pressure can be increased as well as metal clearances when back-up rings are used. For example, O-rings that are recommended for 1500 lbf/in.² (10.3 MPa) can be extended to pressures up to 3000 lbf/in.² (20.7 MPa). Lubrication of the seal caused by entrapment of oil by the back-up rings is another desirable side effect. Common materials used in the construction of back-up rings include leather, hard rubber, stiff elastomers, and some plastics such as Teflon®. Rubber or another elastomer compound is required if stretching of the back-up ring is necessary for installation.

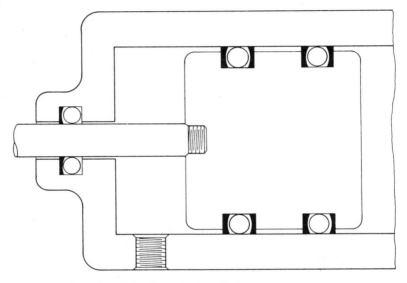

Figure 11-20 Back-up ring installation.

Wear rings are installed in fluid power cylinders to prevent contact between the moving piston and the stationary metal cylinder wall (Fig. 11-21). In addition, seal performance is improved and seal life is extended by reducing the tendency of the cylinder rod to flex from excessive clearance. Close tolerances usually associated with the piston fit in the cylinder bore can be reduced by having the clearance taken up with the wear ring. Dissimilar and incompatible materials, such as aluminum pistons and steel cylinders that have a tendency to skuff and gall during contact, can also be used together with this support method that effectively keeps the piston and cylinder bore separated. Wear rings are made of bronze and other sintered metals, nylon, Teflon®, and other plastics. Requirements of wear rings include some of those associated with seals. Included in these are swell characteristics and resistance to operating temperatures, in addition to bearing strength qualities normally associated with support and wear-resistant materials.

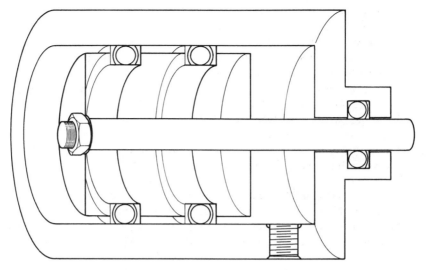

Figure 11-21 Wear-ring installation (*courtesy of Parker Seal Company*).

11-6 ELASTOMER SEAL TESTING

Elastomer seals are subject to physical and chemical property tests, as well as to functional tests. Physical and chemical property tests include such initial properties as durometer hardness, tensile strength, ultimate elongation, modulus of elongation, and specific gravity. Functional tests include changes in physical properties occurring in simulated service testing, as well as hardness change, tensile strength change, elongation change, and volume change. Other functional properties related to actual service include compression set, low temperature characteristics, and heat aging. Most tests are derivations of standards developed by manufacturers with many years of accumulated experience in the development and testing of elastomers.

Care must be exercised in the interpretation of laboratory test results that do not actually simulate or duplicate service conditions as they exist in hydraulic systems in the field. Some materials that fail to pass laboratory tests may be perfectly satisfactory when subjected to actual use in hydraulic systems applications. This is not to reduce the value of laboratory testing of elastomer seals, but to caution that the final test, service in the field, is the authority from which the appropriateness of a seal material for a particular application is determined. Thus, field test data must be the source of information from which many decisions regarding appropriateness can be made.

11-7 SUMMARY

Fluid power seals are used to make systems internally and externally fluid tight. Static applications prevent the passage of fluid from inside the system and seal the spaces between two parts in stationary contact with each other. Dynamic seals

Sec. 11-7   Summary   **275** prevent fluid from leaking between moving parts, either within the system itself, such as between pistons and cylinder bores, or outside the system, such as between moving cylinder rods and stationary rod-end caps. Materials used for fluid power seals must be compatible with system fluid as well as have the physical characteristics that will insure efficient operation under varying conditions of pressure, speed, temperature, and external environment. Seal configuration describing the cross section area of the seal has its origin in attempts to decrease leakage, insure long and trouble-free service life, reduce manufacturing and service costs, and, in general, increase the mechanical efficiency of the system. Common configurations include the square cut, O-ring, Q-ring, U-cup, V-cup packing, lip seal, and flange seal. Other specialty seals are sometimes attached to retaining parts, such as bolt and fitting seals, to insure that they are properly installed and operate effectively. Of all seals, the O-ring is the most common and extensively used. Seals are tested using standard procedures, many of which have been adopted by the American Society for Testing and Materials. Common tests applied to elastomeric materials include hardness, tensile strength, ultimate elongation, modulus of elongation, specific gravity, and other changes occurring during operation or exposure to the effects of heat, pressure, and the hydraulic fluid. These include volume change or swell, compression set, and retraction temperatures.

STUDY QUESTIONS AND PROBLEMS

1. How do static and dynamic seals differ?
2. How much should confined gaskets such as O-rings be compressed to make a proper seal?
3. Compute the groove diameter, width, and outside diameter for a 6-in. close-fitting static male plug seal using an O-ring given an initial squeeze of 20%. Use an O-ring with a free cross section diameter of 0.140 in. (clue: Draw the application first.)
4. A static face seal groove with a mean diameter of 4.000 in. is to be recessed in a flange and covered with a face plate. If the O-ring has a free cross section of 0.125 and is to be given an initial squeeze of 25%, what are the dimensions of the groove (mean, diameter, depth, and width)?
5. A female gland is to accommodate a 10-cm tube with a clearance of 1.5 mm. If the O-ring has a free cross section of 4.76 mm and is given a 20% squeeze, what are the dimensions of the groove (diameter, depth, and width)?
6. What factors influence cylinder friction?
7. Each time a double-acting hydraulic cylinder extends and retracts through an 8-in. stroke cycle, a large drop of oil is seen to drip from the front seal. If the cylinder rod has an O.D. of 2 in., and the oil film thickness is 0.0008 in., what is the volume of this "drop of oil" for one cycle?
8. If the leakage rate in Problem 7 remains constant with the cycle rate which is 25 strokes per minute, how long would it take for the system to lose a gallon of fluid?
9. A front end loader cycles the bucket cylinder at an average rate of 10 cycles/min. If an oil film thickness of 0.0005 in. (five ten thousandths of an inch) lubricates a cylinder rod

with an O.D. of 1.5 in. as it extends and retracts through a 24-in. stroke, and the rod wiper removes the oil when the cylinder retracts, how much fluid would the cylinder lose during an 8-hour shift?

10. What elastomer sealing materials could be used with a petroleum-base fluid if the operating temperature ranges from $-10°F$ ($-23.3°C$) at start-up to $200°F$ ($93.3°C$) when the machine is warm and under load?

11. Which sealing material would have the greatest all-around resistance to high and low temperatures?

12. An elastomer piston seal with no back-up rings must operate at pressures up to 2000 lbf/in.^{2} with a clearance gap of 0.010 in. What Shore A hardness would be recommended to prevent extrusion? What Shore A hardness would be recommended if the pressure were reduced to 1000 lbf/in^2?

13. What are the two main factors that influence compression set in elastomers?

14. What factors are used to compute final squeeze of an elastomer seal?

15. An elastomer O-ring with a free cross section diameter of 0.250 in. is given an initial squeeze at assembly in a static application of 20%. Fluid swell accounts for an additional 15% swell, and compression set accounts for a reduction of 10%. Compute the final available squeeze on the O-ring in inches and as a percent of the original free cross section.

16. An elastomer static seal with an original cross section thickness of 0.250 in. measures only 0.210 in. in the direction of squeeze after it has been removed from service for some time. What is the compression set characteristic of the material?

17. An elastomer O-ring with a nominal cross section diameter of 3/16 in. swells 18% in the hydraulic fluid used in the system and has a characteristic compression set of 8%. If the final squeeze is to be 25%, how deep would the gland have to be?

18. A tensile test of a polymer material with a cross section of 1 cm^2 requires a force of 250 N to separate the material. What is the tensile strength of the material in MPa and lbf/in.^{2}?

19. What is the difference between tensile strength and ultimate elongation?

20. An elastomer seal with a mean diameter of 4.200 in. and free cross section of 0.250 in. must be stretched over a piston with an O.D. of 4.400 in. If the ultimate elongation is given as 20% for the material, and it is assumed that the cross section remains constant, a) can the material be fitted over the piston, and b) at what length would the elastomer separate?

21. In Problem 20, what would the minimum separation length of the material have to be for the piston ring to fit over the piston.

22. List the more common seal configurations and draw their cross sections.

23. Why are back-up rings used?

24. Why are wear rings used on cylinder pistons?

25. How is a durometer hardness reading of A/75/15 interpreted?

26. A durometer hardness test result is reported as A/75/5. What is the force exerted by the indentor rod from an A-scale durometer?

27. Explain what is meant by retraction value.

CHAPTER 12

SYSTEMS
COMPONENTS

12-1 INTRODUCTION

Hydraulic systems components include both single hydraulic devices, such as reservoirs, accumulators, and lubrication devices, as well as others that are compound in nature, including several single devices in one package or hydraulic enclosure. Power units and hydrostatic transmissions, for example, are compound components. Components represent the state of the art in fluid power products. They also represent advances by individual manufacturers to meet specialized needs of the industry. Many fluid components, such as accumulators, are more easily explained and understood using related circuits. Others, such as pressure gauges and lube devices, may be explained and understood just as clearly by themselves.

This chapter presents a representative sampling of fluid power components not covered in detail elsewhere in preceding chapters, including accumulators, intensifiers, conductors, instrumentation and lube devices, hydraulic power packages, and electric controls.

12-2 RESERVOIRS AND HEAT EXCHANGERS

Fluid conditioners consist of reservoirs, heat exchangers, strainers, filters, and filter condition indicator systems. Reservoirs and heat exchangers are explained here.

An appropriately sized industrial reservoir is seen in Fig. 12-1. Several features are apparent. The overall dimensions should enclose the oil with sufficient surface area to allow air bubbles and foam to escape the fluid. The depth must be sufficient to assure that during peak pump demands the oil level will not drop below the suction line. The reservoir should be sized so that fluid contact with the

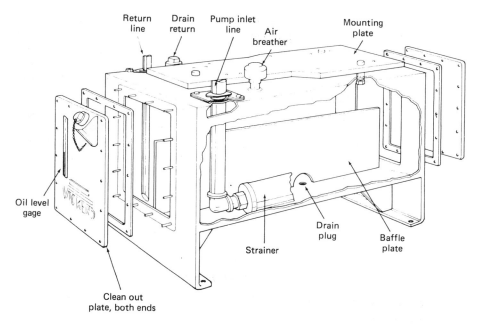

Figure 12-1 Production hydraulic reservoir (*courtesy of Sperry-Vickers*).

tank wall and bottom will permit adequate exposure for cooling the fluid. Baffles are provided to separate system suction and return oil, promote circulation toward the bottom and side walls, and drop out foreign material such as dirt, sludge, and metallic articles. The return line is usually cut at a 45° angle to assist in redirection of the fluid. Notice that a drain is located at the lowest point in the reservoir, and that a cleanout plate is provided for inspection and cleaning purposes. Sight gauges are often included to monitor fluid level. The suction and return lines are sealed where they enter the reservoir to prevent dirt from entering the system. A filtered breather system is included to admit clean make-up air to maintain atmospheric pressure as fluid is pumped out of the reservoir. The strainer prevents recirculation of particulate material back through the system.

Location of the return line below the surface of the fluid reduces the tendency for the oil to mist or form air bubbles as it enters the tank. In some instances, however, the return line is purposely located above the surface level to reduce back pressure or syphoning on the system that might impede flow in gravity return systems.

Heat exchangers are used in hydraulic systems to maintain the operating temperature of the fluid within specified limits. Nominal operating temperatures range between 120°F and 150°F. Excessive temperatures are caused by extreme environments, such as locations near hot metals in steel mills or foundries, excessive pumping of the hydraulic fluid across relief valves between working cycles, and high cycle rates requiring fluid to flow through restrictive valves, plumbing, and fittings which cause undue pressure drop and generate heat. High temperatures promote thinning, oxidation, and breakdown of the hydraulic oil; cause deterioration of seals, packings, and hydraulic hoses; and cause malfunction or inefficiency in components because required clearances between such parts as valve spools and bodies and pump and motor parts cannot be maintained.

The first step in sizing a heat exchanger is to compute the horsepower loss and then convert this to a rise in fluid temperature.

The heat energy generated internally by the hydraulic system equals the power input to the pump minus the power consumed at the output, multiplied by the operation time. This must be in compatible units, usually British thermal units (Btu) or Joules. A Btu is the amount of heat required to raise one pound of water one degree Fahrenheit.

In equation form

$$\text{Energy Loss} = \text{Time (Pump Input Horsepower}$$

$$- \text{Actuator Output Horsepower)}$$

and

$$\text{Energy Loss} = t(Ihp - Ohp) \tag{12-1}$$

where the total time is (t). Using conversion factors to obtain compatible units in the British and SI systems,

$$1\ HP = 2{,}544\ \text{Btu/hr} = 42.4\ \text{Btu/min} = 0.7067\ \text{Btu/sec} \tag{12-2}$$

and

$$1\ HP = 2\ 685\ 600\ \text{J/hr} = 44\ 760\ \text{J/min} = 746\ \text{J/s} \tag{12-3}$$

The energy loss then equals

$$\text{Energy Loss (Btu)} = 2{,}544 \times t(Ihp - Ohp) \text{ where } t \text{ is in hrs} \tag{12-4}$$

and

$$\text{Energy loss (J)} = 746 \times t(Ihp - Ohp) \text{ where } t \text{ is in seconds.} \tag{12-5}$$

Example 1

The hydrostatic transmission in Fig. 12-2 receives 8 Ihp from the engine at the pump. The transmission output delivers 5 hp to the rear wheel output drive. Assuming the difference is lost to heat, compute the total Btu (and Joule) heat loss over a period of 4 hr of operation.

Solution   Substituting in Eqs. 12-4 and 12-5

$$\text{Btu Energy Loss} = 2{,}544 \times t(Ihp - Ohp)$$

$$= 2{,}544 \times 4(8 - 5) = 30{,}528\ \text{Btu}$$

and in SI units

$$\text{Joule Energy Loss} = (746 \times 3{,}600) \times 4(8 - 5) = 32{,}227{,}200\ \text{Joules}$$

The temperature increase resulting from the heat generated in the fluid is computed in degrees Fahrenheit (°F) and degrees Celsius (°C). The heat gener-

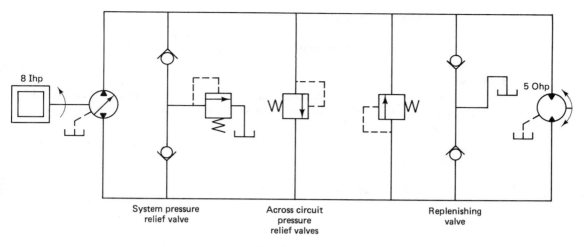

8 Ihp

5 Ohp

System pressure
relief valve

Across circuit
pressure
relief valves

Replenishing
valve

Figure 12-2   Example 1.

ated in fluid power systems is attributed to discharging fluid across such components as pressure relief valves and from the mechanical friction of moving parts. In most hydraulic systems, mechanical friction accounts for less than 20% of the input horsepower.

The formula for computing temperature increase due to heat generation by fluid passing through a restriction is

$$\text{Temperature Increase} = \frac{\text{Heat Generation Rate}}{\text{Oil Specific Heat} \times \text{Oil Mass Flow Rate}} \qquad (12\text{-}6)$$

In compatible units

$$\text{Temperature Increase (°F)} = \frac{\text{(Btu/min)}}{\text{(Btu/lbm/°F)} \times \text{(lbm/min)}} \qquad (12\text{-}7)$$

where the specific heat of oil is given as 0.42 Btu/lbm/°F.

When the flow rate is given in gal/min and the mass of water is taken as 62.4 lbm/ft³, the mass flow rate of oil is computed from

$$\text{Oil Mass Flow Rate (lbm/min)} = \text{gal/min} \times \frac{231}{1728} \times 62.4 \times \text{Oil } Sg$$

and

$$\text{Oil Mass Flow Rate (lbm/min)} = 8.34 \times \text{gal/min} \times \text{Oil } Sg$$

Here it is noticed that because of standard gravitational conditions, the values for the mass (lbm) and weight (lbf) of the oil are numerically equal.

In SI units, the specific heat of oil is given to be 1.75 kJ/kg/°C which is derived from the following conversion of the English units:

(0.42 Btu/lbm/°F)(1.057 kJ/Btu)(2.2 lbm/kg)(1.8°F/°C) = 1.75 kJ/kg/°C and the temperature increase is

$$\text{Temperature Increase (°C)} = \frac{\text{kJ/min}}{(\text{kJ/kg/°C}) \times (\text{kg/min})} \qquad (12\text{-}8)$$

Example 2

The pressure drop across a sticking control valve is observed to be 1000 lbf/in². If the fluid has a Sg of .85 and a flow rate of 3 gal/min, estimate the rise in temperature of the fluid across the control valve.

Solution In English units, the heat generation rate (Btu/min) from the pressure drop across the control valve is

$$\text{Heat Generation Rate} = 42.4 \times \frac{p \times Q}{1714}$$

and

$$\text{Heat Generation Rate} = 42.4 \times \frac{1000 \times 3}{1714} = 74.21 \text{ Btu/min}$$

The mass flow rate of fluid through the valve is computed from

$$\text{Mass Flow Rate} = 8.34 \times \text{gal/min} \times Sg$$

and

$$\text{Mass Flow Rate} = 8.34 \times 3 \times .85 = 21.27 \text{ lbm/min}$$

Finally, solving for the temperature rise in °F of the fluid

$$\text{Temperature Rise °F} = \frac{\text{Heat Generation Rate}}{\text{Oil Specific Heat} \times \text{Oil Mass Flow Rate}}$$

and

$$\text{Temperature Rise °F} = \frac{74.21}{0.42 \times 21.27} = 8.31°F$$

In SI units, the heat generation rate from the pressure drop across the control valve is computed using Eq. 12-3 and Eq. 2-12 (SI fluid horsepower)

$$\text{Heat Generation Rate} = (44\ 760 \text{ J/min}) \frac{(\text{kPa} \times \text{liters/min})}{(44\ 760 \text{ J/min})}$$

$$\text{Heat Generation Rate} = (44\ 760 \text{ J/min}) \frac{(68966 \text{ kPa} \times 11.36 \text{ liters/min})}{(44\ 760 \text{ J/min})}$$

$$\text{Heat Generation Rate} = 78.345 \text{ kJ/min}$$

The mass flow rate of fluid through the valve is computed

$$\text{Mass Flow Rate} = (1 \text{ kg/liter})(11.36 \text{ liters/min})\ (Sg)$$

$$\text{Mass Flow Rate} = (1)(11.36)(0.85) = 9.656 \text{ kg/min}$$

Finally, solving for the temperature rise of the fluid in °C using Eq. 12-8

$$\text{Temperature Rise (°C)} = \frac{(\text{kJ/min})}{(\text{kJ/kg/°C}) \times (\text{kg/min})}$$

and

$$\text{Temperature Rise (°C)} = \frac{(78.345 \text{ kJ/min})}{(1.75 \text{ kJ/kg/°C})(9.656 \text{ kg/min})} = 4.64\text{°C}$$

Heat exchangers relieve the hydraulic fluid of excess heat to lower its operating temperature. In simple terms, the amount of heat to be removed and transferred to the cooling medium equals the difference between the input to the hydraulic pump and the output of all system actuators. This assumes, of course, that the existing ambient temperature is appropriate for system operation and that environmental conditions are not adding or subtracting heat from the fluid. This is seldom the case. In extreme environments, both cold as well as hot, heat exchangers may be employed to counteract primarily environmental circumstances rather than operating conditions to maintain the temperature of the oil within operable limits. For example, in northern climates, heat is frequently added to the fluid to decrease viscosity, whereas locating pumps, transmission lines, and actuators near furnaces requires that heat be subtracted from the fluid to increase viscosity and reduce the temperature. This is a different problem than lowering the temperature of fluids which are heated by the working cycle itself, for example, by being discharged across a relief valve.

Stationary production hydraulics typically use forced air or water circulating heat exchangers because of the low cost and availability of water and other appropriate cooling fluids. Figure 12-3 shows a typical oil cooler that uses circulated air to dissipate the heat. Portable and particularly mobile hydraulics applications use air radiators equipped with a circulating fan. In most large mobile applications,

Symbol

Figure 12-3 Air colled heat exchanger (*courtesy of Young Radiator Company*).

the engine radiator fan serves both radiators or cools a combination radiator. When controlled heating as well as cooling are desirable, the fluid may be piped into a radiator heat exchanger to maintain the temperature of the radiator coolant.

Several types of heat exchangers are available. Those used to add heat to the fluid include the shell and tube-type (see Fig. 12-4), immersed tube-type, and immersed electric coils. Those used to cool the hydraulic fluid include the shell and tube, finned radiator, plate type, and reservoir immersed coils.

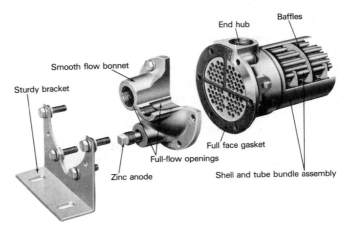

Figure 12-4 Construction features of shell and tube heat exchanger and air cooled radiator heat exchanger (*courtesy of Young Radiator Company*).

12-3 ACCUMULATORS

Hydraulic accumulators use ballast weights, springs, or gas pressure to generate the precharge force against the fluid that is stored under pressure for use in the system.

Weighted accumulator systems use a large ballast over a cylinder with capacities from 1 to 500 gal at pressures to 3000 lbf/in.^{2} (Fig. 12-5.)

The output pressure available from weighted accumulators is a function of the ballast weight and the cross section area of the accumulator and is constant through the delivery of the usable capacity. That is

$$\text{Output Pressure} = \frac{\text{Ballast Force}}{\text{Accumulator Area}} \tag{12-9}$$

where the output pressure (p) is in lbf/in.^{2}, the ballast force (F) is in lbf (N), and the accumulator cross section area (A) is in in.^{2}. If the cross section area of the accumulator is given, the capacity of the accumulator is computed as a function of the stroke. That is,

$$\text{Accumulator Capacity (gal)} = \frac{\text{Area (in.}^2) \times \text{Stroke (in.)}}{231 \text{ (in.}^3/\text{gal)}} \tag{12-10}$$

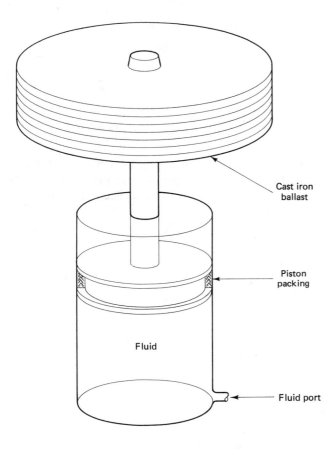

Cast iron
ballast

Piston
packing

Fluid

Fluid port

Figure 12-5 Ballast-type accumulator.

In SI metric units,

$$\text{Accumulator Capacity (liters)} = \frac{\text{Area (cm}^2\text{)} \times \text{Stroke (cm)}}{1000 \text{ (cm}^3\text{/liter)}} \qquad (12\text{-}11)$$

Example 3

(a) Assuming friction is negligible, determine the weight that must be used for ballast to generate 1,500 lbf/in.^{2} from an accumulator with a cross section diameter of 12 in. (b) What must the stroke length be to have a capacity of 50 gal?

Solution

(a) The pressure is computed from Eq. 12-9:

$$\text{Output Pressure} = \frac{\text{Force}}{\text{Area}}$$

$$F = p \times A$$

and

$$F = 1,500 \times \frac{3.14 \times 12^2}{4}$$

$$F = 169,560 \text{ lbf or } 84.78 \text{ tons}$$

(b) The stroke length is computed from Eq. 12-10:

$$\text{Accumulator Capacity} = \frac{\text{Area} \times \text{Stroke}}{231}$$

$$S = \frac{V \times 231}{A}$$

$$S = \frac{50 \times 231}{113.04} = 102.2 \text{ in. or } 8.5 \text{ ft}$$

Spring loaded accumulators function much the same as weighted accumulators except that the force applied to the movable accumulator piston is generated by coil springs rather than ballast weights. Spring loaded accumulators can be mounted in any position. (Fig. 12-6.)

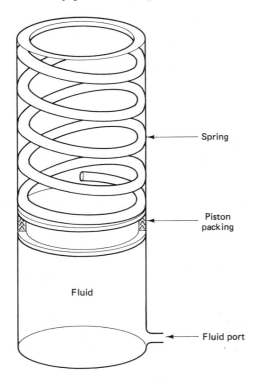

— Spring

— Piston packing

Fluid

— Fluid port

Figure 12-6 Spring-type accumulator.

Gas-charged accumulators use pistons, diaphragms, bladders, or no medium whatsoever to separate the two fluids. Inert gases such as dried nitrogen are used as the precharge. Figure 12-7 illustrates a typical bladder-type accumulator. Provision is made to keep the bladder from damage and rupture through the discharge port when it is precharged by placing a large head poppet valve at the outlet which is closed by the bladder when the accumulator is empty of fluid. Bladder-type accumulators are available in sizes to 80 gal for operation to 6000 lbf/in.² at flow rates of 600 gpm.

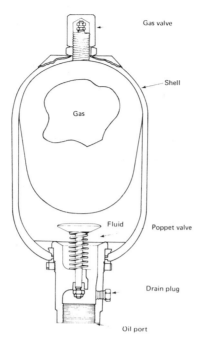

Gas valve

Shell

Gas

Fluid

Poppet valve

Drain plug

Oil port

Figure 12-7 Bladder-type accumulator (*courtesy of Greer Olear Products Division*).

Gas charged accumulators operate by placing the compressible gas over the incompressible hydraulic fluid in a constant volume accumulator. The hydraulic pressure developed and the volume of fluid available to the system are dependent on the precharge pressure and expansion characteristics of the gas.

If the temperature of the gas is kept constant, the pressure change in the gas is inversely proportional to its volume, and the expansion and contraction of the gas is considered to be isothermal (constant temperature). For example, when 100 in.^{3} of gas originally at 1,000 lbf/in.^{2} abs. is compressed to 50 in.^{3}, the pressure will rise to 2,000 lbf/in.^{2}. Conversely, if the volume of gas is increased to 200 in.^{3}, its pressure would reduce to 500 lbf/in.^{2} abs. Where the expansion and contraction of the gas are sufficiently slow to generate little or no heat, performance of the gas can be stated approximately by the formula

$$p_1 V_1 = p_2 V_2 = p_3 V_3 \qquad (12\text{-}12)$$

where (p) and (V) represent the absolute pressure and volume of the gas, and the subscripts represent different pressures and volumes of the gas.

Example 4

What size accumulator is necessary to supply a hydraulic system with 5 gal of oil within the pressure range 2,000–3,000 lbf/in.^{2} gauge, if the accumulator has a precharge pressure of 1,500 lbf/in.^{2} gauge?

Solution Refer to Fig. 12-8. Isothermal contraction and expansion of the gas is computed from

$$p_1 V_1 = p_2 V_2 = p_3 V_3$$

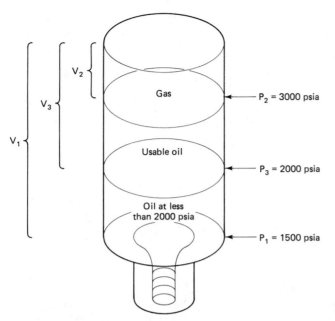

Gas ← P_2 = 3000 psia

Usable oil ← P_3 = 2000 psia

Oil at less than 2000 psia ← P_1 = 1500 psia

Figure 12-8 Examples 4 and 5.

where V_1 is the unknown capacity of the accumulator

$$V_2 = V_3 - 1,155 \qquad (5 \text{ gal} \times 231 \text{ in.}^3/\text{gal})$$

and

$$V_3 = V_2 + 1,155$$

Solving for V_2

$$p_2 V_2 = p_3 V_3$$

$$V_2 = \frac{p_3}{p_2} \times V_3$$

Substituting for V_3

$$V_2 = \frac{p_3}{p_2} \times (V_2 + 1,155)$$

Substituting values

$$V_2 = \frac{2,000 + 14.7}{3,000 + 14.7} \times (V_2 + 1,155)$$

$$V_2 = 0.668 V_2 + 772$$

$$V_2(1.000 - 0.668) = 772$$

and

$$V_2 = 2325 \text{ in.}^3$$

Solving for V_1

$$p_1 V_1 = p_2 V_2$$

$$V_1 = \frac{p_2}{p_1} \times V_2$$

Substituting values

$$V_1 = \frac{3,000 + 14.7}{1,500 + 14.7} \times 2325$$

and

$$V_1 = 4,627 \text{ in.}^3 \text{ or } 20 \text{ gal}$$

When a gas is compressed and expanded quickly, heating and cooling generated cause pressure changes in addition to those occurring strictly as the result of volume changes. If the gas is insulated and no heat is allowed to escape, sudden reduction of 100 in.^{3} of a gas, originally at 1000 lbf/in.^{2} absolute, to 61.2 in.^{3} is sufficient to increase the pressure of the gas to 2,000 lbf/in.^{2} absolute, and the performance of the gas is considered to be adiabatic (no transfer of heat). The general formula that represents adiabatic performance of a gas is

$$p_1 V_1^{1.4} = p_2 V_2^{1.4} = p_3 V_3^{1.4}$$

Example 5

Refer again to Fig. 12-8, and use the problem from Example 4. Adiabatic contraction and expansion of the gas is computed similarly using the volume coefficient 1.4.

Solution Solving for V_2

$$p_2 V_2^{1.4} = p_3 V_3^{1.4}$$

$$V_2^{1.4} = \frac{p_3}{p_2} \times V_3^{1.4}$$

$$V_2 = \left(\frac{p_3}{p_2}\right)^{1/1.4} \times V_3$$

Substituting values

$$V_2 = \left(\frac{2,000 + 14.7}{3,000 + 14.7}\right)^{5/7} \times (V_2 + 1,155)$$

$$V_2 = (.6683)^{5/7} \times (V_2 + 1,155)$$

$$V_2 = (\sqrt[7]{.1333}) \times (V_2 + 1,155)$$

Computing the seventh root by trial calculation

$$V_2 = .7499(V_2 + 1,155)$$

$$V_2 = .7499 V_2 + 866.13$$

$$V_2(1.00 - .7499) = 866.13$$

and

$$V_2 = 3,463 \text{ in.}^3$$

Solving for V_1

$$p_1 V_1^{1.4} = p_2 V_2^{1.4}$$

$$V_1^{1.4} = \frac{p_2}{p_1} \times V_2^{1.4}$$

$$V_1 = \left(\frac{p_2}{p_1}\right)^{1/1.4} \times V_2$$

Substituting values

$$V_1 = \left(\frac{3,000 + 14.7}{1,500 + 14.7}\right)^{5/7} \times 3,463$$

$$V_1 = \sqrt[7]{31.231} \times 3,463$$

Again, computing the seventh root by trial calculation

$$V_1 = 1.635 \times 3,463 = 5,662 \text{ in.}^3 \text{ or } 24 \text{ gal}$$

The percentage of difference in the required size of the accumulator for adiabatic vs. isothermal expansion is

$$\text{Percent difference} = \frac{5,662.22 - 4,627.25}{5,662.22} \times 100 = 18.28\%$$

In most circumstances the performance of the gas is neither isothermal nor adiabatic, but rather somewhere in between—that is, polytropic with a value for the coefficient for volumetric expansion of approximately ($n = 1.25$). Neither adiabatic nor polytropic values take into account the heat dissipated by the accumulator to the surrounding environment, and in actual practice the size of the accumulator is calculated for isothermal expansion (Boyle's law) and then oversized to compensate for increased expansion of the precharge gas due to heat that may be generated during the working cycle and not dissipated through the walls of the accumulator.

Because hydraulic accumulators store pressurized fluid for system use on demand, they can be used to serve a variety of system functions. Typical of these functions are maintaining system pressure, absorbing hydraulic shock, supplementing pump delivery, providing an auxiliary source of power, cushioning loads, dispensing lubricants, and acting as a barrier between dissimilar fluids.

Maintaining system pressure may require that the accumulator compensate for pressure losses due to fluid leakage or increases from thermal expansion of the fluid. Other applications use the accumulator to maintain system pressure while pump delivery is diverted to other work. Emergency fluid is also available should power failure to the pump drive occur. Short-term use of hydraulic components such as clamping devices, lifts, bearing lubricators, and other cylinder actuators is possible using the pressurized fluid stored in the accumulator. Placement of an accumulator in a hydraulic circuit to maintain system pressure is illustrated in Fig. 12-9.

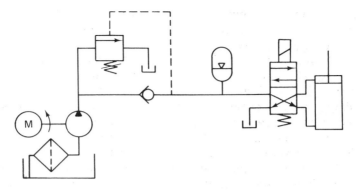

Figure 12-9 Pressure maintenance circuit.

12-4 INTENSIFIERS

Intensifiers or boosters are used to increase the fluid pressure available from the supply source. Typically, they apply a shop air source over oil supplied from a separate reservoir to increase system pressure and add control flexibility not usually available when air is used as the fluid medium to the actuator. They are also employed to reduce costs by eliminating hydraulic pumps and other compo-

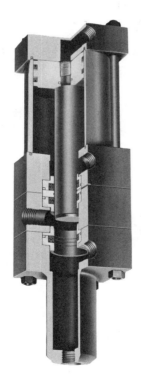

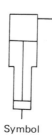

Symbol

Figure 12-10 Air-oil intensifier "booster" (*courtesy of S-P Manufacturing Company*).

nents for single intermittent low volume applications such as bench presses, riveting machines, and others that perform such operations as bending, broaching, cutting, crimping, and forming.

A typical air over oil intensifier is illustrated in Fig. 12-10. Air is directed to the cap end of the air cylinder which forces the hydraulic ram through the lower seal and into the lower hydraulic cylinder. This traps fluid in the lower booster cylinder section developing high pressure. The degree of pressure boost is determined by the area ratios of the air piston to the hydraulic ram. For example, a 5-in. air cylinder driving a 1-in. ram has a ratio of 25 : 1. Given shop air at 80 lbf/in.², the pressure available at the hydraulic output would be (80 × 25) 2,000 lbf/in². The volume output available from the booster cylinder section of the intensifier equals the ram area multiplied by its effective stroke. The effective stroke is less than the actual stroke by as much as 2 in. because of initial travel before the ram passes through the hydraulic seal and traps the fluid in the booster section.

A complete double pressure booster system with air-oil tank return is seen in Fig. 12-11. Systems of this type are used where a cylinder must traverse a substantial distance at low pressure followed by travel through a short distance at high pressure to complete the work part of the cycle. The cylinder must be able to

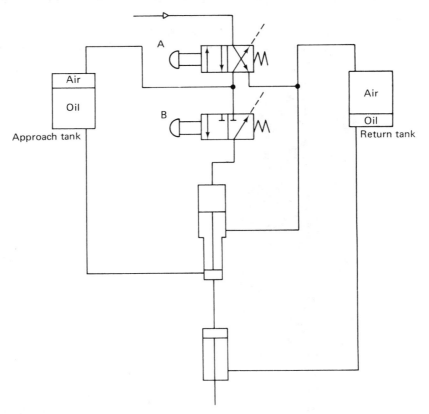

Figure 12-11 Double pressure booster system with air-oil tank return.

extend and retract through the low pressure portions of the cycle using shop air pressure of approximately 100 lbf/in². In operation, valve *A* is used to extend and retract the working cylinder at shop air pressure. Valve *B* applies shop air pressure to the head end of the air cylinder to drive the booster cylinder at high hydraulic pressure. When valve *A* is actuated, the low pressure approach is made by directing air to the approach tank, thereby forcing oil at shop air pressure through the booster cylinder to the head end of the working cylinder. When the working cylinder contacts the workpiece, valve *B* is actuated and the high pressure portion of the cycle is initiated by directing shop air to the head end of the air cylinder which, in turn, intensifies fluid pressure in the booster cylinder. The return passage to the approach tank is closed off by the downward action of the booster cylinder piston, and the working cylinder receives oil at high pressure to perform the designated task. Release of valve *B* blocks shop air and vents the head end of the air cylinder thereby making the high pressure portion of the cycle inactive. Release of valve *A* vents the air in the approach tank and redirects shop air to the return tank. This directs shop air to the rod end of the air cylinder and, at the same time, directs oil from the return tank to the rod end of the working cylinder causing its return. Oil entering the booster cylinder from the working cylinder is returned to the approach tank.

12-5 CONDUCTORS AND CONNECTORS

Hydraulic conductors and connectors contain and distribute fluid throughout the system and include manifolds, pipes, tubing, fittings, and flexible couplings. By objective, they are designed to convey fluid in required amounts with minimum loss due to friction and leakage. Proper design takes into account primarily the operating pressure and flow rate in the system, although other factors are important. These include material, cost, compatibility with the fluid and existing components such as valves, and flexibility where the fluid must be transported to moving components. Leakage that affects the safety of the operator, work spoilage, equipment damage, and the integrity and durability of the system are also taken into account to meet increasingly rigid OSHA Standards in the industry.

Conductors are fabricated from steel pipe, steel tubing, plastic, other rigid synthetic tubing, and flexible hose. Noncorrosive steels are commonly used in corrosive environments. Exterior noncorrosive paints are also used to prevent deterioration. Copper and plating materials such as cadmium and zinc that would react with some hydraulic fluids within the system are avoided. Connectors are machined and fabricated from malleable steel or stainless steel.

The appropriate size diameter conductor is determined from the required flow rate and the continuity equation, maintaining the velocity below 5 ft/s at pressures below 5 in. of Hg for intake lines and to 20 ft/s for pressure lines. Given the flow rate and these acceptable limits, the inside diameter of the conductor can be computed using the formulas

$$\text{Area (in.}^2) = \frac{\text{Flow Rate (gal/min)} \times 0.3208}{\text{Velocity (ft/s)}} \tag{12-13}$$

and in SI units

$$\text{Area (cm}^2) = \frac{\text{Flow Rate (l/min)} \times (0.167)}{\text{Velocity (m/s)}} \tag{12-14}$$

Conductor wall thickness is influenced by the system pressure, tensile strength of the material, outside diameter of the conductor, and the safety factor. For non-critical applications, a safety factor of 4 : 1 is acceptable above 2,500 lbf/in.^{2}; 6 : 1 for pressures between 1,000 and 2,500 lbf/in.^{2}; and 8 : 1 for pressures below 1,000 lbf/in^2. For systems encountering severe pressure shocks and mechanical stress, the safety factor is 10.

Given the inside diameter of the conductor necessary to convey a required flow rate and the maximum system pressure, wall thickness is computed using Barlow's formula. In either English or SI units

Wall Thickness (WT)

$$= \frac{\text{Maximum Pressure } (p \text{ max}) \times \text{Conductor } (OD)}{2 \times \text{Material Tensile Strength } (TS)} \tag{12-15}$$

where the conductor OD equals

$$OD = \text{Conductor } ID + 2 \times WT \tag{12-16}$$

Substituting

$$WT = \frac{(p \text{ max}) \times (ID + 2\,WT)}{2 \times TS}$$

Solving for the wall thickness

$$WT = \frac{(p \text{ max} \times ID) + (2 \times p \text{ max} \times WT)}{2 \times TS}$$

$$(2 \times TS \times WT) - (2 \times p \text{ max} \times WT) = p \text{ max} \times ID$$

$$2\,WT(TS - p \text{ max}) = p \text{ max} \times ID$$

and

$$WT = \frac{p \text{ max} \times ID}{2(TS - p \text{ max})} \tag{12-17}$$

Finally, wall thickness (WT) is multiplied by the appropriate safety factor to ensure the integrity of the conductor.

Example 6

A conductor must carry 20 gal/min with a maximum velocity of 15 ft/sec at a pressure of 2,500 lbf/in^2. Assuming a safety factor of 4, and a conductor made of dead soft cold drawn steel tubing with a tensile strength of 55,000 lbf/in.^{2}, determine the conductor size and wall thickness. Finally, select an appropriate conductor from the schedule of commercially available tubing in Table 12-1.

TABLE 12-1 BASIC DIMENSIONS FOR COMMERCIALLY AVAILABLE TUBING (COURTESY OF MOBIL OIL)

Outside diameter OD(in.)	Wall thickness WT(in.)	Inside diameter ID(in.)
$\frac{1}{8}$	0.028	0.069
	0.032	0.061
	0.035	0.055
$\frac{3}{16}$	0.032	0.1235
	0.035	0.1175
$\frac{1}{4}$	0.035	0.180
	0.042	0.166
	0.049	0.152
	0.058	0.134
	0.065	0.120
$\frac{5}{16}$	0.035	0.2425
	0.042	0.2285
	0.049	0.2145
	0.058	0.1965
	0.065	0.1825
$\frac{3}{8}$	0.035	0.305
	0.042	0.291
	0.049	0.277
	0.058	0.259
	0.065	0.245
$\frac{1}{2}$	0.035	0.430
	0.042	0.416
	0.049	0.402
	0.058	0.384
	0.065	0.370
	0.072	0.358
	0.083	0.334
$\frac{5}{8}$	0.035	0.555
	0.042	0.541
	0.049	0.527
	0.058	0.509
	0.065	0.495
	0.072	0.481
	0.083	0.459
	0.095	0.435
$\frac{3}{4}$	0.049	0.652
	0.058	0.634
	0.065	0.620
	0.072	0.606
	0.083	0.584
	0.095	0.560
	0.109	0.532
$\frac{7}{8}$	0.049	0.777
	0.058	0.759
	0.065	0.745
	0.072	0.731
	0.083	0.709
	0.095	0.685
	0.109	0.657
1	0.049	0.902
	0.058	0.884
	0.065	0.870
	0.072	0.856
	0.083	0.834
	0.095	0.810
	0.109	0.782
	0.120	0.760
$1\frac{1}{4}$	0.049	1.152
	0.058	1.134
	0.065	1.120
	0.072	1.106
	0.083	1.084
	0.095	1.060
	0.109	1.032
	0.120	1.010
$1\frac{1}{2}$	0.065	1.370
	0.072	1.356
	0.083	1.334
	0.095	1.310
	0.109	1.282
	0.120	1.260
$1\frac{3}{4}$	0.065	1.620
	0.072	1.606
	0.083	1.584
	0.095	1.560
	0.109	1.532
	0.120	1.510
	0.134	1.482
2	0.065	1.870
	0.072	1.856
	0.083	1.834
	0.095	1.810
	0.109	1.782
	0.120	1.760
	0.134	1.732

Solution   First, determine the inside diameter of the conductor using Eq. 12-13 or a continuity equation nomograph

$$\text{Area} = \frac{\text{Flow Rate} \times 0.3208}{\text{Velocity}}$$

$$\text{Area} = \frac{20 \times 0.3208}{15} = 0.4277 \text{ in.}^2$$

and the inside diameter is computed from

$$\text{Area} = \frac{\pi \times ID^2}{4}$$

$$ID = \sqrt{\frac{4 \times \text{Area}}{\pi}}$$

$$ID = 2\sqrt{\frac{\text{Area}}{\pi}} = 2\sqrt{\frac{.4277}{3.14}} = 0.738 \text{ in.}^2$$

Second, compute the wall thickness from Eq. 12-17

$$WT = \frac{p \max \times ID}{2(TS - p\max)}$$

$$WT = \frac{2,500 \times 0.738}{2(55,000 - 2,500)}$$

and

$$WT = 0.0176 \text{ in.}$$

Multiplying the wall thickness by a safety factor of 4

$$WT = 0.0176 \times 4 = 0.070 \text{ in.}$$

Finally, looking at the schedule of commercially available tubing in Table 12-1 beside 7/8-in. *OD*, it is observed that the inside diameter and wall thickness required (0.738 *ID* and 0.070 *WT*) lie somewhere between two given values (0.731 *ID* and 0.072 *WT*) and (0.745 *ID* and 0.065 *WT*). Either would be considered appropriate.

Steel pipe with both tapered and straight threads is used extensively to fabricate hydraulic conductors because of its availability and mechanical strength. Major disadvantages associated with its use include bulkiness, weight, and the large number of fittings required. Sealing at joints is effected by tightening and requires sealing compound or tape if fittings and components are to be accurately positioned. Two tapered pipe threads are used extensively: the National or American Standard Taper Pipe Thread (*NPT*) and the Dryseal American National Standard Taper Pipe Thread (*NPTF*), the difference being that no contact occurs at the crown and root of the *NPT* thread. Both pipe threads are interchangeable, but proper assembly requires that male and female dryseal threads be mated for effective sealing. When straight pipe threads are used, a special pipe thread locknut and seal must be used to position and seal the fitting. Pipe is available in nominal sizes for hydraulic systems from 1/8 to 8 in. in four standard wall thicknesses. Figure 12-12 illustrates their relative sizes which are given in Table 12-1.

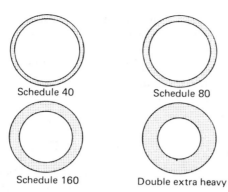

Schedule 40

Schedule 80

Schedule 160

Double extra heavy

Figure 12-12 Relative cross sections of hydraulic pipe.

Steel tubing is widely used for plumbing hydraulic systems because it is easily formed, has fewer fittings, locates a union connection conveniently at every coupling and presents a neat and orderly appearance. The most common tubing in use is dead soft cold drawn SAE 1010 steel with a strength equivalent to schedule 80 steel pipe. Tubing conforms to JIC standards with size increments increasing by 1/16th in *OD* from 1/8–3/8 in., 1/8 in. increments from 1/2–1 in., and 1/4 in. increments above 1 in. *OD*. Table 12-1 lists the basic dimensions for commercially available steel tubing.

Steel tubing is fabricated with bends made to standard radii. In small sizes, bends are made with a portable or bench-mounted hydraulic bending tool. Larger sizes require the use of factory preformed bends that are fabricated using fittings or annealed welds. The radius for a tube bend is a function of the outside diameter of the tube, with the minimum being 2½ to 3 times the *OD*. Table 12-2 lists standard bend radii for steel tubing. Number sizes correspond to the outside diameter in 16ths—that is, 4 equals 4/16 or 1/4-in. outside diameter.

TABLE 12-2 STANDARD TUBE BENDING RADII

Tube size no.	Tube outside diameter	Standard radius R
2	$\frac{1}{8}$	$\frac{3}{8}$
3	$\frac{3}{16}$	$\frac{7}{16}$
4	$\frac{1}{4}$	$\frac{9}{16}$
5	$\frac{5}{16}$	$\frac{11}{16}$
6	$\frac{3}{8}$	$\frac{15}{16}$
8	$\frac{1}{2}$	$1\frac{1}{2}$
10	$\frac{5}{8}$	2
12	$\frac{3}{4}$	$2\frac{1}{2}$
14	$\frac{7}{8}$	3
16	1	$3\frac{1}{2}$

Except for welded connections, all tube connectors seal at two points: at the component port and at the junction of the tube and the fitting. The seal between the component port and the fitting can be made by equipping the fitting with a tapered pipe thread, straight pipe thread and gasket-fitted lock nut, tapered metal-to-metal bottom seat or flanged port connections. A typical tapered pipe (NPTF) dry seal port is illustrated in Fig. 12-13.

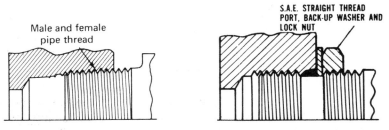

Figure 12-13 Tapered and straight thread port connections (*courtesy of Imperial Eastman Corporation*).

Refer to Fig. 12-14. Fittings with straight threads are assembled by turning the locknut, back-up washer, and O-rings as far back on the fitting as possible and lubricating the O-ring with petrolatum (Vaseline). The fitting is then run down into the port until the back-up washer just contacts the port boss. Then it is positioned by turning it out (up to 359°). The mechanical stability and seal are effected by holding the fitting with a wrench and tightening the lock nut to seat the back-up washer on the face of the port boss. The O-ring is squeezed into the recess to make a leak-free seal.

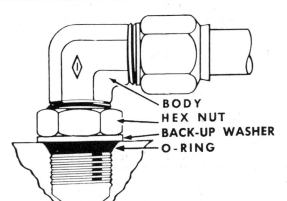

Figure 12-14 Straight thread fitting with O-ring assembly (*courtesy of Imperial Eastman Corporation*).

The seal at the junction of the fitting and the tube is made using welded joints, flare or flareless fittings. Welded connections attach the fitting firmly to the tubing. Assembly requires bolting the fitting to the fluid power component. Socket and butt weld fittings are applicable where excessive temperatures, pressures, or vibration would damage mechanical connections. Flare fittings are available with 45° and 37° angles. Figure 12-15 illustrates a correct 37° angle and

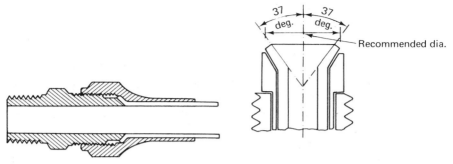

Recommended dia.

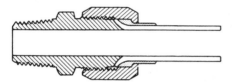

Flared 2 piece assembly is low cost and reusable. Tubing flare is compressed against a mating cone on the fitting by a female tubing nut. Cannot be used with extra heavy wall tubing.

Flared 3-piece assembly uses shorter nut than 2-piece assembly. Allows tube bends closer to the fitting and is less affected by vibration. Tubing nut tightens a sleeve against the flare. Seal is between the flare and the fitting cone. Fittings are reusable.

Figure 12-15 Correct flare angle and typical two- and three-piece flared fitting assemblies (*courtesy of Mobil Oil Corporation*).

typical two and three piece flared fitting assemblies. Flareless fittings seal by compressing a metal ferrule or seal against the tube wall. Both flare and flareless fitting connections are reusable after disassembly.

Where movement is necessary between stationary and movable components, flexible couplings are used. Swivel joints are available in a variety of configurations for use with steel tubing at pressures to 6,000 lbf/in.^{2} and temperatures to 450°F. Seals are selected to be compatible with the working environment, fluid, and mechanical stress to be encountered. Figure 12-16 illustrates a typical

Figure 12-16 Ball bearing swivel fitting (*courtesy of Chiksan Company*).

mechanical swivel fitting assembly. Notice the use of ball bearings in the swivel support.

Flexible hydraulic hose is available for pressures to 10,000 lbf/in². It has wide application where plumbing must connect components between which relative movement must occur. Portable power units that are moved in-plant and mobile applications are excellent examples requiring flexibility of movement between components. Flexible hoses are available with 1-6 plies of reinforcing wire for working pressures in excess of 10,000 lbf/in². Working pressures of 2000-3000 lbf/in.² usually employ 1 and 2 plies of wire around a center tube. Friction layers of elastomeric material are placed between plies as a bonding agent. Wire plies are spiral wound or braided. Figure 12-17 illustrates flexible hose with one to four plies. Of prime importance in the selection of wire flexible hose is compatibility between the hose elastomeric material and system fluid. The appropriateness of a flexible hose material should be compared with manufacturer's data before installation and use since softening or disintegration could result in system contamination and component damage. Common materials used as the center tube include synthetic rubber and Buna-N to be compatible with most available commercial fluids.

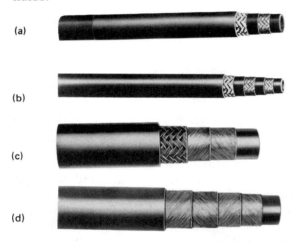

(a)

(b)

(c)

(d)

Figure 12-17 Flexible hose using (a) one-wire brade, (b) two-wire braid, (c) two-spiral wire, (d) four-light spiral wire (*courtesy of Imperial Eastman Corporation*).

Thermoplastic flexible hose reinforced with polyester braid is currently available for hydraulic applications to 3,000 lbf/in.² and above. Seamless heat stabilized thermoplastic hose is commonly used as the center tube. Thermoplastic hose provides excellent mechanical and dialectric properties and is particularly applicable around power line boom equipment because it is a nonconductor. Figure 12-18 illustrates a typical reinforced thermoplastic hose.

Figure 12-18 Thermoplastic hose reinforced with polyester (*courtesy of Imperial Eastman Corporation*).

Flexible hydraulic hose is available in factory-made units, or may be purchased separately with fittings and assembled in the field. The three most important factors requiring attention are 1) hose selection and assembly, 2) engineering minimum bend radii and flexing, and proper installation of the hose with adequate supports to assure trouble free operation, and 3) maximum service life. Hose selection should be made from an assessment of pressure and flow requirements and an examination of manufacturers' product information to assure system and service compatibility. Instructions and a list of required special tools are a part of this literature. Engineering the position of the hose to assure minimum bends takes into account the size of the hose, the number of reinforcing plies, and the amount of flexing required between stationary and moving components. Manufacturers' data also should be consulted for this information. Proper hose installation results from a combination of good workmanship, experience, and common sense, taking into account the mechanical properties of the hose. Figure 12-19

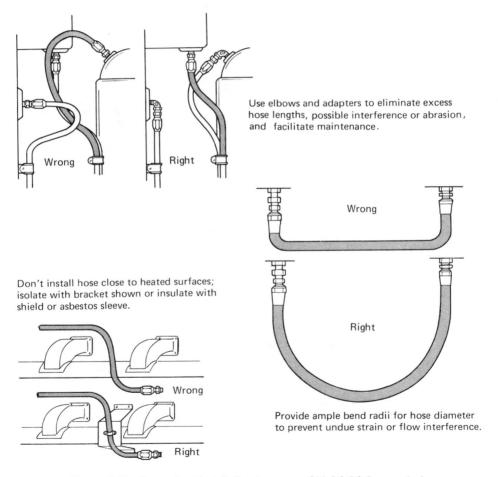

Use elbows and adapters to eliminate excess hose lengths, possible interference or abrasion, and facilitate maintenance.

Don't install hose close to heated surfaces; isolate with bracket shown or insulate with shield or asbestos sleeve.

Provide ample bend radii for hose diameter to prevent undue strain or flow interference.

Figure 12-19 Proper host installation (*courtesy of Mobil Oil Corporation*).

Right

Don't install hoase in twisted position. It weakens the hose, sometimes as much as 90% for a 7″ twist on a large hose, and pressure surges tend to loosen fitting connections.

Wrong

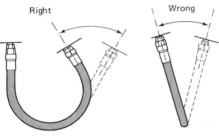

Right   Wrong

When flexing or movement is part of operating characteristics, install hose and fitting so that movement takes place in same plane and minimum bending radii are not exceeded.

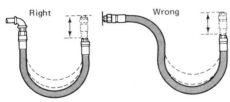

Right   Wrong

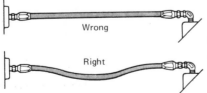

Wrong

Right

Always install hose of correct length. Short hose length will twist on installation. Further pressure can change hose length as much as + 2% to − 6%. Provide slack to compensate for pressure changes.

Figure 12-19 (*continued*)

illustrates several right and wrong installations with notations describing common errors found in the field. Twisting a hose during installation, for example, can reduce its life more than 75%.

Quick-disconnect hose couplings are used where frequent connections are made and broken between components such as inplant portable power units and mobile agricultural equipment. Hose couplings are available in straight-through, one-way shut off, and two-way shut off designs and incorporate rings, balls, and pins in the locking mechanism. Straight-through couplings offer a minimum restriction to the flow of fluid but do not prevent fluid loss from the system when the coupling is disconnected. One-way shut off quick disconnects locate the shut off at the fluid source connection and leave the actuator component unchecked. Leakage is usually not excessive in short runs, but system contamination caused by the entry of dirt in the open end of the fitting is a problem, particularly with mobile equipment sitting at the work site. Two-way shut off quick disconnects stop the flow of fluid at the junction of the coupling when it is disconnected, preventing fluid loss and the entry of dirt and air in the system. Selecting a quick disconnect requires an examination of the volume flow rate and maximum restriction that can be tolerated. While low friction two-way quick disconnects are receiving widespread use, loss at the coupling both with suction and pressure line applications is substantial and must be considered. Oversizing reduces fitting loss, but increases costs associated with the larger flexible hose required in addition to the oversize coupling.

12-6 INSTRUMENTATION

Five parameters are commonly monitored by the fluid power machine operator and maintenance personnel: system pressure, temperature, flow rate, speed, and fluid level.

System pressure is measured at the pump inlet with vacuum gauges to monitor inlet strainers, filters, and line obstructions between the reservoir and the pump. At the pump outlet and other places in the system, pressure is monitored using pressure gauges to check components condition and the pressure value of the load resistance. While a variety of gauges are available to serve these functions, the bourdon pressure gauge shown in Fig. 12-20 is the most common found in use in all pressure ranges because of its simplicity, cost, and robust construction. Gauges are often filled with a dampening fluid such as glycerine or light gauge oil to smooth pulsations that might cause damage to the sensitive gear mechanism. Gauge snubbers, which are adjustable flow valves placed at the gauge connection, are also used for this purpose. Gauge isolating, consisting of placing diaphragms or other similar barriers between the bourdon gauge and system fluid, are used when it is desirable to separate the gauge from the corrosive effects of system fluid. Vacuum gauges are usually scaled in in. or mm of mercury (Hg) for use in intake lines, although scales in lbf/in.^{2} and dual scales are also used extensively for personnel convenience. Conversion from in. of Hg to absolute

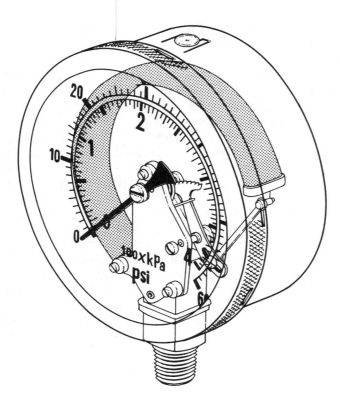

Figure 12-20 Bourden tube pressure gauge.

pressure (lbf/in.^{2} absolute) may be made using Table 12-3. Other variations of the bourdon pressure gauge include compound gauges that measure both pressure and vacuum, dual gauges that use two movements, pressure connections and dual hands to monitor two pressure ranges, and gauges with maximum reading hands to hold system peak pressure readings for later observation and recording.

Fluid temperature is usually measured at the reservoir to monitor heat being generated and stored by the system. Fluid temperature must be kept within acceptable limits, usually between 120°F and 140°F, in order to retain its viscosity and other properties necessary for satisfactory operation for any extended time interval.

Fluid temperature is often measured with dial mounted gauges driven by temperature sensitive fluids that are metal encased and immersed in the hydraulic fluid. A typical direct mounted combination temperature fluid level gauge is illustrated in Fig. 12-21. Extension of the gauge from the reservoir to a remote control panel frequently makes use of fluid filled tubes to sense and transmit the temperature to the gauge, or of thermocouples immersed in the fluid connected by wires to the gauge at the panel.

Fluid flow is measured to determine the volume flow rate in the system from the pump or through an actuator. Determining slippage of components and volumetric efficiency are common purposes for their use. Fluid flow devices can

TABLE 12-3 CONVERSION FROM ABSOLUTE PRESSURE IN lbf/in.² abs TO INCHES OF Hg (*1 lbf/in.² abs = 2.0353741 in. Hg*)

Absolute pressure (in. of Hg)	Absolute gauge pressure (lbs/in.²)
1.0	0.49
2.49	1.0
4.07	2.0
6.11	3.0
8.14	4.0
10.16	5.0
12.21	6.0
14.25	7.0
16.28	8.0
18.31	9.0
20.35	10.0
22.39	11.0
24.42	12.0
26.46	13.0
28.49	14.0
29.92	14.7

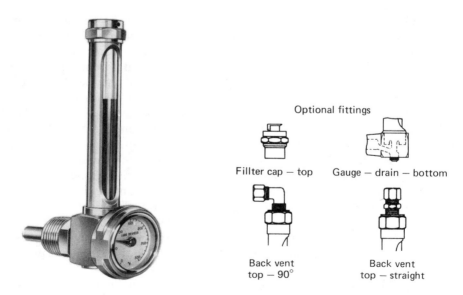

Optional fittings

Fillter cap — top   Gauge — drain — bottom

Back vent top — 90°   Back vent top — straight

Figure 12-21 Combination temperature fluid level gauge (*courtesy of Lube Devices, Inc.*).

measure flow by a simple flow sight indicator such as that in Fig. 12-22, as a percentage of the flow capacity, or they can measure and read out the flow on a gal/min or other convenient scale.

Figure 12-22   Flow sights (*courtesy of W.E. Anderson, Inc.*).

Flow sites are used as a quick indicator that fluid is moving through the system and, with an impeller, the direction of flow can be determined as well. In-line flow meters such as that shown in Fig. 12-23 display the flow rate in gal/min or liters/min at a given temperature. They can be inserted in the line for test purposes, or dedicated to the system. The flow indicator consists of a sharp edge orifice and tapered metering piston which moves in proportion to changes in flow rate. As the flow increases, the pressure difference across the orifice formed by the metering piston and fixed orifice moves the piston against the calibrated spring. Piston movement is directly proportional to flow rate with the sharp edge orifice minimizing the effects of viscosity. The pointer which reads the flow rate on a rotary calibrated scale is coupled to the piston magnetically.

Figure 12-24 illustrates a portable combination–test apparatus that measures pressure, temperature, and flow rate. It has extensive application for in-plant and mobile hydraulic testing applications by maintenance personnel for determining causes of failure and monitoring system work cycles.

System speed indicators are used to determine the operating rpm of pumps and motors. They can be of the direct counter type and measure the number of turns (n) which then must be divided by the time to determine rpm, or they may measure speed (N) directly in rpm.

Fluid level is monitored to indicate the visual condition and level of the fluid in the reservoir. Figure 12-25 illustrates two typical fluid level gauges. Fluid level gauges are frequently scaled or strategically mounted in the reservoir to indicate the level and amount of make-up fluid necessary to fill the system.

Operation

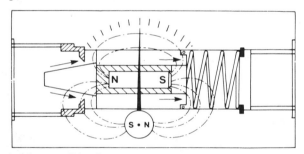

Typical Pressure Drop Curves
Oil Viscosity 25 Centistrokes   100 SSU

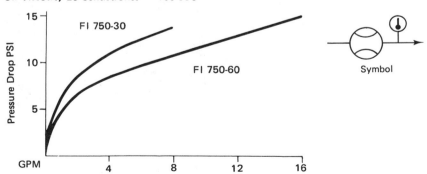

Figure 12-23 Calibrated flowmeter (*courtesy of Webster Instruments, a division of Webster Products*).

Figure 12-24 Portable hydraulic tester (*courtesy of Schroeder Brothers Corporation*).

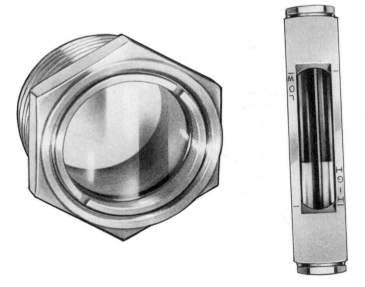

Figure 12-25 Fluid level sight gauges (*courtesy of Lube Devices, Inc.*).

12-7 HYDRAULIC POWER UNITS

Power packages consist of a pump and other components assembled into one unit to supply pressurized fluid. They have been developed from extensive experience by manufacturers of fluid power components to supply a need and market for packaged units conforming to JIC and other fluid power standards, and result in substantial cost savings to the consumer not available with in-plant systems constructed from off-the-shelf components.

Power packages are available as stock units or can be assembled to meet customer specifications, incorporating features peculiar to a particular application. Figure 12-26 illustrates a power package equipped with a pressure gauge, monitoring system, pressure relief, heat exchanger, and sight level gauge. The reservoir conforms to JIC specifications. Units of this type are available in 2-10 hp sizes with reservoirs to 40 gal. Single or double pumps of the gear and vane type are most common, mounted directly to the motor through a flexible coupling.

Figure 12-26 Power unit (*courtesy of Delta Power*).

Flow dividers receive fluid from a single source and accurately meter it in equal amounts of two, three, or more circuits (Fig. 12-27). They provide a simple means to synchronize cylinders and other actuators that receive fluid from a single source. Fluid flow enters the divider and is piped to each of the banked gear units that function much the same as gear motors with the exception that little power is consumed. Flow is divided by the connected gear units that pass the same amount of fluid to each outlet port. If unequal flow division is desired, gear units of different displacements are selected to match the required outlet flow ratios.

Flow dividers can also be used for pressure amplification using appropriate valving to direct one of the divided flows to power the actuator and return the remaining flow to the tank. A part of the flow divider thus becomes a motor,

Figure 12-27 Flow divider (*courtesy of Delta Power*).

driving the flow portion of the flow divider as a pump, increasing the pressure potential in that leg of the circuit.

12-8 HYDROSTATIC TRANSMISSIONS

Hydrostatic transmissions convert a unidirectional variable speed shift input to a bidirectional variable speed output. They are used for a variety of production and mobile applications and provide advantages in power transmission and control not available by other means. These include remote transmission to difficult access areas and constant torque, speed, or brake horsepower output regardless of variation in output load. Precise matching between available power and workload, infinite selection of drive speed from forward to reverse, and dynamic braking all can be incorporated in one unit.

Figure 12-28 illustrates a hydrostatic transmission consisting of a variable displacement pump capable of over-center operation supplying oil to a fixed dis-

Figure 12-28 Two-unit hydrostatic transmission (*courtesy of Abex Corporation, Denison Division*).

placement motor connected in a closed circuit. A typical schematic circuit is shown in Fig. 12-29. The charge package forming part of the variable pump includes the necessary relief, shuttle, and check valves as well as a micrometer-range filter and gear pump for circuit supercharging. Manually operated servo controls are available to match the transmission to various drive requirements. The addition of a heat exchanger and reservoir completes the circuit. In operation, the low pressure precharge pump circuit returns about 65% of its flow capacity to the reservoir via the heat exchanger that cools the closed loop in addition to supplying make-up oil.

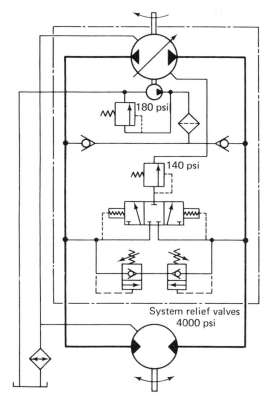

Figure 12-29 Typical circuit for hydrostatic transmission (*courtesy of Webster Electric Company, Inc.*).

Figure 12-30 illustrates an in-line hydrostatic transmission consisting of an over-center lever controlled variable volume axial piston pump that drives a fixed displacement dual rotation axial piston motor. The circuit shown in Fig. 12-31 incorporates either an integral or separately mounted precharge pump to keep the aluminum housing flooded. An oil cooler and filter system are also recommended to keep operating temperatures at less than 150°F and the oil free from contaminants.

Hydrostatic transmissions are also available as complete power packages for industrial production. The packaged unit which is shown in cutaway with its

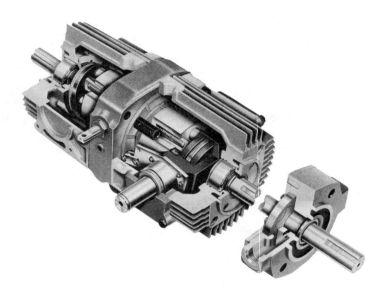

Figure 12-30 In-line single-unit hydrostatic transmission (*courtesy of Webster Electric Company, Inc.*).

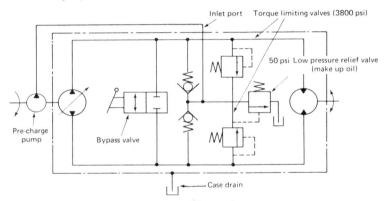

Figure 12-31 Circuit for Fig. 12-30—hydrostatic transmission (*courtesy of Webster Electric Company, Inc.*).

circuit (Fig. 12-32) contains the electric drive motor, variable delivery reversing axial piston pump, charge pump, fixed displacement axial piston motor, and all valving necessary in one complete housing which also serves as the oil reservoir. The fan shown on the pump through-shaft is driven by the electric motor and cools the finned housing and oil.

12-9 ELECTRICAL CONTROLS

Electrical controls are used with fluid power components to designate the operation cycle of the system, usually by switching the power input to the prime mover or to pressure, volume, and flow control valves. Safety devices are incorporated

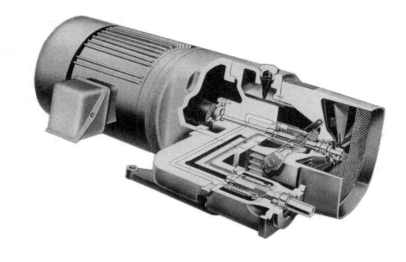

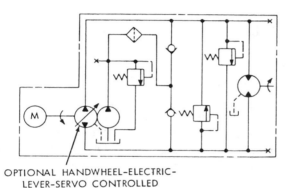

OPTIONAL HANDWHEEL-ELECTRIC-
LEVER-SERVO CONTROLLED

Figure 12-32 Hydrostatic transmission power package (*courtesy of Sperry-Vickers*).

to protect operating personnel and the system from damage. These are signaled by pressure, temperature, current, load, and position sensitive devices. Fundamental electrical devices to close the circuit to the electric motor driving the pump are the switch and the relay. Valves are usually operated by electric solenoids in response to closing of contact relays.

Standard electrical symbols are combined with ANSI fluid power symbols to indicate the location and operation of the electrical circuit. Fluid power system operated circuits such as pressure switches and limit switches are shown on the fluid power diagram to correspond to symbols on the electrical diagram.

Six basic electrical devices are used in the control of fluid power systems: manually operated switches, limit switches, pressure switches, relays, timers, and solenoids. Other devices, such as temperature switches and float switches are used to a lesser extent in control, but perform a valuable function in safeguarding machinery. Electrical diagrams use a ladder format with the power connected to the left side and the ground connected to the right. Components having more than one position are shown in the unactuated position.

Manually operated switches open and close circuits to initiate or stop operation of the system cycle. The symbol for simple switches in the normally open and normally closed positions is seen in Fig. 12-33. Limit switches open and close

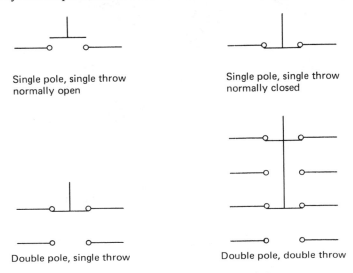

Single pole, single throw
normally open

Single pole, single throw
normally closed

Double pole, single throw

Double pole, double throw

Figure 12-33 Manually operated switches.

circuits in response to the position of linear actuators or the workpiece. The symbol for the limit switch in the open and closed positions is seen in Fig. 12-34. Pressure switches respond to preset limits determined by a spring or other setting in the pressure switch device to open or close an electrical circuit. They are commonly used in sequencing operations and as an overload protection for the system. Symbols for pressure switches in the normally open and normally closed positions are shown in Fig. 12-35. Relays are switches that are actuated electri-

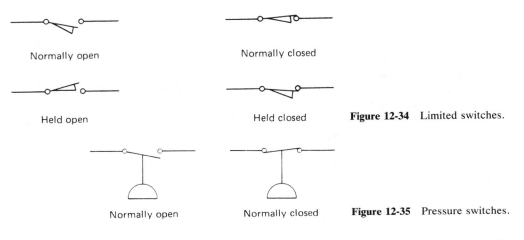

Normally open

Normally closed

Held open

Held closed

Figure 12-34 Limited switches.

Normally open

Normally closed

Figure 12-35 Pressure switches.

cally using an electromagnetic coil. Figure 12-36 illustrates the symbols used to designate the open and closed contacts of the relay switch and the relay coil. Timers are used to delay a sequence either during the work or slack portions of the cycle, for example, at the end of a drilling operation to clear the hole of chips. The symbols for energized and de-energized timers are seen in Fig. 12-37. Solenoids are used to operate directional control and check valves in the system. The symbol for a solenoid is seen in Fig. 12-38.

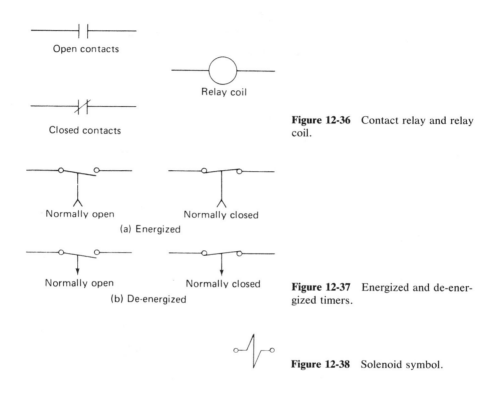

Open contacts

Relay coil

Closed contacts

Figure 12-36   Contact relay and relay coil.

Normally open   Normally closed

(a) Energized

Normally open   Normally closed

(b) De-energized

Figure 12-37   Energized and de-energized timers.

Figure 12-38   Solenoid symbol.

A typical circuit used to extend and then retract a cylinder illustrates the use of electrical controls and symbols in designing fluid power systems. The circuit in Fig. 12-39 uses four manual switches and two limit switches to control the actions of the double-acting hydraulic cylinder. Forward and reverse switches control the direction of the cylinder. Limit switches are used to stop the cylinder rod at both extremes of its travel. Three lights indicate the machine is on and the direction of travel.

When the ''ON'' switch is pushed, relay coil (CR_1) latches the circuit from disconnecting when the switch is released after momentary contact and energizes the hydraulic motor starter transformer. This circuit is fused. Fluid passes freely through the open center spring-returned valve from the pump to the reservoir. Pressing the *FWD* switch energizes relay coils CR_2 and CR_5 unlatching control relay CR_5 that breaks the reverse circuit containing (Sol_2). At the end of the

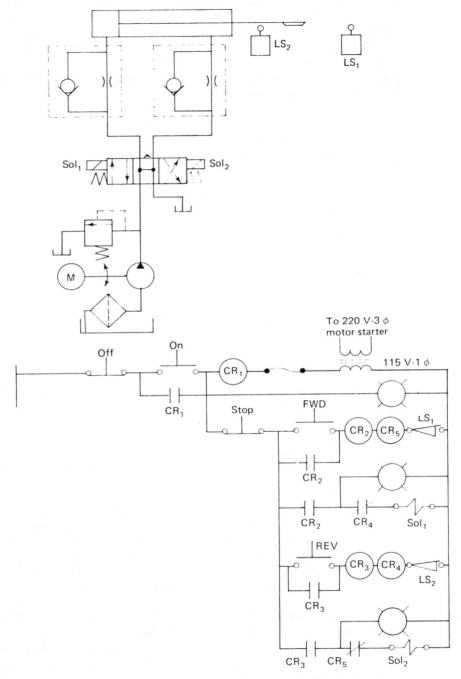

Figure 12-39 Electrically controlled hydraulic circuit.

forward travel, the cylinder rod cams limit switch LS_1 breaking the forward circuit and piston rod travel stops. Depressing the reverse circuit switch REV energizes relay coils CR_3 and CR_4 latching control relay CR_3 which energizes Sol_2 and unlatches control relay CR_4 which breaks the forward circuit containing Sol_1, thereby causing the cylinder rod to be retracted. At the end of the retraction stroke, the cylinder rod cams limit switch LS_2 which opens the reverse circuit and the spring-centered control valve stops cylinder travel.

The cylinder rod can be stopped at any position in its travel by pushing the STOP switch which disconnects both the forward and reverse circuits and allows the spring-centered control valve to assume a center position. Depressing the REV switch when the piston rod is traveling forward will cause the forward solenoid circuit to unlatch and the reverse circuit solenoid (Sol_2) to be energized. Depressing the FWD switch when the piston rod is traveling in the reverse direction will have the opposite effect. Depressing both the FWD and REV switches at the same time will have no effect until one is released, at which time the latter will control the circuit.

The circuit can easily be rearranged to cause the cylinder rod to return automatically when either or both limit switches are cammed. Delay timers and other electrical devices can be added for operator convenience and machine versatility.

12-10 SUMMARY

Hydraulic components represent a wide variety of devices from which fluid power systems may be constructed. Some of these devices include fluid conditioners, accumulators, intensifiers, conductors, instrumentation, hydraulic power units, and electrical controls. Selecting appropriate components must be systematic to assure within-system compatibility, safety to personnel, and minimum initial and maintenance cost. The literature from manufacturers is a major source for determining the availability of appropriate components and their capability. System design must consider purpose, duty, cycle rates, and maintenance in the selection of components, as well as workmanship required to fabricate quality systems requiring minimum maintenance and repair down-time. Workmanship and fabrication skills result from training, common sense, and experience. Pride in workmanship by personnel seems to be a critical element.

STUDY QUESTIONS AND PROBLEMS

1. List the important features to be considered when selecting a reservoir.
2. Show the conversion of Btu's to J/s and J/s to Btu's.
3. A hydrostatic transmission driven by a 10-hp motor delivers 6 hp at the output shaft. Assuming that 80% of the loss can be attributed to heat loss, compute the total Btu heat loss over a 3-hr period.

4. A partially open pressure relief valve discharges 160°F system oil with a Sg of 0.91 at 5 gal/min into the reservoir with a pressure drop of 800 lbf/in². What will be the temperature of the oil as it flows into the reservoir? The specific heat of the oil is 0.42 Btu/lbm/°F.

5. A high-pressure press with a 10 gal/min pump cycles every 8 min. During the 2-min high-pressure portion of the cycle, oil is dumped over a high-pressure relief valve at 2,500 lbf/in². During the 6-min low-pressure portion of the cycle, oil is dumped over a low-pressure relief valve at 500 lbf/in². If the time during which the cylinder is traveling is considered negligible, compute the total Btu heat loss per hr generated by dumping fluid over both relief valves.

6. A sticking control valve spool causes a pressure drop of 700 lbf/in². If the fluid is being pumped across the valve at 5 gal/min and has a specific heat of 0.42 Btu/lbm/°F and a Sg of 0.91, estimate the temperature rise in the fluid.

7. At full capacity, a pump delivers fluid to a wheel motor at 10 liters/minute at a pressure of 10 MPa. If the motor delivers 1.5 hp at the ground and 90% of the loss can be accounted for by internal leakage which generates heat in the fluid, compute the heat generation rate in kJ/min.

8. If the fluid in problem 7 has a Sg of 0.87 and no heat is lost to the surroundings, what temperature rise might be expected in the fluid? The specific heat of the oil is 0.42 Btu/lbm/°F.

9. List the important factors to be considered when selecting a heat exchanger.

10. A 400 in.³ gas-charged accumulator is precharged to 1,500 lbf/in². Fluid is then pumped into the accumulator until a relief valve setting of 3,000 lbf/in². is reached. If a system operating pressure of 2,000 lbf/in². is assumed, how much fluid will be available from the accumulator for useful purposes? (Clue: draw the figure for the problem.)

11. An accumulator is to be sized to supplement pump delivery during peak demand with 10 liters of fluid between the pressures of 14 MPa and 10 MPa. To be sure that fluid is available from the accumulator at 10 MPa, it is precharged to 8 MPa. What size should the accumulator be if atmospheric pressure equals 101 kPa and the process is considered to be isothermal?

12. If the process in Problem 11 is considered to be adiabatic, what size accumulator would be required?

13. Construct a simple hydraulic circuit and indicate appropriate locations for accumulators given the following conditions:
    a) Supplement pump delivery
    b) Independent power supply
    c) Separate two dissimilar fluids
    d) Load shock absorber

14. Use Barlow's formula to compute the inside diameter and wall thickness of a conductor which must carry 10 gpm at 3,000 lbf/in². Use 12 ft/sec as the maximum velocity for the fluid, a safety factor of 8, and conductor tubing which has a tensile strength of 75,000 lbf/in².

15. What minimum commercial size tubing with a wall thickness of 0.95 in. would be required at the inlet and outlet of a 40 gal/min pump if the inlet and outlet velocities are limited to 5 ft/sec and 20 ft/sec, respectively?

16. Assuming a material with a tensile strength of 60,000 lbf/in.², and a SF = 8, select a commercial steel tubing size and weight to deliver 10 gal/min at 3400 lbf/in² without exceeding a velocity of 20 ft/sec.

17. What would be the minimum center-to-center dimension between the ears of a hold down bracket on a No. 8 tube making a return radius bend?

18. List how the five system parameters are commonly monitored by hydraulic machine operators and maintenance personnel.

19. List the advantages commonly associated with hydrostatic transmissions.

20. Define a common function for each of the following electrical devices used in controlling fluid power systems:
    a) Manually operated switch
    b) Limit switch
    c) Pressure switch
    d) Relays
    e) Timers
    f) Solenoids

21. What advantages does a straight port connection have over a tapered thread port connection?

CHAPTER 13

BASIC CIRCUITS AND SIZING HYDRAULICS COMPONENTS

13-1 INTRODUCTION

Systems are composed of circuits. A system is capable of completing one or several operations that constitute a work cycle, and includes the pump drive, pump, reservoir, valves, cylinders, motors, and suitable plumbing to transfer fluid at high pressure. A circuit, on the other hand, is capable of performing one or more specific tasks, but not a complete work cycle. A circuit functions as part of a system. Actuator, control, pump, and other branch circuits are common examples. Circuits are not to be confused with circuit diagrams constructed from standard graphic symbols representing components. Circuit diagrams are drawn of both circuits as well as complete hydraulic systems.

Both systems and circuits are designed to accomplish output objectives. The design phase includes sizing the components and plumbing to meet the output requirements, as well as establishing work cycles using time, flow, pressure, and horsepower calculations. The performance of the machine is then tested against the output requirements to assure that it conforms to the design specifications and that it meets the safety standards established by the industry and OSHA.

The difference between design and analysis can be understood by referring to the two by two table in Fig. 13-1. Circuit as well as system design are functions requiring planning, layout, sometimes modeling, and writing specifications. They are usually performed at a desk-drafting work station. Circuit and system analyses are functions requiring assembly, installation, diagnosis and repair. Typically, they are performed in the shop, laboratory, and at test sites and benches using specialized equipment, tools, and instrumentation. Analysis requires hands-on

	Design	Analysis
Circuits	Planning	Assembly
and	Layout	Installation
Systems	Modeling	Diagnosis
	Writing specs.	Repair

Figure 13-1 Circuit system design and analysis functions.

work. Common to both are basic knowledge of fluid power principles, formulas, and calculations about performance.

Hydraulic systems may be classified as open loop, closed loop, and servo systems. Open-loop systems operate without feedback from the output, except for the operator. Performance is determined by the operational characteristics of individual components. Most industrial circuits are of this type. Closed-loop circuits sample the output and generate a proportional control signal that is used to correct the input command signal. It is worth noting that until recently closed-loop hydrostatic transmissions have not been closed-loop hydraulic systems and the term in this usage described the closed fluid flow path between the pump and motor. Servo systems, on the other hand, feed back control signals to the input command as a result of a change in the mechanical position of the output.[1,2]

The design and analysis of hydraulic systems and circuits is systematic in that it considers several activities in sequence to develop and prove the operation of an objective oriented machine. Typically the following steps are followed:

1. Size actuators from output objectives.
2. Establish work cycles using time, flow, pressure, and horsepower plots.
3. Design the circuit.
4. Size and select components.
5. Assemble the circuit or system.
6. Monitor performance of the machine.
7. Check the machine for safe operation and compliance with Occupational Safety and Health Administration standards.

A preliminary step in designing circuits and systems is to recognize the fluid power symbols which are given in Appendix D. These conform to the standards accepted by the ANSI and NFPA. At this point it would be beneficial to review the symbols for each component.

[1] A lengthy discussion of servo-control systems can be found in James E. Johnson, "Electro-Hydraulic Servo Systems," *Hydraulics and Pneumatics Magazine,* P.O. Box 6197-U/Cleveland, Ohio 44101.

[2] W. J. Thayer, "Remote Control of Hydraulic Equipment," Technical Bulletin 124, Moog Inc., Controls Division. East Aurora, N.Y. 14052, 1975.

13-2 SIZING ACTUATORS FROM OUTPUT OBJECTIVES

The output objective of the hydraulic circuit or system may be described by performance of a task against a load resistance that may be positive, negative, or inertial as with shock loads. The load resistance typically is not constant. That is, the force required from the actuator must vary. Within the timespan of the work cycle, the load resistance is usually started against breakaway friction, accelerated to some level, maintained at some velocity, decelerated, and finally stopped. Overrunning loads often cause the system to act as a brake. Of these, starting the load resistance against breakaway friction requires the greatest effort at the system input, and this is the primary capability against which the system must be matched. Overrunning loads can be limited using valves restricting the flow of fluid rather than applying a resistive force by the prime mover, although fluid power components must be oversized to withstand breaking loads of a greater magnitude than that capable from the system input power source.

Accomplishing the output objective of moving a load resistance through some predictable and consistent work cycle is a two-part problem. While it is given that fluid power will supply the motive force, it cannot be assumed that movement of the load resistance is both supported and powered by the hydraulic drive. This is usually not the case. Car lifts, for example, do support and guide the load resistance through its travel as well as provide the motive force necessary for movement, but most applications do not. Dump body cylinders and fork lift rams, for example, elevate the load resistance, but the mechanical support and guidance are controlled independently. Hydraulic drill presses, turret lathes, and most production hydraulic powered machinery operate similarly. Hydraulics provides the motive force for a machine with built-in pivots, guides, and stops that direct or limit movement patterns. With applications of this type, machine movements and mechanical constraints must be established before placement of hydraulic cylinders and components can be determined.

Recent developments in hydrostatic transmissions and self-supporting wheel motors for such mobile applications as tractors, railroad cars and other transport equipment may start a trend to build supporting machinery around hydraulic capabilities rather than the opposite, which has been the traditional method. Using off-the-shelf self-supported components of this type has many inherent advantages. With wheel motors, for example, the load resistance can be translated directly into pressure levels, flow rates, and ground speeds, and matched directly with available wheel motor actuators and compatible pump packages. Input horsepower, torque, and gear reduction requirements at the prime mover are then determined by maximum output horsepower and the speed range of the pump.

In Fig. 13-2, a load resistance is hoisted some distance, traversed from (a) to (b) and then lowered. In this example, breakaway friction is low because surface friction is minimal. Assuming rolling friction at the overhead trolley is also low, the major force exerted by the system is in hoisting the load resistance vertically. If one pump supplies both vertical and horizontal motors, the load versus time plot would look like that in Fig. 13-3. The open center valve circuit in Fig. 13-4 is

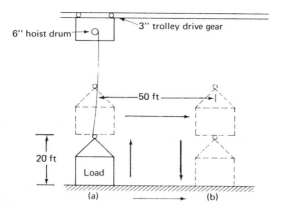

6″ hoist drum

3″ trolley drive gear

—50 ft—

20 ft

Load

(a)

(b)

Figure 13-2 Overhead hoist and trolley.

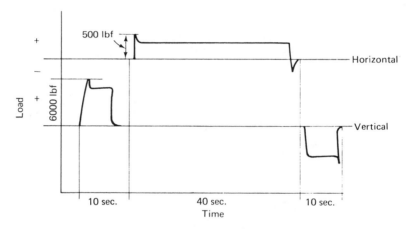

Load

500 lbf

6000 lbf

Horizontal

Vertical

10 sec.

40 sec.

10 sec.

Time

Figure 13-3 Load vs. time plot for Fig. 13-2.

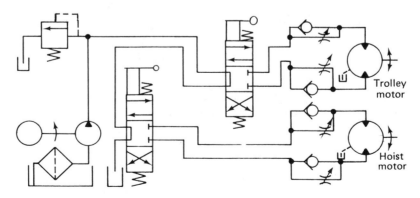

Trolley motor

Hoist motor

Figure 13-4 Hoist and trolley circuit (*exclusive* tandem center).

Sec. 13-2   Sizing Actuators from Output Objectives

used to illustrate this example. The vertical motor first applies sufficient force to break the load resistance away from its position at A causing the first load peak. As the vertical motor continues to apply force, the plot flattens out at the top. When the load resistance reaches the designated vertical position, the vertical motor stops, the load resistance coasts to a stop, and the traverse actuator transports the load resistance from A to B. Breakaway friction causes the force to peak and then to remain flat as the constant speed traverse motor displaces the load from A to B. At the end of the traverse travel, the load overruns slightly causing a small peak. Lowering the load presents the system with a negative rather than positive load resistance requiring braking action. The input from the prime mover is negligible, but hydraulic components must still withstand positive pressures on the actuator side of the control valve when flow is restricted and the hydraulic motors must be provided with an external drain to prevent blowing shaft seals. A peak is experienced at the end, since the load is decelerated before it contacts the ground support at B.

The load versus time plot in Fig. 13-5 can be used to describe several important parameters relating to the output objectives of the system and selection of actuators. These are force, distance, time speed, and system pressure. From these parameters can be computed the output horsepower, actuator torque, or output torque if a reduction unit is used, fluid horsepower, flow rate to the actuator, and, finally, actuator displacement.

First, focusing attention on the hoist drum in Fig. 13-2, observe from the load vs. time plot (Fig. 13-3) that a peak load of 6,000 lbf is encountered by the hoist drum when the load is lifted initially. Using this maximum value to size the hoisting circuit, the output horsepower at the hoist during the 10 sec it operates would be computed from

$$\text{Hoist } Hp = \frac{F \times L}{t \times 33,000} \tag{13-1}$$

$$\text{Hoist } Hp = \frac{6000 \times 20 \times 60}{10 \times 33,000} = 21.82 \text{ hp}$$

If the hoist drum is given a mean diameter of 6 in, the rotational speed can be computed by dividing the number of turns (n) by the time (t) in minutes. That is,

$$\text{Hoist Speed } (N_h) = \frac{n}{t} \tag{13-2}$$

where (n) is computed by dividing the hoisting distance (L) by the mean circumference of the hoisting drum (d_d). That is,

$$n = \frac{L}{\pi \times d_d} \tag{13-3}$$

$$n = \frac{20}{3.14 \times 0.5} = 12.74 \text{ turns}$$

and

$$N_h = \frac{12.74 \times 60}{10} = 76.44 \text{ rpm}$$

The output torque at the hoist drum (T_h), whether the drive is taken directly from the fluid horsepower motor output shaft or from the output shaft of a reduction unit, can be computed from

$$\frac{T_h \times N_h}{5252} = \text{Hoist } H_p \tag{13-4}$$

and

$$T_h = \frac{5252 \times 21.82}{76.44} = 1499 \text{ lbf-ft}$$

If a hoist drive motor speed (N_{hm}) of 1,500 rpm is specified, the reduction between the motor and the hoisting drum is computed from

$$\text{Reduction} = \frac{N_{hm}}{N_h} \tag{13-5}$$

and

$$\text{Reduction} = \frac{1,500}{76.44} = 19.62:1$$

at a drum motor torque reduction equal to (1499/19.62:1) 76.38 lbf-ft (916.82 lbf/in.).

If the operating pressure for the system is specified, for example at 2,000 lbf/in.^{2}, the flow rate to the hoist motor (Q_{hm}) can be computed by equating the system fluid horsepower to the hoist hp. That is,

$$\text{Fhp} = \text{Hoist Hp} \tag{13-6}$$

$$\frac{p \times Q_{hm}}{1,714} = 21.82$$

and

$$Q_{hm} = \frac{1,714 \times 21.82}{2,000} = 18.70 \text{ gal/min}$$

Finally, the displacement of the hoisting motor (V_{hm}) equals the flow rate (Q_{hm}) divided by the motor speed (N_{hm}) expressed in in.^{3} That is

$$V_{hm} = \frac{Q_{hm}}{N_{hm}} \times 231 \tag{13-7}$$

and

$$V_{hm} = \frac{18.70}{1500} \times 231 = 2.88 \text{ in.}^3/\text{rev displacement}$$

Summarizing the specifications of the hoisting motor,

Hoisting motor	
Horsepower	21.82
Speed reduction at 1500 rpm	19.62 : 1
Output torque (before reduction)	76.40 lbf-ft (916.56 lbf-in.)
Flow rate at 1500 rpm	18.70 gal/min
Displacement	2.88 in.^{3}/rev

Computing the same measures for the traverse trolley motor from Eq. 13-1

$$\text{Traverse Hp} = \frac{F \times L}{t \times 33{,}000}$$

and

$$\text{Traverse Hp} = \frac{500 \times 50 \times 60}{40 \times 33{,}000} = 1.14 \text{ hp}$$

With a 3-in. trolley drive gear and 50-ft traverse travel in 40 sec, Eq. 13-3 gives

$$n = \frac{50}{3.14 \times .25} = 63.69 \text{ turns}$$

and Eq. 13-2 gives

$$N_t = \frac{63.69 \times 60}{40} = 95.54 \text{ rpm}$$

Output torque at the trolley gear (T_t) is computed from Eq. 13-4

$$\frac{T_t \times N_t}{5252} = \text{Trolley Hp}$$

and

$$T_t = \frac{5252 \times 1.14}{95.54} = 62.67 \text{ lbf-ft (752 lbf-in.)}$$

If a trolley drive motor speed (N_{tm}) of 1,500 rpm is selected, Eq. 13-5 gives a reduction of

$$\text{Reduction} = \frac{N_{tm}}{N_t}$$

$$\text{Reduction} = \frac{1{,}500}{95.54} = 15.7 : 1$$

will be required with a reduction in torque to (62.67/15.7) 3.99 lbf-ft (47.88 lbf-in.).

If the pressure is also specified at 2,000 lbf/in.^{2}, the flow rate to the motor (Q_m) will be computed from Eq. 13-6

$$\frac{p \times Q}{1,714} = 1.14$$

and

$$Q_{tm} = \frac{1,714 \times 1.14}{2,000} = 0.977 \text{ gal/min}$$

And the displacement of the trolley motor from Eq. 13-7 (V_{tm}) equals

$$V_{tm} = \frac{Q_{tm}}{N_{tm}} \times 231$$

and

$$V_{tm} = \frac{0.977}{1,500} \times 231 = 0.150 \text{ in.}^3/\text{rev}$$

Summarizing the specifications for the trolley motor,

Trolley motor	
Horsepower	1.14
Speed reduction at 1500 rpm	15.70 : 1
Output torque (before reduction)	3.99 lbf-ft (47.88 lbf-in.)
Flow rate at 1500 rpm	0.977 gal/min
Displacement	0.150 in.^{3}/rev

In the previous example, the exclusive tandem center valves allow fluid to be directed to either the hoisting circuit or the traverse trolley circuit, but not simultaneously to both. Maximum delivery, then, occurs when the hoisting circuit is engaged, since it has the largest actuator flow rate (that is, 18.70 gal/min) and the pump is sized accordingly. Flow from the pump (Q_p) and the delivery to the actuators (Q_a) are considered equal. That is,

$$Q_p = Q_a \qquad (13\text{-}8)$$

which in this case is a single actuator hoist motor. If through ports were added to both control valves, as in Fig. 13-5, the system would have the capability of simultaneously hoisting as well as providing traverse movement

$$Q_p = Q_a = Q_{hm} + Q_{tm}$$

and

$$Q_p = 18.70 + 0.977 = 19.68 \text{ gal/min}$$

if leakage were neglected.

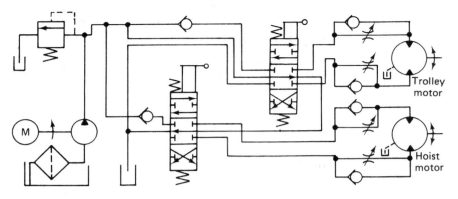

Figure 13-5 Hoist and trolley circuit (*inclusive* tandem center).

Fluid power systems may be classified by the method used to transmit power through the system, as well as by special purpose objectives. As power is transmitted through the system, parameters important to the designer include the system output horsepower, flow rate, system pressure, and system input horsepower, each of which may be constant or variable. Most systems can be classified as

1. Constant flow rate (*C-Q*).
2. Constant pressure (*C-P*).
3. Constant horsepower (*C-Hp*).
4. Load sensing.

13-3 CONSTANT FLOW SYSTEMS

Constant flow systems receive fluid from a fixed displacement pump turning at a relatively stable speed. Figure 13-6 illustrates a simple *C-Q* circuit. Flow control to system actuators can be regulated by the constant volume pump turning at a fixed speed, or by flow control valves appropriately placed in the circuit to meter-in, meter-out, or bleed-off unnecessary flow from the pump. Figure 13-7 illustrates a *C-Q* circuit with a meter-out flow control valve. The circuit in Fig. 13-8 uses an open center valve to route flow to the reservoir in the neutral position and

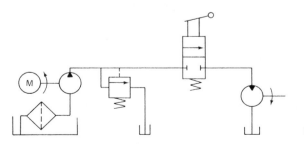

Figure 13-6 Simple C-Q circuit.

Basic Circuits and Sizing Hydraulics Components   Chap. 13

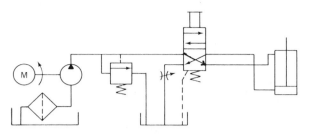

Figure 13-7 C-Q circuit; meter-out control.

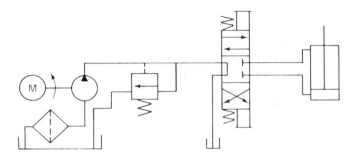

Figure 13-8 C-Q circuit with open center control valve and pilot operated unloading valve.

a pilot-operated relief valve to direct flow to the reservoir when the cylinder becomes fully extended or retracted.

Flow rate from the pump is a function of the pump displacement (V_p) and speed (N_p) and equals

$$Q_p = \frac{V_p \times N_p}{231} \quad \text{in gal/min} \tag{13-9}$$

Ideally, the pump is sized to total actuator demand so that the full capacity of the pump is used at the output against the load resistance. When actuators do not use the full output of the pump, the pressure in the system increases to a predetermined level set by a relief valve or unloading valve.

Since the speed of the pump (N) is approximately independent of the load when constant speed electric motors drive the pump, system input horsepower is proportional to the magnitude of the pressure and varies directly with the load resistance. Accordingly, system pressure fluctuates in the fluid horsepower formula

$$Ihp = Fhp = \frac{p \times Q}{1,714}$$

since (Q) is constant. The input to the electric drive motor measured in amperes, or to governor controlled *I-C* engines measured in lb/hr of hydrocarbon fuel, can also be expected to vary respectively.

Constant operation systems, such as the conveyor circuit in Fig. 13-9, require that flow from the pump at least equal actuator demand plus losses from

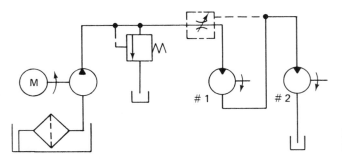

Figure 13-9 C-Q conveyor circuit with flow control valve.

internal leakage and friction. The conveyor or circuit in Fig. 13-9 regulates flow to #1 motor and bleeds excess flow to #2 motor, allowing it to slightly lead #1 keeping the conveyor drive chain tight.

Systems in which the operation is cyclic, such as the press circuit in Fig. 13-10 use appropriately sized accumulators in parallel with the actuator to supply peak demands during the working part of the cycle and to store fluid during the slack part of the cycle. An unloading valve directs delivery to drain when the accumulator is filled. *A spring returned normally open solenoid valve drains the accumulator when the system is turned off* **as a safety precaution to prevent accidental activation** *should the control valve be shifted.*

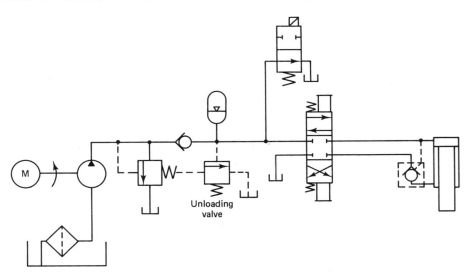

Figure 13-10 C-Q accumulator press circuit.

13-4 CONSTANT PRESSURE SYSTEMS (C-P)

Constant pressure systems keep the pressure fixed while varying the flow rate to one or more actuators. Typically, a pressure-compensated variable displacement piston or vane pump supplies fluid to the actuator through a closed center control valve. Fixed displacement rotary actuators receiving fluid at constant pressure deliver a constant torque output regardless of variations in speed. Pressure surges and drift often encountered when open center systems come to pressure are reduced, since pressure is maintained when the control valve is centered. The circuit in Fig. 13-11 powers two actuators at constant pressure and variable volume. When the rotary actuator control valve is centered, the internal passage connection in the control valve allows the motor to coast to a stop. Cavitation is prevented by the check valve arrangement which will provide additional fluid from the reservoir if necessary. The cylinder actuator, however, is locked in position and will resist overrunning loads, since both ports to the cylinder are blocked. Variable flow demand is accomplished by the variable displacement capability of the pump reacting to changes in pressure. High demand will cause the pressure to drop and the pump to become fully stroked until the preset pressure level is reached. Likewise, reduction in fluid demand will cause the pressure to rise and the pump to shift, thereby reducing output flow. With both valves in the center position, flow through the system is negligible and the pump supplies only make-up oil. The power loss is also small. The use of pressure-compensated variable displacement pumps eliminates the need for both relief and unloading valves, since the pump compensates to limit system pressure. Relief valves or hydraulic fuses are usually installed, however, to protect the system against hydraulic shocks, thermal expansion of the fluid, and possible malfunction of the pressure compensation mechanism in the pump.

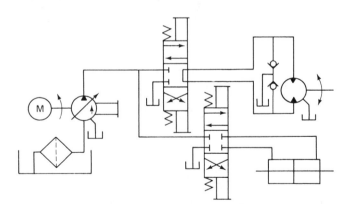

Figure 13-11 C-P circuit with two actuators.

13-5 CONSTANT HORSEPOWER SYSTEMS (C-Hp)

Constant horsepower systems transmit approximately constant power through the system regardless of changes in the magnitude of the load resistance. Whereas constant torque and constant pressure systems fix a constant speed (N) and torque (T), respectively, constant horsepower systems must regulate some combination of the pressure (p) and flow rate (Q) to regulate the speed (N) and torque

(T), so that their product is a constant. That is,

$$\text{Output Horsepower} = \frac{p \times Q}{1,714} = \frac{T \times N}{5252} = \text{Constant}$$

Some circuits that deliver a constant horsepower output use a fixed displacement pump to drive a variable displacement motor to fulfill this function. The winch circuit in Fig. 13-12 uses a constant displacement pump to supply a manually operated over-the-center bidirectional motor at constant pressure. An in-line check prevents the motor from driving the pump when the load resistance is over capacity. A braking valve in the exhaust line is controlled by a manually operated two-way control valve. Setting the motor at maximum displacement will cause it to generate maximum torque at minimum speed. Reducing the displacement will cause an increase in rpm accompanied by a decrease in torque. Further reduction of the displacement will result in further reduction in torque with an accompanying increase in rpm until such time as the torque is not sufficient to sustain rotation. At this displacement, the load resistance will stall the motor.

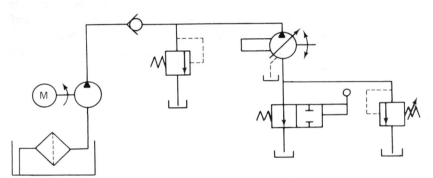

Figure 13-12   C-Hp output circuit.

Other constant horsepower systems have a fixed horsepower input source such as a storage battery bank. Since the output horsepower (Ohp) from the system equals

$$Ohp = \frac{F \times L}{t \times 33,000}$$

increases in the magnitude of the load resistance will result in either an increase in the time required to move the load resistance the prescribed distance or a lesser distance moved in the prescribed time.

An increase in the load resistance (F) will cause an increase in system pressure if the effective area of the actuator is fixed, such as in cylinder actuators and fixed displacement motors. In pressure compensated variable displacement motors, if pressure (p) remains fixed, the effective displacement of the actuator must increase if the load resistance increases.

The *C-Hp* circuit in Fig. 13-13 is used to operate the lift cylinder of a battery powered work lifter. The electrical schematic indicates how the circuit operates.

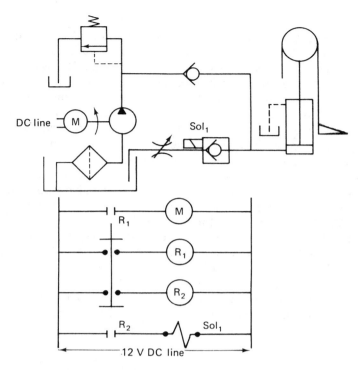

Figure 13-13 C-Hp circuit with electrical controls.

The horsepower input from the storage battery can be computed from

$$Hp = \frac{W}{746} = \frac{E \times I}{746}$$

where (E) is the voltage and (I) is the amperage. If the power from the storage battery remains substantially constant with changes in the magnitude of the load resistance, increases in the current (I) will be accompanied by decreases in the voltage (E) so that $(E \times I)$ remains approximately constant.

In operation, closing the *Lift* switch energizes relay coil #1 and relay #1 which operates the pump motor. Fluid flows through the check valve to the blank end of the lift cylinder raising the load. Opening the *Lift* switch and closing the *Lower* switch opens the motor circuit, energizes relay coil #2 and relay #2 which energizes solenoid #1, thereby shifting the lower check valve open. This allows fluid to return through the variable flow control valve to the reservoir, thereby regulating the lowering velocity of the load.

13-6 LOAD-SENSING SYSTEMS

Load-sensing systems are designed to save energy and improve overall system efficiency. They vary both the pump volume and pressure to suit the demands of the work load. This avoids some of the power losses due to maintaining a constant flow in open center systems, or maintaining a constant high pressure in conventional closed center systems. Load-sensing systems deliver only the amount of flow required by the operation speed, and at a pressure required to move the load. This is accomplished by using a load-sensing compensator valve which unloads the pump if a fixed displacement pump is used, or which actuates the destroking mechanism if a variable displacement pump is used.

In an application that requires a capacity flow of 20 gal/min at 2500 lbf/in.^{2}, for example, some means must be considered to deal with the excess flow and high pressure during the time the system is responding to partial load conditions. In mobile applications and conveyor systems, for example, this is a high percentage of the operating time. If a partial load required a pressure of only 1300 lbf/in.^{2}, and a metered flow rate of 9 gal/min, the flow valve would direct 9 gal/min against the load resistance and the remaining 11 gal/min would flow across the relief valve at 2500 lbf/in.^{2} with a resulting 16 hp loss. This energy would be lost in the form of heat. The traditional closed center system under the same condition would deliver only the required 9 gal/min, but at the destroke pressure of 2500 lbf/in.^{2}, thereby causing a 1200 lbf/in.^{2} pressure drop across the control valve with a resulting 6.3 hp loss in the form of heat.

The load-sensing system which uses a variable displacement pump has the destroking mechanism set at 200 lbf/in.^{2}. This is accomplished by having a sepa-

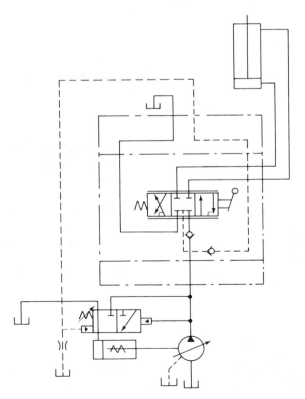

Figure 13-14 Load sensing circuit.

rate pilot line from the work port used to establish the minimum pressure to move the load resistance, plus an amount necessary to provide make-up oil for leakage and serve other pilot and auxiliary needs in the system. In the problem just described, the variable displacement pump would be destroked at 1500 lbf/in.^{2} rather than at 2500 lbf/in.^{2}, and deliver 9 gal/min to the actuator while permitting only 200 lbf/in.^{2} pressure drop across the control valve. The circuit is shown in Fig. 13-14. A load-sensing circuit which uses a fixed displacement pump would sense the load pressure from the work port in the same manner just mentioned, but use an unloading valve operated by a separate pilot to return the unneeded flow to the reservoir. The circuit to accomplish this is shown in Fig. 13-15.

In addition to increased overall system efficiency at partial load, efficiency is also increased when the control valve is in the neutral position. Here, open center

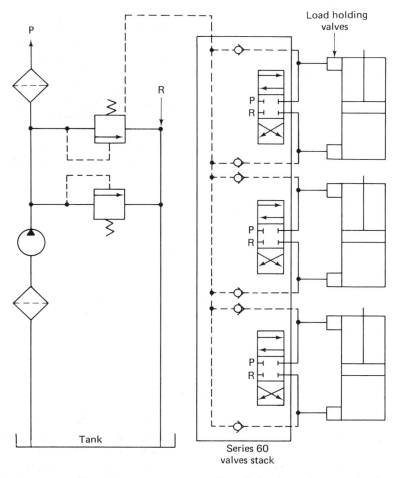

Figure 13-15 Fixed displacement pump with cylinder demand pump unloading valve.

systems deliver the complete flow of the pump through the lines, filter, control valve spool, and heat exchanger, which accumulates to a considerable loss over time. And while the variable delivery system destrokes the pump to reduce flow to a minimum when the valve is in the neutral position, losses are still generated from holding maximum system pressure using conventional pump destroking methods. Load sensing, on the other hand, limits the pressure to approximately 250 lbf/in.^{2} above the load, even in the neutral position. This constant pressure drop results not only in lower leakage and heat losses, but in the very important benefit of improving the metering characteristic of the valve by reducing flow forces which are normally caused by higher pressure drops and flow velocities across the spool.

13-7 BASIC CIRCUITS

Several basic circuits designed to accomplish specific objectives are encountered by the technician who must determine which is most appropriate, and by the mechanic who fabricates, installs, and services them. The most common are unloading, sequencing, regenerative, synchronous, and safety circuits.

Unloading circuits return fluid to the reservoir at low pressure when the system is idle. Their purpose is to reduce operating costs and heat generation and extend pump life and efficiency. Circuits can be unloaded through control valves operated manually, electrically, or remotely by a rise in system pressure.

The open center control valve is probably the simplest example used to unload a circuit at low pressure when the control valve is in the center position (Fig. 13-16).

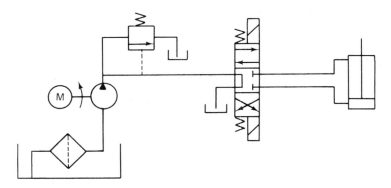

Figure 13-16 Open center circuit with linear actuator.

In Fig. 13-17, the two-way solenoid-operated valve drains the pilot-operated balanced relief valve when it is energized. When the three-way cylinder control valve is centered, the two-way solenoid valve is energized and opens the pilot drain from the relief valve resulting in a pressure imbalance, causing the pilot-operated relief valve to shift open returning flow from the fixed displacement pump to drain.

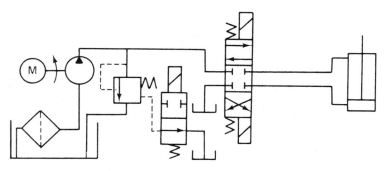

Figure 13-17 Pilot relief valve controlled by solenoid.

Figure 13-18 illustrates a piloted relief valve controlled by a second pilot-operated pressure relief valve. The circuit unloads flow from the pump only when the cylinder is retracted during the slack part of the cycle. When the two-way control valve is in the position shown, system pressure on the rod side of the piston pilots open the pressure relief valve in that leg of the circuit. This drains the pilot-operated pressure relief valve causing it to open. During the time that pressure is applied against the blank end of the piston, the pilot-operated and drained relief valve operates normally to limit system pressure to the relief valve setting.

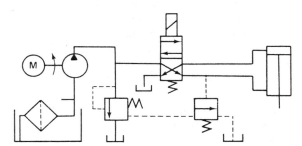

Figure 13-18 Pilot relief valve controlled by pilot valve.

High-low unloading circuits act to isolate one or both pumps from the circuit and return their flow to the reservoir at low pressure. They have application to circuits requiring rapid traverse at low pressure followed by a slow advance at high pressure. Both pumps typically are banked and driven from a common shaft and power source. The circuit in Fig. 13-19 unloads both pumps to drain when the control valve is in the center position. When the valve is shifted to either extreme position, both pumps supply fluid at medium pressure until the cylinder contacts sufficient load resistance to raise the system pressure to a predetermined level. At this point the pilot-operated relief valve unloads the left pump which is then isolated by the check valve. Maximum system pressure is then limited by the pilot-operated relief valve located just upstream of the control valve in the usual manner.

Sequencing circuits order cyclic events, such as the operation of two cylinders, one after the other. Events can be ordered using unequal loading against

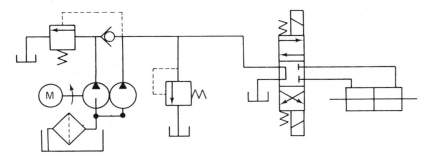

Figure 13-19 High-low unloading circuit.

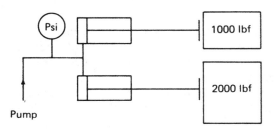

Pump

Figure 13-20 Unequal loading sequencing.

cylinders of equal area as in Fig. 13-20, by using electrical limit switches, or through the use of independent cam operated switches to shift control valves to order cycle operations in the desired way. Figure 13-21 illustrates a clamp and drill circuit in which the sequence is ordered by a pressure step increase across sequence valves S_1 and S_2. When the solenoid valve is shifted, fluid flows to the blank end of both cylinders. The clamp cylinder extends first because the fluid flow is unobstructed. When the clamp contacts the workpiece, the pressure rises shifting sequence valve S_1 and the drilling operation begins. When the solenoid valve reverses the direction of flow, first the drill retracts, sequence valve S_2 is shifted and then the clamp releases the work piece. Notice that integral checks in both sequence valves allow free flow in one direction. Also the pressure at the

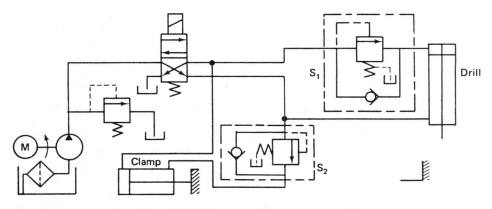

Figure 13-21 Presence step sequence circuit.

blank end of the clamp cylinder and the rod end of the drill cylinder is higher because of pressure drops across sequence valves S_1 and S_2.

Regenerative circuits make use of the differential areas between the blank and rod ends of cylinder actuators to increase cylinder velocity without increasing the flow rate from the pump. In the circuit in Fig. 13-22, the return position of the control valve pipes fluid to both the blank and rod ends of the cylinder simultaneously. Fluid from the rod end joins fluid being piped to the blank end as the cylinder extends.

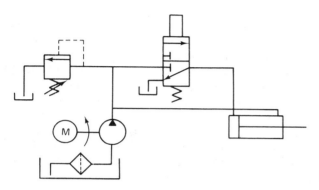

Figure 13-22   Regenerative circuit.

The fluid piped to a cylinder exerts a force

$$F = p \times A$$

where the area (A) may be at either the blank or rod end of the cylinder. At the blank end of the cylinder, the force (F_b) equals the pressure times the area of the bore. That is,

$$F_b = p \times \frac{\pi D_b^2}{4}$$

At the rod end of the cylinder, the force equals the pressure times the difference in area between the bore and the rod. That is,

$$F_r = p \times \frac{\pi (D_b^2 - D_r^2)}{4}$$

where D_b and D_r equal the diameters of the bore and rod cross section, respectively.

If fluid is applied to both ends of the cylinder simultaneously, the differential force between the blank and rod ends of the cylinder equals the force applied to the blank end of the cylinder minus the force applied to the rod end of the cylinder. That is,

$$F_d = F_b - F_r$$

$$F_d = p \times \frac{\pi D_b^2}{4} - p \times \frac{\pi (D_b^2 - D_r^2)}{4}$$

Sec. 13-7   Basic Circuits

Factoring

$$F_d = \frac{\pi p}{4} (D_b^2 - D_b^2 - D_r^2)$$

$$F_d = \frac{\pi p}{4} \times D_r^2$$

and

$$F_d = p \times A_r$$

This says, in effect, that when pressure is applied to both ends of a single-end rod cylinder simultaneously, that the differential force at the blank end of the cylinder equals the pressure multiplied by the cross-section area of the cylinder rod. If in Fig. 13-22 the cylinder rod is sized to have an area equal to one-half that of the cylinder bore, the velocity and force of the cylinder will be equal in both directions, since the flow from the rod end of the cylinder is added to that entering the blank end of the cylinder on extension.

Synchronous circuits allow two or more linear or rotary actuators to operate in unison regardless of differences in the magnitude of the load resistance. One method used is to mechanically link both cylinders or rotary actuators together. Another method is to meter an equal amount of fluid to each circuit by using two fixed displacement fluid motors of equal volume or a flow divider. Assuming the displacement of each segment of the flow divider and respective cylinders receiving fluid is equal, deviation from synchronization can be attributed primarily to leakage. The circuit in Fig. 13-23 uses a fixed displacement pump to supply fluid to two cylinders of equal displacement through a flow divider. The circuit is controlled by a four-way three-position open center valve. On extension, fluid is fed to the flow divider which meters equal amounts to cylinders 1 and 2 causing them to extend in synchronization. Centering the control valve locks the cylinders. Shifting the control valve to retract both cylinders causes fluid to be fed to the rod ends of each cylinder and the return fluid from the blank ends of each cylinder to be metered in equal amounts through the flow divider. While the flow

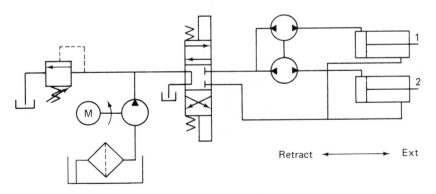

Figure 13-23 Synchronous circuit using flow divider.

divider meters equal amounts of fluid in and out of each cylinder, there is no compensation made for leakage and fluid bypassing the cylinder pistons.

Another circuit that synchronizes two cylinders is illustrated in Fig. 13-24. Here two double end rod cylinders of equal bore are connected in series. Fluid from the first cylinder is used to power the second. As in Fig. 13-24, variations in leakage past piston seals will cause variation in stroke between the two cylinders. Since the deviation is cumulative, the system can be expected to run out of synchronization after repeated cycling.

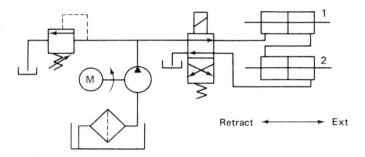

Figure 13-24 Synchronous circuit using equal area cylinders.

Correction for minor variation in synchronization at the end of each cycle can be made by using a replenishing circuit to supply make-up oil for that lost due to leakage. The circuit in Fig. 13-25 uses a four-way, three-position valve to extend and retract cylinders 1 and 2, which are connected in series. Solenoid valve A replenishes the circuit. On extension, if cylinder 1 bottoms first, it contacts a limit switch energizing solenoid valve A that supplies additional fluid to fully extend cylinder 2. On retraction, if cylinder 2 bottoms first, it contacts a limit switch energizing solenoid valve A that supplies additional fluid to retract

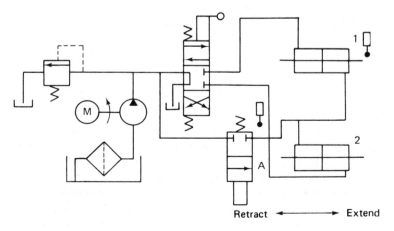

Figure 13-25 Replenishing synchronous circuit.

cylinder 1. In this manner, the strokes of both cylinders can be expected to begin and end when both cylinders are in the same position.

Synchronizing fluid power motors on conveyors and roller feed systems is critical where slack in the system can affect production quality and machinery breakdown. Figure 13-26 illustrates a conveyor fluid motor circuit controlled by a metering pump and two pilot-operated control valves. Fluid is supplied to the system from a common manifold to which each fluid motor is connected. Each fluid motor in turn drives an auxiliary pump connected in series, and supplies fluid to pilot operate normally open two-way control valves that connect the fluid supply from the manifold to each conveyor drive motor. The pilot circuit terminates at the variable displacement metering pump that regulates the speed of the system. When the system is started, pressurized fluid from the manifold is supplied to conveyor drive motors a and b which drive the auxiliary pumps. Metering fluid passes through the auxiliary pumps and metering pump. Fluid motor speed continues to increase until the fluid through the auxiliary pumps and metering pump are equal. An increase in speed of either motor causes pressure to build in the circuit piloting the two-way control valves toward the closed position which reduces flow from the manifold to the respective motor. A decrease in speed of either motor causes pressure to be reduced at the respective auxiliary pump output, and the pilot-operated control valve opens. The metering valves in the circuit are used to dampen surges and natural harmonics generated by the system.

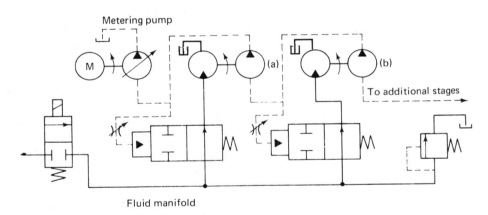

Figure 13-26 Conveyor fluid motor control circuit.

Safety circuits are designed to prevent damage to equipment and injury to operating and maintenance personnel. Typically, they prevent the system from accidentally falling on the operator or maintenance men, from overloading, and usually require two-hand operation of controls to prevent the operator's hands from becoming trapped in the machine. Press circuits are examples that incorporate all of these potential hazards and necessary safety measures.

Figure 13-27 illustrates a safety circuit with a fail safe device to prevent the ram from accidentally falling should a hydraulic line rupture or personnel inadver-

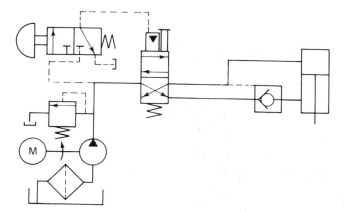

Figure 13-27 Fail safe circuit.

tently operate the manual override on the control valve when the pump is not operating. To lower the ram, pilot pressure from the blank end of the piston must pilot open the check valve at the rod end to permit oil to return through the control valve to the reservoir. The pilot-operated control valve permits free flow in the opposite direction to return the ram when the control valve is reversed.

The circuit in Fig. 13-28 incorporates overload protection to safeguard the system components. The pilot-operated main control valve 2 is controlled by a manually operated three-way pilot valve. The overload valve 3 connects the pilot to drain. Should the ram encounter undue resistance on the downward stroke, sequence valve 4 pilot operates the overload valve 3. This drains the pilot to the spring returned main control valve 2, causing the circuit to reverse. Operating the three-way manual control valve will have no effect on the circuit unless the overload valve 3 is reset manually.

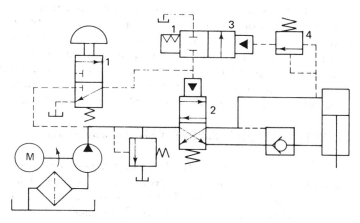

Figure 13-28 Fail safe circuit with overload protection.

Figure 13-29 illustrates a fail safe circuit with overload protection and an interlock to protect operating personnel. The interlock requires that both palm valves be depressed for the cylinder ram to be retracted. This prevents the opera-

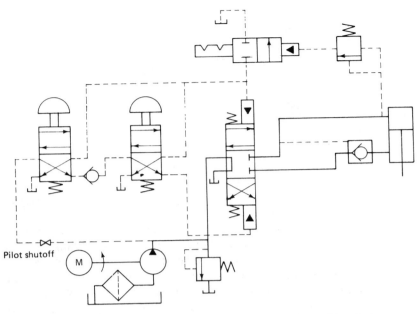

Figure 13-29 Fail safe circuit with overload protection and interlock.

tor from tying one valve down, thus releasing one hand that may become injured in the press during the work portion of the cycle. When the pump is started, fluid initially flows through the four-way open center control valve and the pilot circuit is closed. This allows the circuit to warm up with the pump isolated from the rest of the system. Opening the pilot shut-off valve sends a pilot signal through both four-way interlock valves to operate the main control valve retracting the cylinder. When both four-way palm valves are depressed, the main control valve is shifted to supply fluid to the blank end of the cylinder. Releasing both palm valves causes the press ram to be returned. The fail safe and overload portions of the circuit operate as previously described preventing accidental dropping of the ram and venting the pilot signal to the main control valve to return the ram should an unanticipated obstruction be encountered on the downward stroke.

13-8 OPEN CENTER VERSUS CLOSED CENTER CIRCUITS

Open center circuits connect the fixed displacement pump to the reservoir when the control valve is in the center position. The control valve outlet ports to the actuator may or may not be blocked. Power losses between cycles are small because the discharge from the pump is drained at low pressure. The system comes to an operating pressure, depending on the value of the load resistance, each time the control valve is shifted to pipe fluid to the actuator. Maximum pressure is set by the relief or unloading valve. Actuators are sized to accept the maximum flow of the pump and the velocity at the output is determined by the displacement of the actuator or metering valves. Figure 13-30 shows an open

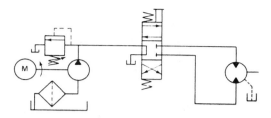

Figure 13-30 Open center circuit with rotary actuator.

center circuit where the velocity is controlled by the displacement of the actuator. Figure 13-31 illustrates a meter-in circuit that regulates flow to the actuator from the control valve and is appropriate for resistive loads. The circuit in Fig. 13-32 is a meter-out circuit used to accommodate overrunning loads. The bleed-off circuit in Fig. 13-33 returns unused fluid to the reservoir. A filter and heat exchanger are added to condition the excess flow.

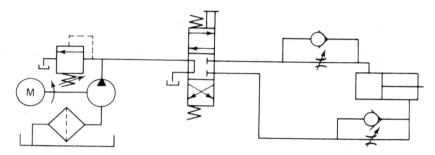

Figure 13-31 Open center meter-in circuit.

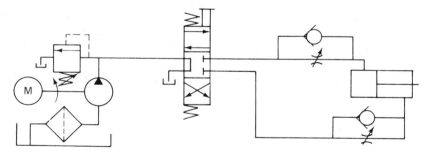

Figure 13-32 Open center meter-out circuit.

Closed center circuits do not connect the pump to the reservoir when the control valve is in the center position. The pump may be blocked or piped to one or both actuator ports to return, replenish, or regenerate the circuit. Closed center circuits that use fixed displacement pumps sometimes incorporate accumulators to divert flow during the slack part of the cycle for use during the work portion of the cycle. Total flow per cycle and cycle rate are used to size the pump

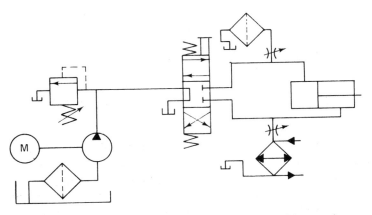

Figure 13-33 Open center bleed-off circuit with filter and heat exchanger.

and accumulator such that unused flow from the pump returned across the relief valve to the reservoir is minimized. Figure 13-34 shows a closed center circuit using a fixed displacement pump to charge an accumulator. During the slack part of the cycle, the fixed displacement pump charges the accumulator. The check valve holds system pressure preventing the motor from being driven by system pressure when the system is idle. The two-way safety solenoid valve drains the accumulator when the system is turned off. Maximum pressure is set by the pilot-operated pressure relief valve which returns unused flow to the reservoir. Also notice the use of an unloading pilot to the relief valve to prevent system overpressure by the load resistance or heat and a check valve that prevents the pump from being motorized when the system is shut down. Figure 13-35 illustrates a closed center circuit operating from a pressure compensated variable displacement pump. A relief valve is used as an added safety precaution. Both linear actuators receive fluid at constant pressure and variable flow depending on their load resistance and the throttling capability built into the control valve.

Open center systems allow the designer to use a few simple and relatively inexpensive components to fabricate an efficient, robust, and reliable system with operating pressures to 3000 lbf/in.^{2} if single or tandem actuators drive a positive

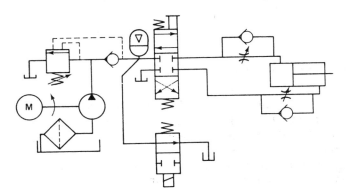

Figure 13-34 Closed center accumulator circuit.

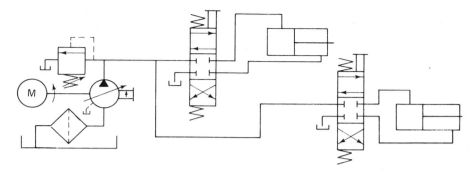

Figure 13-35 Closed center circuit with variable displacement pressure compensated pump.

load resistance under ideal conditions. Preferably, the time between operating cycles is substantial, the load resistance does not have a tendency to overrun, the velocity is not critical, and force rather than fine accuracy is the main consideration. Many production and mobile applications use open center systems. System efficiency is high so long as excessive pressure drop associated with metering is kept to a minimum. This requires that considerable attention be given to sizing pump delivery with actuator demand to provide the desired velocity at the output.

Closed center circuits offer greater flexibility to the designer because system pressure can be maintained regardless of delivery. In addition to the inherent fast response characteristic, this capability allows the use of pilot-operated circuits as well as pressurizing multiple circuits from one pump at varying delivery rates. The cost, however, is greater.

If a fixed displacement pump is used, an accumulator circuit can store fluid for later use and reduces inefficiency caused by discharging fluid at system pressure over a relief valve. In this capacity, small pumps might also supply circuits requiring large deliveries for short periods when connected in parallel with an accumulator. Fixed displacement tandem pump high-low systems accomplish the same task but unload the larger low-pressure pump with the smaller high-pressure circuit.

If initial cost is secondary, a pressure compensated variable displacement pump can be used to respond to all these varying system demands at constant pressure. Some disadvantages resulting from operating the control valve under pressure, such as creep, valve spool seizing, and wear, can be reduced using pressure balanced and more expensive close-toleranced components that make allowance for this contingency. Fine filtration also is required to prevent lapping wear from silt.

13-9 PUMP SELECTION

Selection of the appropriate pump is made primarily from an examination of the work cycle with consideration being given to other factors such as cost, duty rating, expected life, size, porting, and appearance.

Actuator selection is made from force, distance, and time criteria which give the designer system pressure and flow requirements as the basis for selecting input and control components. The work cycle profile can be used to describe pressure and flow requirements with respect to time. This reveals, for example, if the cycle is intermittent or continuous, uses single or multiple branched circuits, and has a long or short slack time in the cycle. This information projects a general indication whether open or closed center circuits are more appropriate using fixed or variable displacement pumps, and whether accumulators would conserve energy and increase system efficiency.

With the exception of accumulator circuits that augment pump flow, pressure and flow requirements determine input horsepower to the pump shaft. That is,

$$Ihp = \frac{p \times Q}{1,714}$$

When pump overall efficiency (e_{po}), consisting of volumetric and mechanical efficiency factors are accounted for

$$Ihp = \frac{p \times Q}{1,714} \times \frac{1}{e_{po}} \tag{13-10}$$

This raises the required input horsepower value by 15–25%, depending on whether piston, vane, or gear pumps are used, with piston pumps usually being the most efficient and gear pumps being the least efficient.

For a given horsepower at some pressure (p), the envelope size is influenced by the operating speed of the pump since

$$Q = \frac{V_p \times N}{231} \quad \text{in gal/min}$$

As operating speed is increased, displacement (V_p) can be decreased with a reduction in pump size. Thus, aircraft and aerospace pumps are made smaller, lighter, and occupy less space than industrial pumps by increasing the rpm. Mobile pumps fall somewhere in between with respect to size, cost, and other factors, since a decrease in size (and usually an increase in pressure) requires more closely controlled tolerances.

Drive rpm is also determined not only by the size of the pump and cost considerations, but by the prime mover to pump drive ratio. When electric motors are used in industrial applications, one of the synchronous speeds (1,200 or 1,750) is selected and the motor and pump shafts are coupled directly to each other. Mobile applications driving from the engine may be coupled directly or driven by belts, chains, or gears at some ratio of engine rpm. Maximum pump drive rpm in this case must be equated using some increase or reduction to a speed of 2,000–4,000 rpm for gasoline or *L-P* fuel engines, and 1,500–2,500 rpm for diesel engines. At the low end of the operating range, the flow from the pump must be sufficient to meet pressure and flow rate demands, and at the high end, not damage the pump from excessive rpm.

For a given required horsepower input and rpm, the required drive torque at the pump can be computed from

$$Ihp = \frac{T_p \times N}{5,252}$$

and

$$T_p = \frac{Ihp \times 5,252}{N}$$

For a given pump displacement (V_p), required input torque from the prime mover may also be computed from

$$T = \frac{V_p \times p}{2 \times \pi} \qquad \text{in lbf-ft} \qquad (13\text{-}15)$$

Cost frequently is the prime consideration when selecting a pump and this determines the design of the system. In general, piston pumps are the most expensive. Vane and gear pumps are less expensive. Variable displacement pumps have a higher initial cost than do fixed displacement pumps, but eliminate other pressure and flow control components. Variable displacement pumps are available in piston and vane types.

Sometimes it may be less expensive to use high-low tandem mounted fixed displacement pumps in an application than to use one large variable displacement pump. While piston pumps are the most expensive, axial piston check valve pumps have the capability to split the pump circuit into two or more independent pressurized supply sources. This offers many possibilities to the designer to eliminate pressure and flow regulating components in branch circuits while still maintaining independent control to several actuators.

Duty rating and pump life describe the severity of operation and how long the unit can be expected to last before failure. Intermittent operation is less severe than continuous operation. Similarly, heavy duty pumps can be expected to be more robust, more forgiving of abuse, and longer lasting than light duty units. Expected life under given conditions is often specified by the manufacturer using the B-10 rating. As with seals, bearings, and other wear-prone parts, this is the time in service hours that units can be expected to perform satisfactorily before 10% of their number will fail.

13-10 MATCHING COMPONENTS

The basis for matching system components can be defined from the pressure and flow rate necessary to accomplish specified objectives at the system output. Pressure drop across components also influences selection since sufficient power must be available after losses to operate the actuator against the load resistance. Other considerations include seals and fluid compatibility, component porting, the quality specified for each component application, cost, expected maintenance, availability of service parts, and testing facilities available by the factory to insure component integrity.

Pressure losses across each component vary with the applied pressure and flow rate. How the circuit is connected to the system also affects its performance and expected pressure drop. If pressure losses and mechanical friction are substantial, the heat generated will have to be dissipated using a heat exchanger. Most manufacturers list these specifications in their literature, or will make the same available for their components, given the specific application from which to work. Pressure drop and heat generated can be verified on the hydraulic machine itself by making provision for instrumentation, such as pressure, flow, and temperature gauges, when the system is fabricated. This provision also can be used to collect data during periodic maintenance checks and reduce diagnosis time should failure occur at some later time.

Porting and plumbing should be specified early in the design. If threaded connections are to be used, sufficient room must be provided to assemble and disassemble plumbing to remove components without taking them apart. Flanged and bolted connections allow a smaller space to be used because components can be grouped closer together and cause less leakage. Where space is at a premium, this can markedly affect the design. Pre-assembled components with integrated circuitry to perform specified cycles within one envelope further reduce size and plumbing complexity. In some applications, flexible tubing can be used to an advantage to connect components, particularly where testing and development are desirable prior to ''hard plumbing'' the system.

13-11 PERFORMANCE TESTING

Performance testing is part of the initial operation of the machine as well as for later periodic maintenance checks and diagnosis after failure. Operating standards for the machine must be established in advance and recorded to make data from later observation meaningful. Once initial operating specifications have been verified, maintenance personnel will have at their disposal standards against which to compare observed performance. The cycle plot is the basis for establishing these standards.

The cycle plot is a graphic picture of the work performed by the machine during a specified time. In recordable data terms, pressure, flow rate, and time can be observed while the specified load is applied to the system. If these verified data are plotted against the external load for each operation the machine performs, tests can then be established and written to isolate the cause of failure by sequentially eliminating circuit components that are operating properly. This will reduce the time necessary to find the failure.

Cycle rates determine work production. For operating systems, cycle time establishes the bounds within which the specific task must be completed, for example, raising a load resistance a specified distance using a cylinder, or completing the work and slack portions of a press cycle. Cycle rates, in turn, depend on limit switch or similar valve settings, system pressure, flow rate, and the value of the load resistance. If the value of the load resistance is specified and the travel limits are established, then cycle time depends on system pressure and flow rate.

The presence of excessive pressures indicates sticking linkage, machine binding, or seizing in the actuator. For near normal cycle rates, below expected pressures will not be observed. Low pressures that are indicated by extended cycle rate times are probably caused by loss of fluid, low pressure settings, faulty pressure sensitive valves (sequencing, unloading, etc.), sticking control valves, or fluid bypassing within the pump, control valve, or actuator.

Assuming pressures are near normal, cycle time is a function of the flow rate. Simple calculations can compare the effective flow consumed at the actuator with the theoretical flow supplied by the pump to give an indication of system volumetric efficiency. Effective actuator output is computed from the observed cycle rate and the known displacement of components. That is, for a 2-in. bore cylinder traveling at 6 in./sec, the effective flow on extension equals

$$\text{Effective Flow} = \frac{3.14 \times 2^2}{4} \times \frac{6 \times 60}{231} = 4.89 \text{ gal/min}$$

Similarly, if the pump has a displacement of 0.750 in.^{3} and is driven at 1,750 rpm, the theoretical flow equals

$$\text{Theoretical Flow} = \frac{0.750 \times 1,750}{231} = 5.68 \text{ gal/min}$$

and the volumetric efficiency for the system is

$$\text{Volumetric Efficiency} = \frac{4.89}{5.68} \times 100 = 86\%$$

The cause of deviations from established system volumetric efficiency standards within the normal pressure range can be isolated by observing the flow rate in specified portions of the work cycle. This establishes whether discrepancies can be attributed to the flow at the actuator or flow from the pump. Higher than expected flow rates can result from excessive pump speeds or other than normal operation of flow control valves. Normal flow and low cycle rates indicate fluid is bypassing in the actuator. Below expected flow sometimes can be attributed to slippage in the pump, low drive speed, overloading, or improper control valve operation.

Other causes of system malfunction include cavitation from high suction lifts, entrapped air from loose or broken lines, low fluid level, and thinning of the fluid from excessive heat build-up. These problems are usually accompanied by excessive pump noise, motor or cylinder chatter, erratic operation, discoloration, or abnormal fluid appearance and high temperatures.[3]

---

[3] An extensive discussion of noise including trouble shooting hydraulic systems by noise can be found in ''Reducing the Operating Noise In Hydraulic Systems'' (PL-1), Parker Hannifin Corporation, 17325 Euclid Ave., Cleveland, Ohio 44112.

13-12 COMPLIANCE WITH SAFETY RECOMMENDATIONS AND STANDARDS

Safe practices and standards consist of rules for avoidance of hazards proven by research and experience to be harmful to personal safety and health. The Joint Industry Conference (JIC) Standards for Industrial Equipment provides recommendations that are helpful to the fluid power designer, technician, mechanic, and operating personnel at the place of employment in meeting this objective. The Occupational Safety and Health Administration under the Department of Labor provides and enforces standards at the place of employment pursuant to regulations written from the Occupational Safety and Health Act of 1970 (PL 91-596). Familiarity with information available from both sources is advisable to establish the safest possible working conditions with and around fluid power machinery.

The general purpose of the JIC Standards For Hydraulic Equipment is to recommend and promote safety to personnel, provide uninterrupted production, lengthen equipment life, and reduce maintenance costs. Safety recommendations are listed that relate to these objectives, particularly to those that would result in injury to operating personnel and others in the vicinity.[4]

OSHA standards of particular interest to the fluid power industry include: Subpart-G, Occupational Health and Environmental Control; Subpart-M, Compressed Gas and Compressed Air Equipment; Subpart-N, Materials Handling and Storage; and Subpart-O, Machinery and Machine Guarding.[5]

Representative sources of additional information include: Compressed Gas Association, Inc., 500 Fifth Avenue, New York, N.Y. 10036; American Society of Mechanical Engineers, Inc., United Engineering Center, 3452 E. 47th Street, New York, N.Y. 10017; National Fire Protection Association, Inc., 60 Batterymarch Street, Boston, Massachusetts 02110; and the American National Standards Institute, 1430 Broadway, New York, N.Y. 10018.

OSHA Standards requirements are developed by evaluating a hazard potential from a base point, the workplace or workstation, and then considering additional potential hazards as other production elements are added. The categories for standards requirements are:[6]

1. Workplace Standards.
2. Machines and Equipment Standards.
3. Materials Standards.
4. Employee Standards.
5. Power Source Standards.
6. Process Standards.
7. Administrative Regulations.

[4] For complete information re JIC Standards write to: 7901 Westpark Drive, McLean, Virginia 22101.

[5] For more information consult the *Federal Register,* Wednesday, October 18, 1972, Vol. 37, No 202, Part II.

[6] For more information, consult OSHA publication 2072, *General Industry Guide for Applying Safety and Health Standards,* 29 CFR 1910.

Representative standards for power sources, including fluid power, are indexed for the appropriate Subpart and Section from Part 1910 of Title 29 of the Code of Federal Regulations (CFR), as revised.

There are certain basic safety and health standards that apply strictly to the workplace, a building, or other work location. These include safety of floors, or other working surfaces, protection of floor and wall openings, access and exit requirements, sanitation, and fire and emergency protection.

When machines and equipment are added to the workplace, new elements of risk come into play. Standards are included for risks involving machine guarding, operational techniques, special safety devices, inspection and maintenance, mounting, anchoring, grounding, and other protection.

Materials that are used, processed or applied on the job add to the hazards. There are standards covering materials yielding dangerous or toxic fumes or mists, ignitable and/or explosive dusts, and other atmospheric contaminants. There are also standards for safe storage and handling of compressed gases and flammable and combustible liquids as well as more stable materials used in production processes.

When the employee is added to the workplace, other variables and related technical standards become important. What medical and first aid services are required? What personal protective equipment and devices must be provided? Are licenses or other accreditation documents required? What about special training or educational requirements?

The power source used creates additional hazards. Electrical, pneumatic, hydraulic, steam, explosive actuated, and other sources of power have standards.

Some standards cover a special process or a special industry. Welding, cutting and brazing, spray finishing, abrasive blasting, and use of dip tanks are hazardous processes. Special standards exist for sawmills, pulpwood logging, and agriculture.

In addition to the safety and health standards, there are administrative requirements for all employers whatever the size of the work establishment. Every employer must display an OSHA poster stating the rights and obligations of employees and employers; keep injury, illness, and exposure records; report fatalities and multiple hospital injury cases (5 or more); and post an annual summary of injuries and illnesses.

13-13 SUMMARY

Systems design and circuit analysis are used to systematically develop and test hydraulic machines to accomplish specific work tasks. At every level, personnel are involved, from the engineer to the repair mechanic. They work as a team to lay out model prototypes, write specifications, and then to assemble, install, diagnose, and repair system component assemblies.

Systems capabilities are developed to match output objectives—that is, the performance of a task against a specified load resistance. These, in turn, are used to size actuators and pumps with matching components. Work versus time plots are used to analyze the force necessary to move the load the required distance in the time specified and determine output horsepower. Appropriate circuits are then designed to accommodate the characteristics of the load resistance to maximize the use of the available power input and efficiency of the system.

Constant flow systems use fixed displacement pumps turning at relatively constant rpm to power output actuators at a constant linear velocity or rpm. Flow control is regulated by the pump itself or flow control valves appropriately placed in the circuit. Circuits designed to regulate the flow from fixed displacement pumps meter-in, meter-out, or bleed-off excess fluids in amounts necessary to fix the velocity of the actuator within specified limits. Basic constant flow systems are relatively inexpensive to assemble because of simplicity and low component cost.

Constant pressure systems vary flow rate keeping the pressure constant. Typically, they use pressure compensated variable displacement vane and piston pumps to supply fluid to closed center circuits. Where open center circuits return fluid to the reservoir when the control valve is in the center neutral position, closed center valves block the flow of fluid and require the flow to be discharged to the reservoir by other means such as unloading or relief valves, or use a variable displacement pump to compensate, reducing flow automatically. Higher efficiency associated with closed center systems may be accompanied by higher pump and component costs.

Pumps are selected to match the requirements of the load resistance and the circuit design appropriate for the application. Input and pump horsepower must at least equal the output through the complete work cycle for the system to sustain operation. Accumulator circuits are sometimes used, for example on press circuits, to reduce the size of the pump when extended intervals are available during the slack part of the cycle.

System performance is monitored using flow meters, pressure gauges, cycle times, and fluid temperature to compare system output with required performance and determine operating efficiency of components including actuators, valves and pumps. Overall system efficiency is the prime consideration.

Industry and government safety recommendations have been adopted and are being enforced to protect personnel at the place of employment from health and injury hazards. Recent legislation creating the Occupational Health and Safety Administration, under the Department of Labor, specifies standards requirements for general industry and other segments of employment. Healthful and safe work conditions must be incorporated into the hydraulic system to assure compliance with these federal regulations.

STUDY QUESTIONS AND PROBLEMS

1. After reading the introduction to Chapter 13, and reviewing Fig. 13-1, list a specific task which might be delegated for each of the four functions explained—that is,
    a) Circuit design.
    b) System design.

c) Circuit analysis.

d) System analysis.

2. What are the differences between open-loop and closed-loop hydraulic systems?

3. List and explain briefly the steps typically followed in the design and analysis of a hydraulic system. (*Clue:* Select a simple application to use as an example to give continuity to your explanation.)

4. In a constant pressure hydraulic system, what occurs when the load resistance decreases? What can be said for the torque of fixed displacement actuators? Does the system deliver constant horsepower at the output?

5. In a constant horsepower hydraulic system, what happens as the load resistance increases in each of the following situations:

   a) Fixed displacement pump; fixed displacement motor.

   b) Fixed displacement pump; variable displacement motor.

6. List the purpose of each of the following circuits:

   a) Unloading.

   b) Sequencing.

   c) Regenerative.

   d) Synchronous.

   e) Safety circuits.

7. Explain briefly the differences between open center and closed center hydraulic systems. (*Clue:* Make two columns listing the operational characteristics of each.)

8. Construct a circuit drawing of a hydraulic dump lift using an open center circuit.

9. Construct a drawing of an open center two-cylinder double-acting circuit.

10. When the directional control valve in Fig. 13-36 returns to the center position, the cylinder rod is observed to move. Determine first whether the cylinder rod extends or retracts, and then compute the force and velocity.

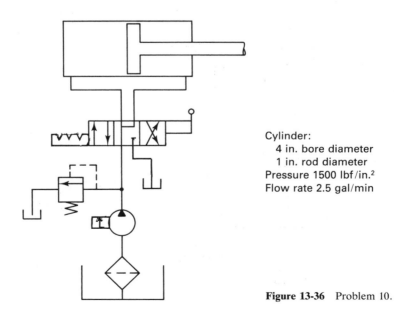

Cylinder:
   4 in. bore diameter
   1 in. rod diameter
   Pressure 1500 lbf/in.²
   Flow rate 2.5 gal/min

Figure 13-36   Problem 10.

11. A fruit grower wants to construct a hydraulic cider press with a $2:1$ ratio ram having a 24-in. stroke. Costs are to be kept to a minimum. The ram should advance the first 12 inches in 5 seconds, followed by a slower squeeze through the final 12 inches. The squeezing action requires a 12-ton force. Since the press portion of the cycle is short compared to the standby portion, the circuit should route the fluid at low pressure through a return line filter to the reservoir in the stand-by position. Before sizing the components for the circuit, construct the circuit diagram. The basic components are given next and shown in Fig. 13-37. Basic components:

a) Fixed displacement pump operating at 2500 lbf/in.^{2} max.

b) Pressure relief valve.

c) Four-way control valve.

d) Double check valve.

e) By-pass (sequence) valve.

f) Hydraulic ram.

g) Low pressure filter and inlet strainer.

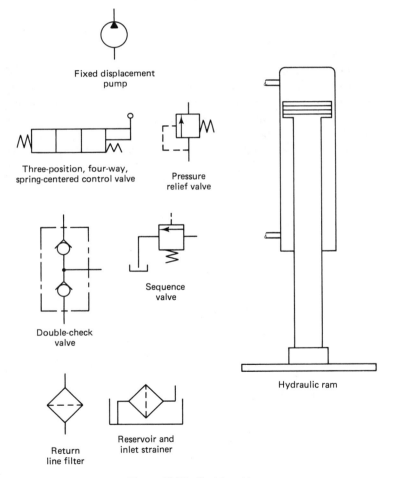

Fixed displacement
pump

Three-position, four-way,
spring-centered control valve

Pressure
relief valve

Double-check
valve

Sequence
valve

Return
line filter

Reservoir and
inlet strainer

Hydraulic ram

Figure 13-37   Problem 11.

h) Reservoir.

i) Connecting lines.

12. For Problem 11 calculate the following for the system and component specifications:

    a) High pressure ram force @ 2500 lbf/in.^{2}_____ .

    b) Ram bore diameter_____in., area_____in^2.

    c) Rod diameter_____in., area_____in^2.

    d) Rod displacement per ft of travel_____in.^{3}/ft.

    e) Pump displacement_____in.^{3}/rev., _____gal/min at 1725 rpm.

    f) Ram velocity:

        i) Advance rate_____in./sec.

        ii) Press rate_____in./sec.

        iii) Return rate_____in./sec.

    g) Pressure relief valve maximum flow rate_____gal/min.

    h) Double check valve maximum flow rate_____gal/min.

    i) Control valve maximum flow rate_____gal/min.

    j) Size of the reservoir_____gal.

13. A simple hydraulic lift circuit is to be drawn that uses a fixed displacement pump to run continuously with the components that are given next. The circuit should free-flow through the check valve while extending the cylinder with the two-position three-way valve, and be retracted with the manual operated shut-off valve. Locate the low pressure filter to clean the fluid when the circuit is in the stand-by position. The basic components are given next and in Fig. 13-38.

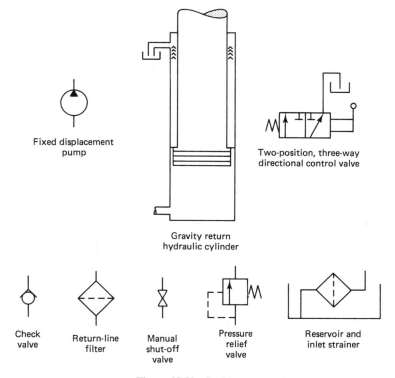

Fixed displacement
pump

Two-position, three-way
directional control valve

Gravity return
hydraulic cylinder

Check
valve

Return-line
filter

Manual
shut-off
valve

Pressure
relief
valve

Reservoir and
inlet strainer

Figure 13-38 Problem 13.

a) Fixed displacement gear pump.

b) Pressure relief valve.

c) Two-position, three-way direction control valve.

d) Gravity return hydraulic cylinder.

e) Manually operated shut-off valve.

f) Check valve.

g) Reservoir, return line filter, inlet strainer and connecting lines.

14. For Problem 13, size the cylinder to lift 10 tons at a system pressure of 2000 lbf/in.^{2}, and size the pump to extend the cylinder 24 in. in 30 secs.

15. Draw an open center circuit that uses a double-end rod hydraulic cylinder to extend

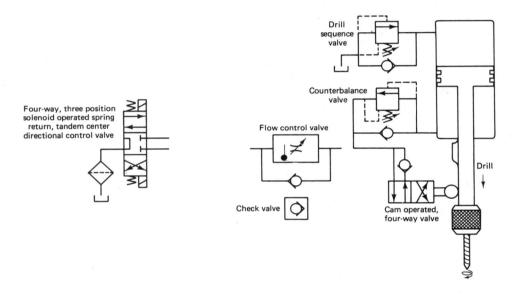

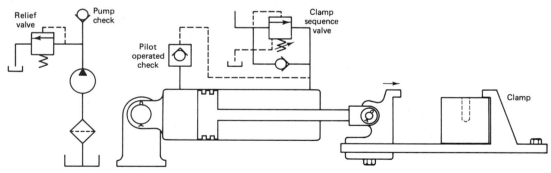

Figure 13-39 Problem 17.

and retract a large workpiece with equal velocity and equal force in both directions. Lock the cylinder in the center position to prevent drift.

16. For Problem 15, if a force of 3800 lbf is delivered through a 1-in. diameter cylinder rod and a rod velocity of 0.75 ft/sec is required, compute the bore of the cylinder and the pump delivery at a pressure of 1000 lbf/in.^{2}.

17. A pressure sequence clamp-drill circuit is to be designed to exert a clamping force on the workpiece, followed by the advance of the drill. A counterbalance valve is to be installed at the rod end of the cylinder to prevent chatter and lunging. The drill portion of the cycle should have a rapid regenerative approach, followed by a variable slow advance, sequenced by a cam-operated four-way valve in series with the flow control valve. The circuit should unload when the control valve is in the center position. On the return portion of the cycle, the drill should retract first, followed by opening of the clamp. A pilot-operated check should be installed at the cap end of the clamp cylinder to prevent the clamp from opening as the drill is retracted. The list of components to construct the circuit follows. They are also shown in Fig. 13-39.

a) Two double-acting cylinders with a 2-in. diameter bore and 1-3/8 in. diameter rod.
b) Fixed displacement pump.
c) Two sequence valves with reverse free flow checks.
d) Drill cylinder counterbalance valve to prevent chatter and lunging.
e) Drill cylinder flow control valve with reverse free flow check.
f) Three position, four-way, double acting, tandem center, solenoid operated, spring return, directional control valve.
g) Relief valve to protect the system.
h) Pilot-operated check valve (blank end of the clamp cylinder).
i) Pump check valve.
j) Check valve (regenerative portion of the circuit).
k) Cam operated four-way valve with check, used as a three-way valve (rapid to slow advance).
l) Return oil filter, inlet strainer, and connecting lines.

Given system conditions	Calculate
a) A minimum clamp force of 1500 lbf	a) Drill sequence valve pressure setting_____ lbf/in.^{2}
b) Clamp closing velocity of 5 ft/min	bi) Pump delivery_____ gal/min
	bii) Pump displacement_____ in.^{3}/rev with 1200 rev/min motor
	biii) Rapid advance velocity on drill_____ in./sec
	biv) Drill and clamp return velocity_____ in./sec
c) Slow advance of 6 in./min	c) Flow control valve setting _____ gal/min
d) Maximum drill force of 750 lbf	d) Drill cylinder rod end counterbalance valve pressure setting_____ lbf/in.^{2} (assume no interaction with the flow control valve)
e) Minimum return force on drill cylinder of 1000 lbf	e) Clamp sequence valve setting_____ lbf/in.^{2}
f) Maximum return force on clamp cylinder of 1500 lbf	f) Pressure relief valve setting_____ lbf/in.^{2}

18. For the circuit in Problem 17, make the calculations for the given system conditions.

19. A traverse feed for a table grinder is to be driven by a hydraulic cylinder with an infinite velocity adjustment within the speed range of the machine. The grinder head is to move with the same velocity in both directions and once the feed is started it should continue to cycle automatically, reversing each time the grinder head reaches the adjustable cam-operated valves which establish the limits of its travel. Use the components listed next to construct the circuit. They are also given in Fig. 13-40. Components for Problem 19:

a) Double-end rod cylinder.

b) Two position, four-way pilot-operated directional control valve.

c) Two, two position, three-way, cam operated, spring return, directional control valves.

d) Flow control valve.

e) Manual shut-off valve.

f) Fixed displacement pump.

g) Pressure relief valve.

h) Reservoir with return line filter and inlet strainer.

i) Connecting lines.

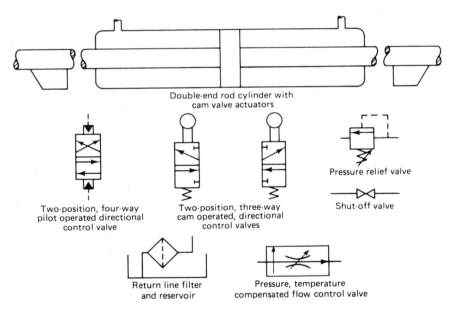

Figure 13-40   Problem 19.

20. In Problem 19, if losses are ignored, the relief valve pressure is set at 100 lbf/in.², and the cylinder rod for structural reasons has an OD of 0.75 in:

a) What inch bore cylinder would be required for the rod to exert a force of 50 lbf in each direction?

b) What pump delivery in gal/min would be required to cycle the cylinder through a 24-in. stroke at 60 cycles/min?

c) What in.^{3}/rev pump displacement would be required if the drive turns at 1200 rev/min?

21. List and explain briefly several factors considered when selecting a pump for a hydraulic system.

22. How are components matched when designing a hydraulic system?

23. How is the performance of a hydraulic system monitored?

24. Cite and explain two safety standards established for hydraulic systems. (*Clue:* Consult 29 CFR 1910, as amended.)

25. Cite a specific standard which applies to the fluid power industry under each of the seven categories established by the Occupational Safety and Health Administration.

CHAPTER *14*

TROUBLESHOOTING HYDRAULIC SYSTEMS

14-1 INTRODUCTION

Troubleshooting means finding the problem. In fluid power systems, problems are first determined to be one or a combination of the five general types, and then through a logical procedure, problem statements are written and tested to determine which specific component or part of the system is at fault. The five general types of problems are

1. Pressure.
2. Flow.
3. Leakage.
4. Heat.
5. Noise and vibration.

The procedure by which specific problems are identified consists of organizing and writing problem statements and then proving or rejecting them through inspection, making simple calculations, conducting tests on the system, and finally doing the teardown and disassembly for visual verification.

Some repairs may have to be made to find the problem, but troubleshooting does not mean repairing the machine. Neither does it mean performing routine maintenance, replacing components, or redesigning the circuit. Rather, troubleshooting means performing the diagnosis and then outlining a course of action that will bring the machine to satisfactory operation. The point is that while several related tasks may be performed to bring the diagnosis to a conclusion, actually fixing the machine is not to be confused with testing probable causes to find out exactly what the problem is. Essentially this means that problems and their causes are best found through a logical systematic diagnosis, and this special way of thinking about problems guides the actions of the troubleshooter.

The factory manual and service history of the machine are important because most problems have a root cause. Of course parts become loose and result in machine failure, but the cause of this kind of problem will be obvious and not require troubleshooting. More often, problems that require troubleshooting have their root cause in the design, age, assembly, operation, maintenance schedule, or even the most recent repairs made to the machine. So without the factory manual and service history of the machine, the diagnosis of a failure and its root cause could be very time consuming. Before troubleshooting a machine, secure the factory manual and service history.

The factory manual and service history are used when the initial inspection of the machine is made. This is the time to find obvious problems and get a general feel for the condition of the machine. And if the service history of the machine is vague, the owner, operator, or maintenance personnel should be asked to provide the necessary information about the age, maintenance schedule, and repairs made to the machine when the inspection is made. The manual will help locate the more obvious components on the system such as pumps and valves, as well as those that are mounted behind other machine parts or even in the reservoir. Isolation valves, check valves, strainers, filters, and electrical controls are often hidden from view. Look at the condition of the fluid first, and then inspect line connections and places where dirt might enter the reservoir. Mentally trace the circuit while looking for leaks and collapsed inlet lines. Check connections at each valve and component, as well as loose housing and mounting bolts. Finally, check shafts and cylinder rods. At each stage of the inspection, note any signs of abuse from improper operation, maintenance, or repair of the machine. This would include the more obvious signs of damage, such as nicks and dents in cylinder rods, as well as less obvious damage from improper repairs, such as pipe wrench marks, overtightened fittings, and misaligned components. Finally, the machine is started, warmed up, and cycled so the symptoms of the problem can be described by the operator.

After making the initial inspection, the symptoms are listed. For example,

1. Hydraulic cylinder moves too slow.
2. Operation is erratic.
3. Machine makes strange noises.
4. Machine runs hot.
5. Pumps have short service life.

Every problem has at least one symptom, but the symptom is not the problem. Rather, the symptom is used to find the problem. Troubleshooting relates known symptoms with the unknown problems. But because some symptoms can lead to more than one problem, the process of elimination must be used to decide which problems are most likely, and those are put in rank order for testing, starting with those that require the least time and expense.

Basically, then, the process of troubleshooting goes like this: With the service manual, service history of the machine, and list of symptoms, a few basic calculations are made to verify that the problem can be classified as one of five general types. Then a list of specific problem statements that can be tested are written down. Those that test positive are proven and accepted, while those that test negative are checked off and rejected. Then to be absolutely sure, those problems that test positive are verified by further testing and tearing down the machine for visual inspection, measurement, and comparison with manufacturer's specifications. The troubleshooting process is shown in Fig. 14-1.

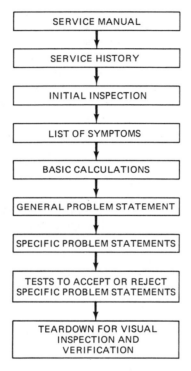

Figure 14-1 Troubleshooting process.

Problem statements can be original, but this is time consuming and a number of problems can be overlooked simply because the list is too long to remember. The accepted practice is to review procedure charts and match the symptoms with those given under one of the basic types of problems. Then the specific problem is fitted to one of the statements given for these symptoms. The problem statements are then listed and tested to find out if they identify the problem.

To be a bona fide problem, then, requires testing, and if the test proves to be positive, that the cause can be verified visually when the machine is torn down. Thus, the written problem statements must have the quality of being tested, because if they cannot, there is no basis for finding out what is wrong with the system. For example, the symptom given in Fig. 14-2 that the hydraulic cylinder moves too slow is classified first as a flow problem. And then a number of specific causes for low flow are written as problem statements. Notice that each one can

Symptom	General problem	Specific problem statements
Hydraulic cylinder moves too slow	Flow (low)	1. Restriction in the pressure line 2. Fluid level low 3. Oil viscosity too high 4. Inlet filter or line plugged 5. Air leak in the suction line 6. Pump running too slow 7. Oil bypassing relief valve 8. Low or high pressure 9. Cylinder bypassing fluid 10. Worn pump

Figure 14-2   How to organize problem statements.

be tested, and that they are put in the order of easiest to most difficult to test. The order is important because testing takes time, and tearing down the machine is not only expensive, but stops the diagnosis until it is put back together. Thus problem statements are listed in the order that will require the least time and expense, and if at all possible without disabling the machine.

Notice that the first five problem statements can be tested by careful inspection and listening when the machine is running, for the tell-tale whine that occurs when air is drawn into an inlet line. And even though one would have to ask when the inlet filter was last changed, replacing it normally would not involve an extended time or expense if it were suspected to be the problem, and the machine would still be operable. Number 6 could be checked with a strobe light tachometer, as could the coupling and shaft keys. Number 7 could be checked by feeling the pressure relief valve for heat generated when fluid flows across it and listening when the machine is stalled for changes in the sound. This does not eliminate the pressure relief valve from the list, but it is enough for a first test. A pressure gauge could also be inserted in the line to determine if the load resistance is binding. And that leaves Number 8, Number 9, and Number 10 to be tested to determine whether the problem is in the cylinder or in the pump. At this point some basic calculations should be made before breaking the connections at the cylinder or pump to insert test equipment, and both of these should be done before the machine is torn down for visual verification because this step will take an extended period of time and disable the machine.

When connections are broken or the machine is torn down, care should be taken not to introduce dirt in the system. Mining, quarry, mobile, and production equipment gathers dirt from the surroundings, and unless special precautions are taken, the system will contain more dirt after it is worked on than before. For small components, wiping fittings and lines in the area is a must, followed by cleaning with an aerosol solvent and blowing clean and dry with air. Larger teardown will require cleaning with steam or pressure washer equipment before air cleaning and drying, but it must be remembered that if solvent or water is used, the machine must be dried before tearing it down. Solvents and water are just as damaging as dirt when introduced in the system.

Finally, remember that "quick fix" solutions usually don't last. Avoid the inclination to do unnecessary work such as removing components or tearing the machine down just to put the system back in service. This is the wrong approach. Not only can this be time consuming and expensive, but more harm than good may result. Replacing a failed pump, for example, could damage the new unit if the fluid were contaminated with debris from the failed unit, even if the pump is a major part of the problem. Instead, test all the probable problem statements to gain the total picture of what the problem is and how it should be corrected.

14-2 WRITING PROBLEM STATEMENTS

Problem statements have three distinct qualities. They are

1. Concise.
2. Relate to a specific function.
3. They can be measured and compared with manufacturer's specifications.

Concise problem statements are short and to the point. They contain at least a subject and verb. The subject names the component, function, or system output, and the verb describes the condition indicated by the symptom. For example, under the general problem area of pressure, the following specific problem statements could be written:

1. System pressure is low.
2. Pressure fluctuates.
3. Pressure spikes when control valve is shifted.
4. Pressure drops when machine warms up.

Notice that each of the problem statements is concise, relates to a specific function, which in this case is pressure, and can be measured with a pressure gauge and compared to specifications given by the manufacturer or calculated by knowing the value of the load resistance.

14-3 PROBLEM STATEMENTS FOR THE SYSTEM

The time required to put a list of problem statements together can be reduced by looking over reference problem statements and selecting those that best fit the symptoms the present system has. For convenience, these are grouped under the five general types of problems. Each of the general categories of problems statements gives an overall view of what can cause this type of problem. And when the problem statements are looked at separately, they pinpoint exactly where in the system such a problem can be found. Following, then, is a list of problem statements under each of the five basic general types of problems that occur in hydraulic systems:

Pressure Related Problem Statements

NO PRESSURE

1. Faulty pressure gauge.
2. Complete pump failure.
3. Circuit open to the reservoir.
4. Motor or pump coupling failure.

LOW PRESSURE

1. Inaccurate gauge.
2. Load resistance less than expected.
3. Pressure relief valve set low or leaking.
4. Unloading valve set too low or leaking.
5. Pressure reducing valve set too low or bypassing.
6. Worn pump.
7. Variable displacement pump compensator yoke mechanism inoperative.

HIGH PRESSURE

1. Strained pressure gauge.
2. Load resistance binding.
3. Load resistance higher than expected.
4. Undersized actuators.
5. Pressure relief valve set too high.
6. Unloading valve set too high.
7. Pressure reducing valve set too high.
8. High-low valve inoperative.
9. Pressure compensator setting on variable displacement pump set too high.
10. Variable displacement pump compensator yoke mechanism inoperative.
11. Restriction in the line between the pump and pressure relief valve.

Flow Related Problem Statements

NO FLOW (FROM THE PUMP)

1. Low fluid level.
2. Plugged inlet filter.
3. Inlet line broken.
4. Pump drive coupling sheared.
5. Pump rotation reversed.
6. Pump mechanical failure.

7. Pump not primed.
8. Broken directional control valve.
9. Flow bypassing at the relief valve.

LOW FLOW

1. Low fluid level.
2. Partially clogged inlet filter.
3. Clogged inlet vent.
4. Inlet line leaking.
5. Low flow control valve setting.
6. Relief valve not fully closing.
7. Internal leak in the system.
8. Low setting on variable displacement pump.
9. Variable displacement pump yoke not shifting.
10. Pilot pressure to variable displacement pump missing or low.
11. Low pump speed.
12. Worn pump.
13. Control valve not fully shifting.
14. Restriction in the line between pump and actuator.

EXCESSIVE FLOW

1. Oversized pump.
2. Undersized actuators.
3. High setting on flow control valve.
4. Overstroking of the yoke mechanism on variable displacement pump.
5. High pump speed.

Leakage Problem Statements

LINES AND FITTINGS

1. Broken line.
2. Cracked port or fittings.
3. Fatigue from hydraulic shock or vibration.
4. SAE/ISO straight thread O-ring fitting parts improperly assembled or damaged.
5. Fittings too loose or overtightened.
6. Improperly paired fittings.
7. Tapered fittings don't have sealing tape or compound.
8. O-ring seal pinched, rolled, cut, or nibbled.

1. O-ring seal pinched, rolled, cut, or nibbled.
2. Sealing surface damaged.
3. Gland misaligned or overtightened.
4. Seal has hardened and set, losing resilience.
5. Misaligned flange does not contain the seal.

DYNAMIC SEALS

1. Seal fails from normal wear.
2. V-ring (gland) needs adjusted.
3. Pressure activated seals hardened and set.
4. Incorrect seal for application.
5. Garter ring broken on lip seal.
6. Seal undersized and extruded.
7. Seal in backwards (not facing pressure).
8. Seal improperly installed without thimble or driving tool.
9. Shaft wobbling from wear, runout, or bend.
10. Shaft worn, notched, or has rough finish.
11. External drain plugged on pump/motor, blows seal.

Excessive Heat Problem Statements

FLUID

1. Pressure relief or unloading valve pressure set too high.
2. Fixed displacement pump too large for the application.
3. Circuit dumps excessive fluid over the pressure relief valve.
4. Pump or motor overloaded.
5. Excessive pump, motor, or cylinder slippage.
6. Low fluid level.
7. Air in the fluid.
8. Fluid viscosity too high.
9. Transmission lines undersized.
10. Heat exchanger undersized or restricted.

COMPONENTS

1. High viscosity fluid.
2. Excessive pressure.
3. Excessive slippage caused by wear.
4. Cavitation (fluid starvation).
5. Excessive speed.

6. Mechanical interference (metal to metal contact).
7. Misalignment.
8. Warn or damaged bearing.

Noise and Vibration Problem Statements

HYDRAULIC NOISES

1. Pump starved for fluid.
2. Cavitation.
3. Air leak in the suction side of the pump.
4. Pressure relief valve or component sticking.
5. Fluid noise across broken or notched valve seat.
6. Air in the fluid.
7. Fluid noise across restriction.
8. Fluid to viscous (or cold).

SYSTEM NOISES

1. Pump or motor is failing.
2. Hydraulic transmission lines rattle.
3. Pump/motor coupling slaps from slack or being loose.
4. Electric motor chatters or whines.
5. Fan housing chatters.
6. Power supply mounting transmits noise.
7. Broken valve spring allows chatter.

SYSTEM VIBRATION

1. Hydraulic transmission lines vibrate or pound.
2. Electric motor coupling vibrates.
3. Electric motor vibrates.
4. Cooling fan out of balance.
5. Pump/motor coupling out of alignment.
6. Electric motor/pump shaft bent.

14-4 PROBLEM STATEMENTS FOR COMPONENTS

In addition to problem statements that describe problems associated with pressure, flow, heat, leakage, noise, and vibration, there are a number of problems that can be associated with specific components. These failures are peculiar to the components themselves and are listed with a number of problem statements for each that describe the probable cause of the problem.

Pumps

PUMP WON'T TURN

1. Bearing seized.
2. Drive motor seized.
3. Internal parts seized or broken.
4. Varnish build-up between parts.

NO PRESSURE (ALSO SEE PRESSURE SECTION)

1. Improper assembly.
2. Pump shaft sheared inside case.
3. Broken internal parts.
4. Vanes stuck in slots.
5. Fluid supply obstructed.
6. Pump turning dry.
7. Yoke mechanism in wrong position.
8. Pump turning in wrong direction.

LOW PRESSURE (ALSO SEE PRESSURE SECTION)

1. No load resistance.
2. Improper assembly.
3. Internal parts damaged from running dry.
4. Excessive internal wear.

NO FLOW (ALSO SEE FLOW SECTION)

1. Pump shaft sheared inside case.
2. Complete pump failure.
3. Pump loses prime.
4. Pump turning in wrong direction.
5. Pump turning dry.
6. Improper assembly.
7. High inlet head.
8. Broken internal parts.

LOW FLOW (ALSO SEE FLOW SECTION)

1. Excessive internal wear.
2. Pressure port connected to inlet or drain.
3. Inlet or outlet restrictions.
4. Variable displacement pump setting low.
5. Faulty pilot operator or stroking mechanism.

LEAKS (ALSO SEE LEAK SECTION)

1. Loose case bolts.
2. Static seals set, dissolved (incompatible with fluid).
3. Normal seal wear.
4. Bearing failure causes seal wear.
5. Seal installed backward or incorrectly (doesn't face pressure).
6. Shaft damaged or subject to abrasion.
7. Case drain obstructed.
8. Cracked case.
9. Misaligned flange.

BEARING FAILS

1. Normal wear.
2. Installed incorrectly.
3. Coupling misalignment.
4. Applied end force through coupling.
5. Overhung load.

Motors

MOTOR WON'T TURN

1. Load resistance seized.
2. Internal parts broken.
3. Return line restricted (quick disconnected uncoupled).

LOW TORQUE FROM THE MOTOR

1. Pressure setting low.
2. Load resistance too high for motor size.
3. Misalignment causes binding.
4. Undersized plumbing causing pressure drops.
5. Excessive wear allows internal leakage.

LOW SPEED FROM THE MOTOR

1. Flow control setting too low.
2. Fixed displacement pump too small for application.
3. Motor displacement too large for application.
4. Variable displacement mechanism setting too low or blocked.
5. Undersized plumbing reduces flow from pump.
6. Excessive wear allows internal leakage.
7. Pump speed too slow.

Valves

PRESSURE RELIEF/PRESSURE UNLOADING/PRESSURE REDUCING

1. Spool or poppet sticks open or closed on debris.
2. Pressure gauge giving inaccurate information.
3. Pilot or return spring broken or bent.
4. Valve set incorrectly.
5. Valve assembled incorrectly.
6. Valve body or parts damaged.

CHECK VALVES

1. Valve check reversed.
2. Valve check stuck open.
3. Incorrect valve size or spring for application.
4. Hydraulic shock has damaged internal parts.
5. Valve leaks from normal wear.
6. Worn plug, ball, spool, or poppet prevents leak tight seal due to misalignment.

DIRECTIONAL CONTROL VALVES (FAILS TO SHIFT, SHIFTS SLOW)

1. Burr or debris blocking valve spool.
2. Valve spool is silted.
3. Valve solenoid malfunctions or is burned out.
4. Oil viscosity is too high.
5. Ports connected incorrectly.
6. Valve assembled incorrectly.
7. Valve spool reversed.
8. Pilot pressure is low.
9. Pilot drain is blocked.
10. Faulty electric solenoid control.
11. Solenoid overheats (high spool force or internal short).

Cylinders

WON'T EXTEND OR RETURN LOAD RESISTANCE

1. Load resistance too great.
2. Load resistance binding.
3. Cylinder undersized.
4. Cylinder rod overextended (cocked).
5. Cylinder barrel bent, binding piston.
6. Blown piston seal.

MOVEMENT TOO SLOW

1. Dirty filter/reservoir vent.
2. Oversize cylinder bore.
3. Undersize pump.
4. Incorrect flow control setting.
5. Restriction in the line.
6. Directional control valve not shifting completely.
7. Fluid viscosity too low (from selection or overheating).
8. Inappropriate circuit.

MOVEMENT TOO FAST

1. Undersize cylinder bore.
2. Oversize pump.
3. Incorrect flow control settings.
4. Overrunning load resistance.
5. Inappropriate circuit.

ERRATIC MOVEMENT

1. Dirty filter/reservoir vent.
2. Air in the oil causing sponginess.
3. Load resistance binding/releasing.
4. Directional control valve chatters.
5. Low pressure relief valve setting.
6. Rod seal ingests air on return stroke (load returned system)

DRIFT

1. Internal leak past piston.
2. External leak at rod seal, fittings, or lines.
3. Leak past directional control valve spool.
4. Directional control valve not centering.
5. Relief valve leaking across seat.

14-5 BASIC CALCULATIONS

Basic calculations verify which of the five general types the problem is. For example, if the symptom is that the actuator will not move the load resistance, there appears to be a pressure problem. And further, the pressure could be too low in the system, or correct in the system but the load resistance too high at the output. To verify what is happening, the pressure will ultimately have to be measured with a gauge, but before this step is taken, some basic calculations should be made to establish what the expected pressures should be. This step gives a better understanding of the problem.

The manual should be consulted to see if this is given by the manufacturer, but in many cases what is listed will be the load capacity, displacement of the pumps, actuators, and other components, and the pressure setting of the relief valve.

The easiest calculations to perform are those for pressure, flow rate, and leakage; but just as important are those for temperature rise in the system and noise and vibration. Noise and vibration are usually isolated and measured directly.

Pressure calculations are derived from the design force placed on linear acting cylinders, or the torque expected from rotating motors. The bore diameter and load, for example, can be used to calculate expected pressures to move the load resistance. The same is true for motors, except that torque rather than linear force is used to make the calculation.

For cylinders, the relationship of pressure to force is given by $p = \dfrac{F}{A}$; whereas for torque, the basic equation comes from the torque-horsepower formula:

$$T = \frac{\text{FHP} \times 5252}{N} = \frac{p \times Q}{1714} \times \frac{5252}{N}$$

and

$$T = 3.06 \times p \times \frac{Q}{N} \tag{14-1}$$

where T = torque (lbf-ft)

p = pressure (lbf/in.^{2})

Q = flow rate (gal/min)

N = pump or motor speed (rev/min)

and the constant 3.06 converts the units of pressure, flow, and speed to torque in the units lbf-ft. This is the basic equation for a rotating unit. If the pump or motor is not rotating, stall or breakaway torque is derived by incorporating the relationship

$$\frac{Q}{N} = \frac{V_d \times N}{N} = \frac{V_d(\text{in.}^3/\text{rev}) \times 1/231(\text{gal/in.}^3) \times N}{N}$$

so that

$$\frac{Q}{N} = \frac{V_d}{231}$$

Thus, when the displacement of the pump is measured in in.³/rev., Eq. 14-1 becomes

$$T = 3.06 \times p \times \frac{V_d}{231}$$

$$T = 0.013 \times p \times V_d \qquad (14\text{-}2)$$

Theoretically, the effects of flow loss are ignored in the calculation, and torque is a function only of the pressure and displacement V_d. In practice the torque curve is lower than the computed value, from breakaway through the first 100 rev/min, and then remains nearly flat until the speed reaches the limits of volumetric efficiency. Here the torque curve droops because flow losses reduce the pressure. Figure 14-3 illustrates torque curves for three typical motors.

Gerotor motors are an exception to the general rule because of the torque multiplication which is inherent to the design. The gerotor ring always has one more lobe than the gerotor. Thus, a 6-lobed gerotor ring paired with a 5-lobed gerotor would produce five times more torque than a conventional motor at one-fifth the speed. Gerotors are particularly well adapted to power steering applications.

Most high-speed motors are oversized to compensate for the loss in torque efficiency at breakaway, and this can be worked through the calculation to account for expected losses. As a rule of thumb, stall torque efficiency is given as 15%–20% below the mechanical efficiency for the particular type of motor (gear, gerotor, vane, or piston). Finally, the computed value is used to evaluate the

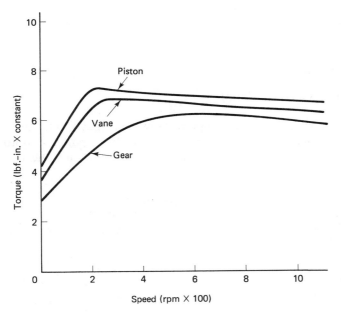

Figure 14-3 Relative torque curves for gear, vane and piston motors of same displacement at slow speeds.

Troubleshooting Hydraulic Systems   Chap. 14

condition of the motor, compared with the torque value necessary to move the load resistance at breakaway. Or, by rearranging Eq. 14-2, the value of the required pressure can be computed for a motor when the torque to start the load is known. Again, typical torque curves for gear, vane and piston motors are given in Fig. 14-3.

Example 1

A gear motor with a displacement of 1.5 in.^{3}/rev with the pressure set at 1800 lbf/in.^{2} stalls sometimes. The mechanic has a torque arm that can be fitted to the coupling, but before breaking the coupling to make the test, compute the value that might be expected.

Solution Fig. 14-4 illustrates the proposed test set-up. The expected breakaway torque from the motor is computed from Eq. 14-2.

$$T = 0.013 \times p \times V_d = (0.013)(1800 \text{ lbf/in.}^2)(1.5 \text{ in.}^3/\text{rev}) = 35.1 \text{ lbf-ft}$$

If stall torque efficiency is assumed to be 20% less than mechanical efficiency, which for a gear motor can be estimated at 90%, the expected torque from the test would probably be no less than 70% of the computed value using Eq. 14-2, or approximately 24.6 lbf-ft.

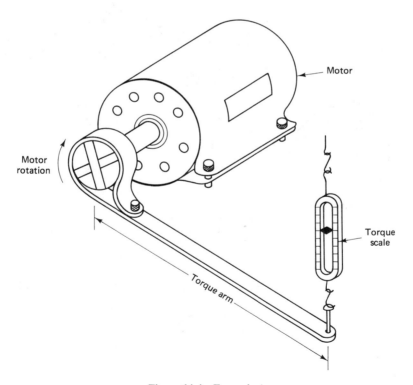

Figure 14-4 Example 1.

Example 2

The chain driven conveyor illustrated in Fig. 14-5 requires 80 lbf-ft of torque to start and bring the system to operating speed. If the drive comes from a hydraulic motor with a displacement of 3.5 in.^{3}/rev and a mechanical efficiency of 95%, what minimum pressure at the motor would develop the required torque?

Solution   From Eq. 14-2,

$$p = \frac{T}{0.013 \times V_d} \times \frac{1}{75\%} = \frac{80}{0.013 \times 3.5} \times \frac{1}{0.75} = 2344 \text{ lbf/in.}^2$$

Similar values for both torque and pressure can be read from Table 14-1.

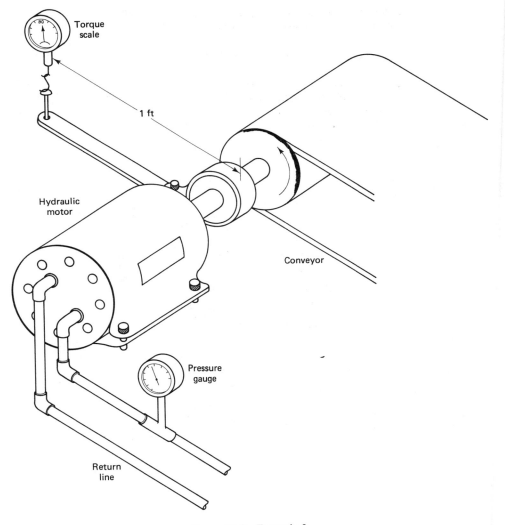

Torque scale

1 ft

Hydraulic motor

Conveyor

Pressure gauge

Return line

Figure 14-5   Example 2.

TABLE 14-1 TORQUE OUTPUT IN lbf-ft FOR VARIOUS DISPLACEMENT MOTORS AT GIVEN FLOW RATES AND PRESSURES

Flow @ 1200 rev/min Gal/min (l/min)	Displacement in.³/rev (cm³/rev)	Pressure levels							
		500 lbf/in.²	750 lbf/in.²	1000 lbf/in.²	1250 lbf/in.²	1500 lbf/in.²	2000 lbf/in.²	2500 lbf/in.²	3000 lbf/in.²
2.5 (9.46)	0.481 (7.88)	3.18	4.78	6.37	7.96	9.56	12.74	15.92	19.12
3.0 (11.36)	0.578 (9.46)	3.83	5.74	7.66	9.57	11.48	15.31	19.14	22.96
4.0 (15.14)	0.770 (12.62)	5.10	7.65	10.20	12.75	15.30	20.40	25.50	30.60
5.0 (18.93)	0.963 (15.78)	6.38	9.57	12.76	15.95	19.13	25.51	31.89	38.27
8.0 (30.28)	1.54 (25.24)	10.20	15.30	20.40	25.50	30.60	40.80	51.00	61.20
10 (37.85)	1.93 (31.54)	12.74	19.13	25.50	31.88	38.25	51.00	63.75	76.50
12.5 (47.32)	2.41 (39.43)	15.94	23.90	31.87	39.84	47.81	53.74	79.68	95.62
15 (56.78)	2.89 (47.32)	19.14	28.71	38.28	47.85	57.42	76.57	95.71	114.85
17.5 (66.24)	3.37 (55.20)	22.32	33.48	44.64	55.80	66.96	89.28	111.60	133.92
20 (75.71)	3.85 (63.09)	25.50	38.25	51.00	63.75	76.50	102.00	127.50	153.00
25 (94.64)	4.81 (78.86)	31.86	47.79	63.72	79.65	95.58	127.43	159.29	191.15
40 (151.42)	7.70 (126.18)	51.0	76.5	102.0	127.5	153.0	204.0	255.0	306.0
50 (189.27)	9.63 (157.72)	63.8	95.7	127.6	159.5	191.3	255.1	318.9	382.7
75 (283.91)	14.40 (236.59)	95.4	143.1	190.8	238.4	286.1	381.5	476.9	572.3
100 (378.54)	19.25 (315.4)	127.5	191.3	300.0	318.8	382.5	510.0	637.5	765.0

For extending and retracting cylinders, the basic calculation for flow rate is made from

$$Q_{cyl} = \frac{\text{Cyl Area} \times \text{Rod Velocity} \times 60 \text{ sec/min}}{231 \text{ in.}^3/\text{gal}} = 0.26 \times A_c \times v_c$$

where Q_{cyl} = flow rate (gal/min)

v_c = cylinder rod velocity (in./sec)

and 0.26 converts seconds to minutes and in.^{3} to gallons.

For a rotating pump or motor,

$$Q_{motor/pump} = \frac{V_d \times N}{231}$$

where Q = flow rate (gal/min)

V_d = pump or motor displacement (in.^{3}/rev)

N = speed of rotation (rev/min)

and the constant 231 converts in.^{3} to gallons.

Whether the flow rate calculation is for theoretical flow Q_t or actual flow Q_a depends upon the direction of flow. As fluid flows from the pump to the actuator, there are losses because of internal leakage, and less fluid powers the actuator than was displaced by rotation of the pump. Thus, the flow rate Q_t calculated from the product of the pump displacement and speed of rotation will be more than the actual flow rate Q_a calculated from the product of the motor displacement and speed of rotation, with the difference between the two attributed to internal leakage. That is,

$$\text{Leakage} = Q_t - Q_a$$

Or, expressed in terms of volumetric efficiency e_v,

$$e_v = \frac{Q_a}{Q_t} \times 100$$

Example 3

A hydraulic pump with a displacement of 0.75 in.^{3}/rev is used to drive a hydraulic motor with a displacement of 0.57 in.^{3}/rev. If the pump turns at 1200 rev/min and internal leakage through the pump, motor, and control valve accounts for 0.50 gal/min, (a) what is the expected speed of the motor, and (b) what is the volumetric efficiency of the system?

Solution (a) The theoretical flow from the pump is computed from

$$Q_{pump} = \frac{V_d \times N_{pump}}{231} = \frac{0.75 \text{ in.}^3/\text{rev} \times 1200 \text{ rev/min}}{231 \text{ in.}^3/\text{gal}} = 3.9 \text{ gal/min}$$

And the actual speed of the motor equals

$$N_{motor} = \frac{231 \times Q_{pump}}{V_d} = \frac{231 \text{ in.}^3/\text{gal} \times (3.9 \text{ gal/min} - 0.50 \text{ gal/min})}{0.57 \text{ in.}^3/\text{rev}}$$

$$= 1378 \text{ rev/min}$$

(b) The volumetric efficiency of the system is computed from

$$e_v = \frac{Q_a \times 100}{Q_t} = \frac{(3.9 \text{ gal/min} - 0.50 \text{ gal/min})}{3.9 \text{ gal/min}} \times 100 = 87.2\%$$

Example 4

If in Example 3 the volumetric efficiency remains the same, how fast could the pump be expected to extend a single-end rod hydraulic cylinder with a 2.5-in. bore diameter?

Solution From Example 3, $e_v = 87.2\%$ so that the effective flow rate Q_a driving the cylinder would be 87.2% of the theoretical flow from the pump:

$$Q_a = 87.2\% \times Q_t = 0.872 \times 3.9 \text{ gal/min} = 3.4 \text{ gal/min}$$

This would extend the cylinder rod with a velocity equal to

$$v_{rod} = \frac{Q_a}{0.26 \times A_c} = \frac{3.4 \text{ gal/min}}{0.26 \times 4.9 \text{ in.}^2} = 2.67 \text{ in./sec}$$

The basic calculations for heat determine first how much heat the system is generating. And, then given conditions as they exist, how much additional capacity must be added to lower the temperature to acceptable limits, usually between 130°–180°F (54°–82°C). The same situation exists for cold temperature operation except that heat must be added rather than taken away from the fluid. As a rule of thumb the temperature of a gallon of fluid can be raised 1°F per hour for each watt of power added (1hp = 746 watts).

As much as 25% of the power driving the pump is converted to heat. This shows up as a fluid temperature rise in the reservoir. About half the heat is caused by fluid dumping across the pressure relief valve, with the remainder generated by pump and motor friction and flow through the directional control and pressure reducing valves. The plumbing dissipates more heat than it generates and should not cause a heat problem unless it is undersized. Finally the fluid power at the output does not generate heat because it is converted to a mechanical output.

One way to check the system is to calculate the cooling capacity of the reservoir as if it were a heat exchanger and, if it doesn't equal 25% of the power output, then additional cooling capacity should be added. If the reservoir will dissipate 25% of the power input and the fluid still runs hot, then either the reservoir isn't actually cooling the fluid or more cooling capacity is needed because the system is generating additional heat. Before adding cooling capacity, however, an effort should be made to increase the cooling efficiency of the reservoir and lower the heat generated by the system.

The easiest method to increase cooling efficiency is to lower the relief valve pressure but this is limited by the working pressure, with enough margin to prevent pressure surges from cracking the relief valve unnecessarily, which will cause the system to malfunction.

The cooling capacity of the reservoir can be increased by cleaning it inside and out. Cleaning the inside will remove any sludge build-up that acts as an insulator, and raising the bottom at least 6 in. above the mounting surface will allow air circulation underneath. Finally, a circulation fan should be added where poor air movement around the power unit can reduce cooling capacity by as much as 50%.

The cooling capacity of steel oil reservoirs can be estimated from

$$Hp_{cooling} = 0.001 \times A \times (T_{max} - T_{amb}) \qquad (14\text{-}3)$$

where $Hp_{cooling}$ = cooling capacity of the reservoir (hp)

A = area of the reservoir

T_{max} = °F temperature of the oil in the reservoir

T_{amb} = °F temperature of the ambient (surrounding) air

And since 1 hp = 2545 Btu/hr, an estimate of the heat value (q) of the cooling capacity is also available in the form

$$q = 2.545 \times A \times (T_{max} - T_{amb}) \qquad (14\text{-}4)$$

These two formulas are useful to approximate (1) the horsepower and heat generated by the system, (2) the expected temperature rise in the fluid, and (3) the sizing requirement for the heat exchanger to add or remove the additional heat.

Example 5

What area reservoir would be required to reduce the temperature 75°F for a system having a 10 gal/min pump operating at 1500 lbf/in.^{2}?

Solution   If it can be assumed that the reservoir is engineered to dissipate 25% of the fluid power input

$$\text{FHP} = \frac{P \times Q}{1714} = \frac{(1500 \text{ lbf/in.}^2) \times (10 \text{ gal/min})}{1714} \times 0.25 = 2.2 \text{ hp}$$

and from Eq. 14-3,

$$A = \frac{Hp_{cooling}}{0.001 \times 75°F} = \frac{2.2 \text{ hp}}{0.001 \times 75°F} = 29.3 \text{ ft}^2$$

For a rectangular reservoir with an end cross section 32 in. × 32 in., the length would be about 4 ft.

Example 6

A 5-hp power pack is mounted on a reservoir with a surface area of 15 ft^2. If 25% of the power returns to the fluid in the form of heat, and the ambient temperature is 75°F, at what temperature would the fluid in the reservoir be expected to operate?

Solution   Rearranging Eq. 14-3,

$$(T_{max} - T_{amb}) = \frac{Hp_{cooling}}{0.001 \times A}$$

$$T_{max} = \frac{Hp_{cooling}}{0.001 \times A} + T_{amb} = \frac{(0.25 \times 5 \text{ hp})}{0.001 \times 15 \text{ ft}^2} + 75°F = 158.3°F$$

When the fluid in the reservoir runs too hot even after the system has been checked, the common practice is to add a heat exchanger to take care of the additional heat that surface radiation from cylinders, components, plumbing, and the reservoir cannot dissipate. With the system running hot, the operating temperature is measured and compared with an ideal operating temperature. The difference between the two is then used to calculate the heat value that must be handled by the heat exchanger.

Example 7

When a hydraulic system comes to operating temperature, the fluid in the reservoir which has a 100 ft^2 surface area is running at 180°F. How much heat would an add-on heat exchanger have to dissipate to reduce the fluid temperature to 140°F?

Solution   Using the difference between the high temperature of 180°F and low temperature of 140°F in Eq. 14-4, the additional heat to be dissipated is approximately

$$q = 2.545 \times A \times \Delta T$$

$$q = 2.545 \times 100 \text{ ft}^2 \times 40°F = 10,180 \text{ Btu/hr}$$

or the average generation rate of 4 hp.

Under the Occupational Safety and Health Act, noise exposure levels for workers must be kept under 90 dB measured for an eight-hour shift. Measurements are taken in accordance with NFPA T3 9.70.12 using the A scale at a distance three feet from the system with a precision sound level instrument and recorder such as that shown in Fig. 14-6.

Noise consists of airborne sound waves generated by pulsations in the system. The sound given off has a frequency (pitch) and an amplitude (loudness). While it is the sound pressure due to the amplitude which causes damage to hearing, higher pitch sounds are more irritating than are lower pitch sounds at the same amplitude. The decibel A scale filters and then registers the amplitude of the sound within a range of frequencies received by the human ear. Most people experience irritation in the presence of sound levels above 90 dB, with discomfort increasing as the pitch of the sound increases. A number of common sounds and their noise levels are given in Fig. 14-7.

Most of the noise in a hydraulic system is generated by the pump and transmitted or sometimes amplified by the reservoir, hydraulic tubing, or resonant machine parts such as shrouds. The speed of the pump accounts for about half the noise from a given pump with the displacement and pressure level contributing equally to the remaining half. The noise can be reduced by increasing the displacement of the pump, thereby reducing the speed, as well as by lowering the system pressure. However, before attempting changes of this magnitude in the system, a simple checklist of those items that contribute to the sound problem should be considered first.

The checklist to isolate noise must consider that noise is generated both by the action of the fluid moving through the pump and system, for example by cavitation, as well as by mechanical generation and sound transmission from the reservoir, vibration, misaligned or loose couplings, and undamped mountings.

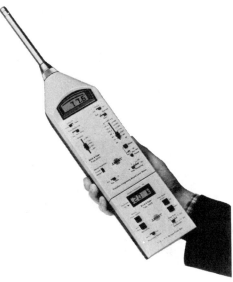

Figure 14-6 Precision Sound Level Instrument and Recorder (*courtesy of Bruel and Kjaer Instruments, Inc.*).

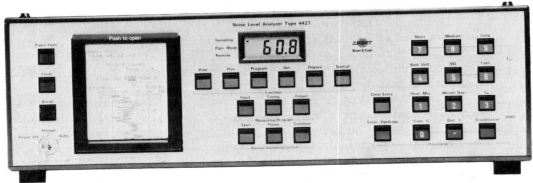

Loud noises are the easiest to identify. They also contribute the most to the additive effect of noise on the Db (A) scale. The source of lower amplitude noises should be isolated and remedied the same as others, but it should be recognized at the outset that the loudest noises contribute a disproportionate share to the amplitude reading.

The major and ever present noise in any hydraulic system is caused by the normal action of the pump as it moves fluid from the low pressure inlet to the high pressure outlet. Here the amplitude of the noise is caused by the step up in pressure, with the amplitude influenced by how abrupt the transition from low to high pressure is. One characteristic of gear pumps that use machined stock, for example, is a high noise level caused in part by unevenness in the pressure transition associated with gear teeth. Figure 14-8 lists the major pump types with the range of noise levels. Here it is seen that gear and vane pumps are the loudest, whereas the screw pump is the quietest. The wide range of noise levels is attributed both to the pressure level and the age of the pump. At the high end, the

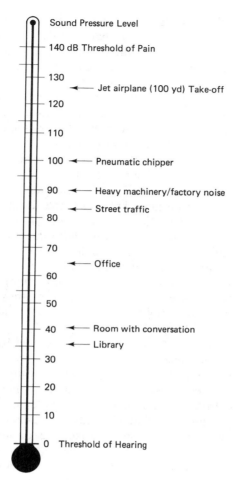

Sound Pressure Level

— 140 dB Threshold of Pain

— 130
← Jet airplane (100 yd) Take-off

— 120

— 110

— 100 ← Pneumatic chipper

— 90 ← Heavy machinery/factory noise
← Street traffic
— 80

— 70

← Office
— 60

— 50

— 40 ← Room with conversation
← Library
— 30

— 20

— 10

— 0 Threshold of Hearing

Figure 14-7 Common sounds and noise levels.

sound level is associated not only with the pressure limitation of the pump—for example 3000 lbf/in.² —but with the age and design of the particular pump in question. What has occurred over the last several years is that sound-conscious manufacturers now distribute pumps of the same type but with sound levels 15%–25% lower than their earlier counterparts. Thus, solutions to the problem of an

Pump type	Pressure range (lbf/in.²)	Noise level (dB-A scale)
Screw Type	500–1000	55–80
Vane (industrial)	1000–2000	60–80
Vane (mobile)	1000–2000	70–90
Axial Piston	1000–3000	65–85
External Gear	1000–2500	65–100

Figure 14-8 Noise levels associated with pump types.

inherently loud pump would consider not only the pressure level at which it operates, but its age, condition, and sound characteristics compared with currently available models. Assuming this is the problem, the solution may require replacing an aging pump and design with a current version that operates more quietly.

The noise from pump cavitation is caused by the violent formation and then collapse of gas bubbles in the fluid near the pump inlet. An air leak will generate a similar sound but the two phenomena are different. Cavitation can be caused by starving the inlet with the source of gas coming from the 5%–10% air entrained normally in the fluid. An air leak, on the other hand, is caused by a loose connection or break in the inlet above the fluid level. Both will destroy the pump, however, so this cause of noise should be isolated and fixed first.

Fluid related noises can also be caused by rough and abrupt transitions in the pump or plumbing as well as by restrictions that choke the flow. These should be examined closely so as not to confuse their origin with the sounds generated or transmitted by the structural members or the plumbing because of pressure ripples or hammer from quick shifting directional control valves.

Cooling fans, shrouds, and housings near hydraulic pumps and motors vibrate if they are not bolted securely. And in some cases they will have to be damped with sound deadening material to prevent vibration and noise transmission.

The alignment of the pump with the drive motor should be checked with a dial indicator. Noise from the metallic coupling can also result from looseness and variation in the angular velocity caused by rapid pressure changes. A coupling with a nonmetallic connecting segment can eliminate this source of noise. Wear under the pump or drive motor mounts is a telltale sign of misalignment problems. Tightening the mount bolts will help only if the alignment problem has been remedied. Both pump and misalignment noises are amplified by the metal surfaces of the hydraulic reservoir.

Hydraulic tubing must be mounted securely in clamped cushion supports to isolate pump and line noises from the building structure. Another common method used to isolate pump and line noises is to insert short sections of hose in the line to break the transmission medium. Where surging line pulsations are present, desurgers are inserted to reduce the source of noise. Accumulators can be used to absorb low-frequency line surges, but a desurger is usually required to absorb pump pulsations.

14-6 TEE TEST FOR HYDRAULIC CIRCUITS

The tee test is a basic procedure used for trial test problem statements. It simulates the performance of various components in the circuit while the machine is in operation and has the advantage of pinpointing the problem without tearing down the machine.

Test connections (see Figs. 14-9 and 14-10) can be made at A, B, C, or D. When the connection is made at C or D, the tester is in series with the control valve and cylinder to check individual components.

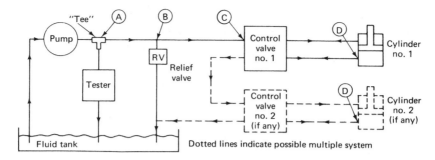

Figure 14-9 Tee test circuit.

First, basic calculations determine how much fluid should be circulating through the circuit. Then by controlling the pressure with the load valve on the tester, the amount of fluid available through each component can be determined. If the test indicates an insufficient flow when the pressure is raised to specifications for the relief valve, the cause can be pinpointed to components by verifying problem statements such as

1. Pump slipping.
2. Fluid flowing over a faulty relief valve.
3. Fluid leaking past control valve spools to the reservoir.
4. Compensation mechanism malfunctioning (pressure or flow).
5. Fluid leaking past pump or motor parts directly to return without power transfer (low volumetric efficiency).

Do not exceed relief valve pressure to prevent damage to the system. Be sure to operate the system long enough to bring the temperature of the fluid within

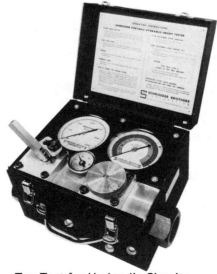

Figure 14-10 Portable hydraulic tester (*courtesy of Schroeder Brothers Corporation*).

the operating range. If the system is driven by an internal combustion engine, tests should be run at near constant rpm.

The tee test checks overall system volumetric efficiency:

Tee test	
1. Connect the tester at point A as shown in Figure 14-9	OK ___
2. Run the system until the oil temperature is within the operating range.	OK ___
3. Position the control valve to direct fluid to extend the cylinder.	OK ___
4. Close the load valve just enough to extend the cylinder.	OK ___
5. Release the load valve and record the total flow through the tester flowing at no load and low pressure to the reservoir.	___ gpm
6. Close the load valve to bring the pressure to 90% of the relief valve pressure.	___ lbf/in.^{2}
7. Record the flow rate at 90% relief valve pressure.	___ gpm
8. Compute the volumetric efficiency (e) from	

$$e_0 = \frac{\text{Flow at 90\% load}}{\text{Flow at no load}} \times 100$$

9. If the test is inconclusive, repeat the procedure applying load pressure in 500 lbf/in.^{2} increments. Do not exceed manufacturer's specifications or relief valve setting.

14-7 SUMMARY

Finding and solving problems in hydraulic systems is quicker and less costly when a logical procedure is used. The approach taken here uses the symptoms to classify the problem as one of five general types, and then pinpoint the trouble with specific problem statements that can be verified through tests and visual inspection.

To be sure the symptoms point toward one of the five general problems, basic calculations are made that compare the observed performance of the machine with the manufacturer's specifications. Once the general problem has been verified, problem statements are written and put in logical order for testing. This saves time and reduces costs associated with tearing down the machine prematurely. Those statements that test positive are accepted for further verification, while those that test negative are rejected. Finally, those problem statements that test positive are used to direct disassembly of the machine so the problem can be verified through visual inspection.

STUDY QUESTIONS AND PROBLEMS

1. List two sources of written information needed to troubleshoot a hydraulic machine problem.

2. List the six-step procedure to troubleshoot a hydraulic machine problem.

3. If the symptom is: "hydraulic cylinder will not move the load resistance," classify the problem and write four problem statements to be tested.

4. If the symptom is: "operation is erratic and pump whines," classify the problem and write four problem statements that can be tested.

5. After warming up, the fluid in a hydraulic system operates at 190°F (87.8°C). Write five problem statements, and put them in order for testing using the rules given in the text to establish their priority.

6. What basic calculations would be made for the general problem statement: "cylinder pressure insufficient to lift load resistance"?

7. What basic calculations would be made to verify the general problem statement: "lift cylinder extends too slow"?

8. A torque motor with a displacement of 4.81 in.^{3}/rev that operates at 1500 lbf/in.^{2} has been removed from a conveyor drive because it stalls at breakaway. Before disassembling or replacing the motor, it has been decided to turn the conveyor shaft coupling manually with a torque wrench to see if it is binding. If the manufacturer's specifications indicate the torque motor should develop 75% of theoretical operating speed torque at breakaway, what should be the maximum torque required to turn the conveyor shaft coupling?

9. If a wheel torque motor with a displacement of 9.62 in.^{3}/rev stalls when the load reaches 150 lbf-ft, how much would the pressure have to be increased to provide the necessary torque to turn the wheel and provide an additional 25% torque margin?

10. A pump with a displacement of 0.351 in.^{3}/rev turning at 1200 rev/min cycles a 2-in. bore cylinder with a 1-in. diameter bore rod through a 12-in. stroke 5 times each minute. What is the volumetric efficiency of the system?

11. After warming up, a hydraulic gear motor with a displacement of 1.75 in.^{3}/rev loses speed from 1000 rev/min to 875 rev/min. If the flow rate to the motor is 10 gal/min, how much fluid is bypassing through the motor, and what is the loss in volumetric efficiency?

12. If the ambient temperature is 75°F, and the oil temperature is 140°F, what is the horsepower cooling capacity of a steel reservoir with an end section 24 in. × 20 in., 50 in. long, with all sides exposed to open circulation?

13. If a hydraulic reservoir is approximately cubic and dissipates 25% of the fluid power from a 2-hp powerpack with a 50°F temperature rise through all but the top, how much fluid would the reservoir require if it were filled three-fourths full?

14. If the ambient temperature is 80°F, what temperature could the fluid be expected to reach if 30% of the horsepower from a 7.5-hp pump is returned as heat to a reservoir, 4 ft. long, with a square end section of 4 ft^2?

15. If it is determined that as much as 5 hp will be returned to the reservoir as heat, and the reservoir has the dimensions 30 in. × 30 in. × 60 in. long, what °F and °C temperature rise can be expected in the fluid?

16. How much heat in Btu can a hydraulic reservoir fully exposed to cooling on all sides and with a surface area of 50-ft^2 be expected to dissipate through a 50°F temperature rise?

17. If a 200-gal. steel hydraulic reservoir twice as long as its square end section is wide and overheats the fluid with a temperature rise of 100°F, how large would the reservoir have to be to reduce the temperature rise to 70°F?

18. When a double-acting, vertical mounted lift cylinder is stopped in mid-stroke, it slowly

drifts back, but there is no leak at the fittings or rod seal. How is it determined if the fault lies with the cylinder?

19. The oil in a steel reservoir with a surface area of 16 ft² normally operates with a 50°F temperature rise. If half the temperature rise is attributed to fluid friction losses through a 3 hp gear pump, how much additional friction loss at the pump would result in an 80°F temperature rise?

20. A three-year old forklift has been lifting slow. The circuit uses a 6 gal./min pump driven directly from the fork motor turning at 2200 rev/min and a 3-in. bore double-acting cylinder connected directly to the forklift mechanism. A manually operated tandem center control valve directs fluid through a pilot operated check valve to lift and hold the cylinder. Reversing the control valve pilot operates the check to retract the cylinder. Out of curiosity, the operator timed the lift and found it to be 1.75 in./sec when the fork was loaded. The operator claims that as the machine warms up the fork lifts faster, but never as fast as when the machine was new. Other than being slow, the machine operation is normal. Maintenance replaced the pump and changed the fluid recently in an effort to correct the problem but performance was improved only slightly. Before changing additional components, it has been decided to find out exactly what the problem is by using the following procedure:

(a) From the discussion of the problem given, make a list of the symptoms.

(b) Perform the basic calculations to verify which of the five general types the problem is.

(c) From the basic calculations that have been made, how should the *general* problem statement be written:

General Problem Statement:_____

(d) From the general problem statement, write at least five *specific* problem statements that can be tested.

(e) For each specific problem statement written in d., list one test that can be made to accept or reject them.

(f) Finally, list the order in which the machine would be disassembled to verify the cause of the problem.

CHAPTER 15

BASICS
OF PNEUMATICS

15-1 INTRODUCTION

Pneumatics provides industry with valuable and economic power for production and control. Unlike the relatively incompressible nature of hydraulic fluid, air used in pneumatics is compressible and, with only minor variations, obeys the laws for a perfect gas. Applications take into account the compressible nature of air by providing additional means to stabilize actuator movement to accurately position the load resistance. Bleed-out circuits, mechanical stops, gear reduction units driven by air-motors, and hydropneumatic systems are often used for this purpose.

Pneumatics has a variety of uses in industry. Air at 80–125 lbf/in.^{2} (550–862 kPa) is provided at construction sites and plants. In construction, it is an indispensable source of power for air drills, hammers, wrenches, and even air cushion supported structures. Also remember that vehicles use air suspension, braking, and pneumatic tires.

In manufacturing, air is used to power high-speed clamping, drilling, grinding, and assembly using pneumatic wrenches and riveting machines. Plant air is also used to power hoists and cushion supports to transport loads throughout the plant.

Operational characteristics make air a desirable medium for energy transfer. Because air can be inducted and exhausted into the atmosphere, a return line is not necessary. Air is also safe in hazardous environments from ignition and burning should leaks occur around sparks or open flames.

Because air contains oxygen (about 20%) and is not sufficient alone to provide adequate lubrication of moving parts and seals, oil is usually introduced into the air stream near the actuator to prevent excessive wear and oxidation.

15-2 STANDARD AIR

Air has mass and exerts pressure on the surface of the earth. A barometer, consisting of an inverted tube closed at the top, will support a column of mercury (Hg) at exactly 760 mm (29.92 in.) at sea level when measuring standard conditions at 32°F. Since Hg has a Sg of 13.5951, this is the equivalent of (29.92 × 13.5951/12) 33.897 ft of water or (0.433 × 33.897115) 14.677 lbf/in². Values for standard atmospheric pressure are usually accepted as 34 ft of water and 14.7 lbf/in.² abs. Figure 15-1 illustrates the construction of a simplified mercury barometer and lists common units for standard atmospheric conditions. Gauge pressures above atmospheric conditions are read as positive gauge pressures, whereas pressures below atmospheric conditions are read with a negative sign. By convention it is understood that positive gauge pressures do not include local atmospheric conditions, and when absolute pressure is used to solve problems, it is signified by p abs.

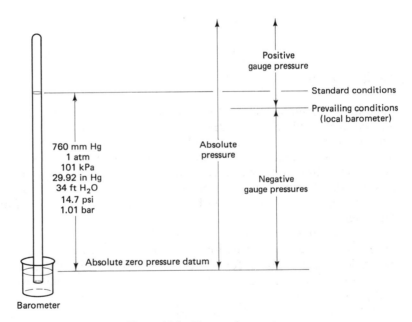

Figure 15-1 Mercury barometer.

Pressures above 14.7 lbf/in.² are positive, whereas pressures below 14.7 lbf/in.² abs cause a vacuum to be formed. Both positive pressures and vacuum pressures have useful purposes in pneumatics. Vacuum measurement is usually given in in. of Hg and then converted into a holding force for such devices as suction pads and cylinders with a specific diameter. A vacuum less than 1-in. Hg is measured in microns, where

1 micron = 0.000 000 1 meter = 0.001 mm

and

$$0.001 \text{ mm} = \frac{0.001 \text{ mm} \times 13.5951 \times 0.433 \text{ lbf/in.}^2 \text{ ft}}{25.4 \text{ mm/in.} \times 12 \text{ in./ft}} = 0.000 \ 019 \ 3 \text{ lbf/in.}^2$$

Pressure in gases is thought to result from activity between atoms. As the temperature decreases, molecular activity also decreases until at absolute zero, $-460°F$ (zero degrees Rankine) or $-273°C$ (zero degrees Kelvin), it is thought to cease altogether. For calculations involving gases, absolute temperature can be defined as

$$\text{Absolute Temperature} \ °R = °F + 460 \tag{15-1}$$

or

$$\text{Absolute Temperature} \ °K = °C + 273 \tag{15-2}$$

The temperature conversion from °F to °C is accomplished with the formula

$$°C = \frac{5}{9} \ (°F - 32) \tag{15-3}$$

and from °C to °F with

$$°F = \frac{9}{5} \ °C + 32 \tag{15-4}$$

While barometric changes caused by altitude, weather conditions, and relative humidity affect atmospheric conditions slightly (about 2½% per 1000 ft), this too must be considered when computing the air consumption of components and the necessary delivery from the compressor. Standard air and air flow (std ft³/min) are defined at a temperature of 68°F, a pressure of 14.7 lbf/in.² abs, and a relative humidity of 36% (0.0750 density). This is in agreement with definitions adopted by ASME, although in the gas industries the temperature of standard air is usually given as 60°F.

15-3 THE GAS LAWS

The compressible nature of air is governed by the laws for a perfect gas. While air itself, composed of about 80% nitrogen and 20% oxygen by volume, is not a perfect gas, deviations do not affect computed results much.

Boyle's Law

Boyle's Law relates the absolute pressure (p) and volume (V) of a given quantity of gas held at constant temperature such that

$$p_1 V_1 = p_2 V_2 \tag{15-5}$$

Where the desired quantity is (p_2) or (V_2), Eq. 15-5 is rearranged.

Example 1

If an empty 20 gal (75.7 l) water system tank on which the pressure gauge initially reads 20 lbf/in.^{2} (137.9 kPa) is half filled with water, such as that shown in Fig. 15-2, what will be the pressure reading on a gauge attached to the tank?

Solution From Boyle's law, Eq. 15-5

$$p_2 = \frac{p_1 V_1}{V_2}$$

and

$$p_2 = \frac{(20 \text{ lbf/in.}^2 + 14.7 \text{ lbf/in.}^2) \times (20 \text{ gal} \times 231 \text{ in.}^3/\text{gal})}{(10 \text{ gal} \times 231 \text{ in.}^3/\text{gal})} = 69.4 \text{ lbf/in.}^2 \text{ abs}$$

The gauge pressure would be

$$p_2 = 69.4 - 14.7 = 54.7 \text{ lbf/in.}^2$$

In SI units

$$p_2 = \frac{(137.9 \text{ kN/m}^2 + 101 \text{ kN/m}^2) \times (75.7 \text{ l} \times 10^{-3} \text{ m}^3/\text{l})}{(37.85 \text{ l} \times 10^{-3} \text{ m}^3/\text{l})}$$

$$= 478 \text{ kPa abs (377 kPa gauge)}$$

Charles' Law

Charles's law relates the absolute temperature (T) and volume (V) of a given quantity of gas held at constant pressure such that

$$\frac{T_1}{T_2} = \frac{V_1}{V_2}$$

and

$$T_1 V_2 = T_2 V_1 \tag{15-6}$$

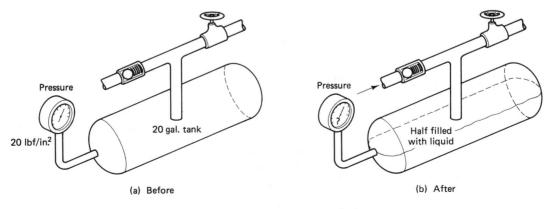

(a) Before   (b) After

Figure 15-2 Example 1.

Where the desired quantity is usually (T_2) or (V_2), and the temperature is degrees absolute—that is, (°F + 460) or (°C + 273).

Example 2

If an accumulator using a dead weight ballast against an initial volume of 1500 in.^{3} of a gas is heated from 80°F to 200°F, what volume will the heated gas occupy?

Solution   Refer to Fig. 15-3. From Charles' law, Eq. 15-6

$$V_2 = \frac{T_2 V_1}{T_1}$$

and

$$V_2 = \frac{(200°F + 460) \times (1,500 \text{ in.}^3)}{(80°F + 460)} = 1833.3 \text{ in.}^3$$

Using the Celcius temperature scale

$$V_2 = \frac{(93.3°C + 273)(1500 \text{ in.}^3 \times 16.39 \text{ cm}^3/\text{in.}^3)}{(26.67°C + 273)} = 30\ 051 \text{ cm}^3$$

Gay Lussac's Law

Gay Lussac's law relates the absolute pressure and temperature of a given quantity of gas held at constant volume such that

$$\frac{p_1}{p_2} = \frac{T_1}{T_2}$$

and

$$p_1 T_2 = p_2 T_1 \tag{15-7}$$

Where the desired quantity is usually (p_2) or (T_2).

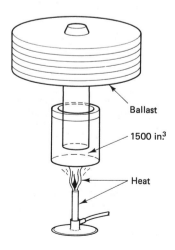

Figure 15-3   Example 2.

Example 3

The constant volume gas pressure vessel shown in Fig. 15-4 on which the pressure gauge reads 2,000 lbf/in.^{2} is heated from 80°F to 250°F. What will the gauge read?

Solution From Gay Lussac's law, Eq. 15-7

$$p_2 = \frac{p_1 T_2}{T_1}$$

$$p_2 = \frac{(2,000 \text{ lbf/in.}^2 + 14.7 \text{ lbf/in.}^2) \times (250°F + 460)}{(80°F + 460)} = 2,649 \text{ lbf/in.}^2 \text{ abs}$$

the gauge pressure is

$$p_2 = 2,649 \text{ lbf/in.}^2 - 14.7 \text{ lbf/in.}^2 = 2,634 \text{ lbf/in.}^2$$

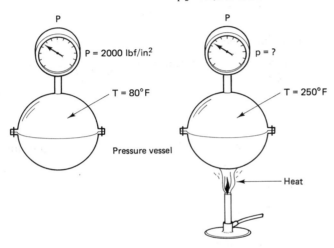

Figure 15-4 Example 3.

General Gas Law

Boyle's, Charles', and Gay Lussac's laws are combined in the general gas law and written as

$$\frac{p_1 V_1}{T_1} = \frac{p_2 V_2}{T_2} \qquad (15\text{-}8)$$

Example 4

Gas in a 1,500-in.^{3} cylinder at 2,000 lbf/in.^{2} gauge is reduced in volume to 1,000 in.^{3} while heated from 75°F to 250°F. What is the final gauge pressure in the cylinder?

Solution Refer to Fig. 15-5. From the general gas law, Eq. 15-8

$$p_2 = \frac{p_1 V_1 T_2}{T_2 V_2}$$

$$p_2 = \frac{(2,000 \text{ lbf/in.}^2 + 14.7 \text{ lbf/in.}^2) \times (1,500 \text{ in.}^3) \times (250°F + 460)}{(75°F + 460) \times (1,000 \text{ in.}^3)}$$

$$= 4,011 \text{ lbf/in.}^2 \text{ abs}$$

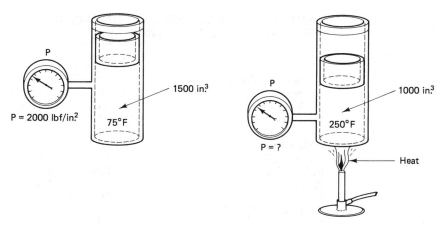

Figure 15-5 Example 4.

or

$$p_2 = 4{,}011 \text{ lbf/in.}^2 - 14.7 \text{ lbf/in.}^2 = 3{,}996 \text{ lbf/in.}^2 \text{ gauge}$$

In SI units

$$p_2 = \frac{(13\ 793 \text{ kN/m}^2 + 101 \text{ kN/m}^2)}{(23.89°C + 273)}$$

$$\times \frac{(1500 \text{ in.}^3 \times 16.39 \text{ cm}^3/\text{in.}^3 \times 10^{-6}\text{m}^3/\text{cm}^3)(121°C + 273)}{(1000 \text{ in.}^3 \times 16.39 \text{ cm}^3/\text{in.}^3 \times 10^{-6}\text{m}^3/\text{cm}^3)}$$

and

$$p_2 = \frac{(13\ 894 \text{ kN/m}^2)(0.0246 \text{ m}^3)(394°K)}{(297°K)(0.0164 \text{ m}^3)} = 27\ 648 \text{ kPa abs} (27\ 547 \text{ kPa gauge})$$

$$= 27.5 \text{ MPa gauge}$$

15-4 APPLICATION OF THE GAS LAWS

The gas laws are applied to pneumatic systems as air is compressed, directed through components, and then exhausted into the atmosphere. While air enters the compressor at atmospheric pressure and room temperature, components operate at 75–150 lbf/in.^{2} with temperatures ranging from below 70°F to 200°F. Thus, changes in pressure, temperature, and volume must be figured in to arrive at the free air displacement of the compressor. Table 15-1 lists the air requirements for various pneumatic tools at 800–125 lbf/in^2.

Because the pressure is relatively constant at the actuator—for example, through the entire stroke of a pneumatic cylinder—the air consumption at operating pressure can be computed the same as fluid consumption in hydraulic systems. The free air calculation is made after consumption at operating pressure is determined by using Boyle's Law to derive approximate values, or the general gas

TABLE 15-1 AIR REQUIREMENTS FOR VARIOUS PNEUMATIC TOOLS AT 80–125 lbf/in.^{2}

	ft^3/min	m^3/min
Air hoist (cylinder)	4	0.11
Air hammer	16	0.45
Air hoist (motor)	4	0.11
Rotary drills $\frac{1}{16}"-\frac{5}{8}"$	7	0.20
Rotary drills $\frac{1}{4}"$	20	0.57
Rotary drills $\frac{3}{8}"$	40	1.13
Rotary drills $\frac{1}{2}"-\frac{3}{4}"$	70	1.98
Rotary drills $\frac{7}{8}"-1"$	80	2.27
Piston drills $\frac{1}{2}"-1\frac{1}{4}"$	45	1.27
Piston drills $\frac{7}{8}"-1\frac{1}{4}"$	80	2.27
Piston drills $1\frac{1}{2}"-2"$	90	2.55
Piston drills $2"-3"$	110	3.11
Grinders vertical and horizontal	20	0.57
1 hp air motor	10	0.28
2 hp air motor	15	0.42
3 hp air motor	20	0.57

law if conversion to standard conditions is specified. The following example is used to clarify the calculations.

Example 5

A single-acting cylinder with a 2-in. (5.08 cm) bore and 10-in (25.4 cm) stroke operates at 80 lbf/in.^{2} (552 kPa) and cycles at a rate of 40 cycles/min. (See Fig. 15-6.) Compute the air consumption in ft^3/min of free air.

Solution The volume per minute (V_1) at 80 lbf/in.^{2} equals

$$V_1 = \frac{\text{Area} \times \text{Stroke} \times \text{Cycle rate}}{1,728}$$

$$V_1 = \frac{3.14 \times (2 \text{ in.})^2 \times 10 \text{ in./cycle} \times 40 \text{ cycles/min}}{4 \times 1728 \text{ in.}^3/\text{ft}^3} = 0.727 \text{ ft}^3/\text{min @ 80 lbf/in.}^2$$

From Boyle's law (Eq. 15-5) at constant temperature, free air consumption equals

$$p_1 V_1 = p_2 V_2$$

and

$$V_2 = \frac{p_1 V_1}{p_2} = \frac{(80 + 14.7)(0.727)}{14.7} = 4.68 \text{ ft}^3/\text{min free air}$$

If temperature correction is desirable, for example if air at the work station is delivered at 75°F and correction to 68°F is necessary, the general gas law (Eq. 15-8) is used

$$\frac{p_1 V_1}{T_1} = \frac{p_2 V_2}{T_2}$$

$$V_2 = \frac{p_1 V_1}{p_2} \times \frac{T_2}{T_1}$$

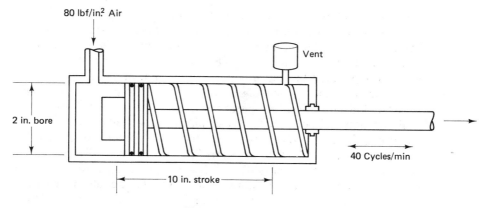

80 lbf/in² Air

Vent

2 in. bore

40 Cycles/min

10 in. stroke

Figure 15-6 Example 5.

and the correction for temperature can be made using absolute temperature units (°R)

$$V_2 = 4.68 \times \frac{T_2}{T_1}$$

$$V_2 = 4.68 \times \frac{(68 + 460)}{(75 + 460)}$$

and

$$V_2 = 4.68 \times 0.9869 = 4.62 \text{ ft}^3/\text{min free air @ 68°F}$$

In SI units

$$V_1 = (20.3 \text{ cm}^2)(25.4 \text{ cm})(40 \text{ cycles/min})(10^{-6} \text{ cm}^3/\text{m}^3) = 0.0206 \text{ m}^3/\text{min}$$

From Boyle's law (Eq. 15-5) for isothermal conditions, free air consumption equals

$$V_2 = \frac{p_1 V_1}{p_2} = \frac{(552 \text{ kN/m}^2 + 101 \text{ kN/m}^2)(0.0206 \text{ m}^3/\text{min})}{(101 \text{ kN/m}^2)}$$

and

$$V_2 = 0.133 \text{ m/min free air at } 23.9°C$$

Correcting for temperature from 23.9°C (75°F) to 20°C (68°F)

$$V_2 = (0.133 \text{ m}^3/\text{min}) \frac{(20°C + 273)}{(23.9°C + 273)} = 0.131 \text{ m}^3/\text{min at } 20°C$$

15-5 MOISTURE AND DEW POINT

Air from the atmosphere contains moisture to varying degrees depending on atmospheric conditions. On humid days, more moisture is present. The amount of moisture present in air is measured in terms of the relative humidity, which is the ratio of the amount of water actually contained in the air to the amount of water the air contains when saturated. The point of saturation is termed the dew point and is defined as that temperature at which moisture begins to condense out of the air. The lower the dew point, the lower the temperature necessary before free moisture will condense and separate out of the air stream, causing damage to transmission lines, hoses, and system components. When air is compressed, for example to 100 lbf/in.², about eight times the amount of moisture [(100 + 14.7) × 1 ÷ 14.7] is pumped through the compressor and into the system.

Figure 15-7 shows the weight in lbf/1000 ft³ of moisture carried in suspension in saturated air at various temperatures. For example, air at 80°F contains about 1.58 lbf of moisture per 1000 ft³ of free air. Compressing the air to 100 lbf/in.² and then cooling it back to 80°F lowers the moisture content to approximately 0.2 lbf which removes about 1.38 lbf or nearly 90% of the water. Thus it can be seen that compressing air (which heats it) and then cooling it back to its original temperature is an effective means to remove moisture which could be harmful to components downstream. It should be noticed from the chart, however, that continued cooling removes proportionally less water, even with an increase in pressure. If for the example given the temperature was further reduced to 60°F, the moisture content would still be 0.1 lbf/1000 ft³, only an additional 6%, but at a much higher energy cost for cooling. As a result, the rule of thumb for cooling is to maintain a 15°F temperature difference between the cooling medium and the air at the inlet of the after-cooler.

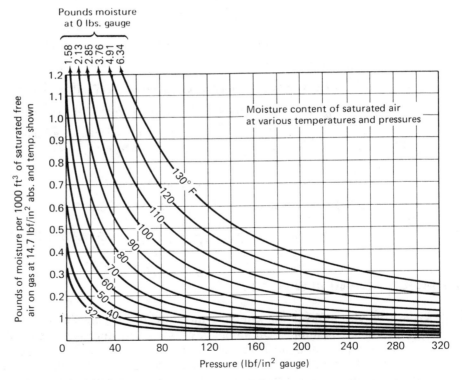

Figure 15-7 Moisture content of saturated air at various temperatures and pressures (reprinted from *The Compressed Air and Gas Handbook*).

Example 6

Saturated air at atmospheric pressure and 90°F is drawn into the inlet of a compressor and compressed to 120 lbf/in.^{2}. How much water could be expected to be removed from the compressed air if the temperature is restored to 90°F?

Solution   Look at Fig. 15-7. At the top left of the graph, saturated air at 90°F contains about 2.13 lbf/1000 ft^3. To arrive at the amount of moisture remaining after it is compressed to 120 lbf/in.^{2} and the temperature restored, the 90°F temperature curve is followed down and to the right until it intersects with 120 lbf/in.^{2} on the vertical axis where the remaining moisture is estimated to be 0.225 lbf/1000 ft^3. Thus the moisture removed would be 2.13 lbf − 0.225 lbf = 1.9 lbf/1000 ft^3 or about 89%.

15-6 CONDITIONING AIR

Conditioning compressed air makes it more acceptable as a medium to transmit power. It also makes it safer for operating personnel. Fluid conditioners for compressed air systems consist of after-coolers, moisture separators, air dryers, air receivers, filters, regulators, lubricators, and mufflers. Compressed air that has not been conditioned contains both moisture and contaminants. Moisture rusts the plumbing system and washes away lubricants in cylinders, motors, and air tools, causing erratic action and increasing wear. It also sticks solenoid valves and freezes in cold weather. All of these result in failure, downtime, and costly repairs. Airborne particles in compressed air and high noise levels caused by exhausting components directly into the atmosphere also endanger operating personnel with possible impairment to vision and hearing.

In practice, after-coolers are placed immediately downstream of the compressor and remove about 85% of the moisture drawn in and pumped through the compressor. They are a necessity for plant air systems. While several designs are available, the shell and tube-type water cooler receive widespread use because of low operating and maintenance costs. Automatic moisture traps are sometimes located between the after-cooler and the receiver or in transmission lines near drops to separate and discharge this moisture from the system before it enters the receiver.

If dry air is required by a pneumatic system—that is, air that does not release moisture at the temperature and pressures encountered by the system—dryers are installed downstream of the after-cooler and moisture trap. To remove moisture, the air may be refrigerated, heated in a furnace, or passed through desiccants such as silica gel, activated alumina, or molecular sieves which remove oil and water vapors.

Refrigerated dryers remove moisture by lowering the temperature of the pressurized air to 35–38°F, causing moisture vapor to condense. If the air piped throughout the system stays above this temperature, condensation will not occur and the remaining humidity in the airstream will pass through components without the harmful effects associated with moisture accumulation. In the refrigerated drying cycle shown in Fig. 15-8, hot, wet air from the compressor enters the air-to-air heat exchanger (1) for precooling, creating initial condensation of vapor.

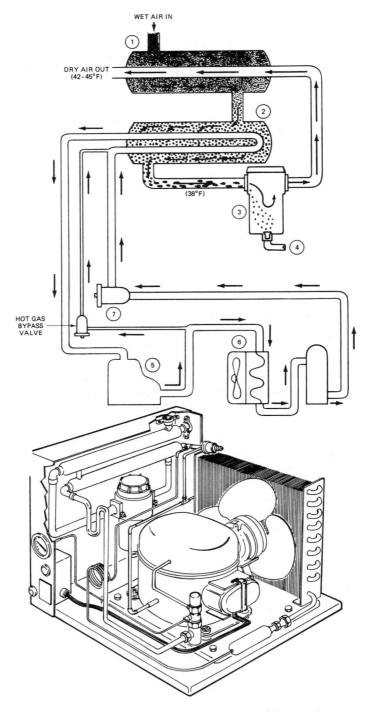

WET AIR IN

DRY AIR OUT
(42–45°F)

(38°F)

HOT GAS
BYPASS
VALVE

Figure 15-8 Refrigerated dryer cycle (*courtesy of Van-Air, Inc.*).

Final cooling is achieved in the refrigerant-to-air heat exchanger (2) as the heat from the compressed air is transferred to the cool refrigerant. The water vapor condensed by the cooling process is collected in the moisture separator (3) and discharged through an automatic drain (4). The cool dry air is then returned to the air-to-air heat exchanger where it is reheated by the hot inlet air. Reheating the air restores its original volume and eliminates pipe sweating. In the refrigeration cycle, the compressor (5) converts cool, low-pressure gas into hot, high-pressure gas. It then is pumped to the condenser (6) which cools the gas and changes it to a high-pressure liquid. Prior to the refrigerant-to-air heat exchanger, the expansion valve (7) throttles the liquid, reducing it to a low pressure. As the liquid absorbs heat from the compressed air, it gradually changes to a gas and is returned to the compressor for recycling.

Air filters are located in the compressor intake, in the pressure line before components, and immediately preceding power tools. They protect against catastrophic failure from large particles as well as gradual failure from small particles that promote wear by lapping action using dirt and other fine particulate matter as an abrasive agent. Care must be exercised that the filter does not become clogged, causing higher than normal suction at the inlet. This would starve the compressor and result in reduced volumetric and pumping efficiency. In-line filters, such as that in Fig. 15-9, which incorporate a cleanable feature initiated by turning the upper T-handle are suitable for removing dirt, rust particles, and other foreign airborne matter in the line. A bypass check valve is incorporated to route air around the filter should the element become clogged. Filters of this type remove particles down to about 25 μ from the air stream.

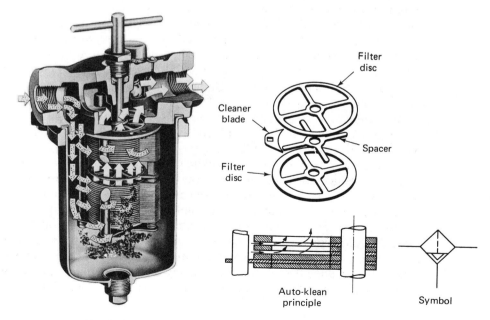

Figure 15-9 In-line cleanable air filter (*courtesy of AMF Cuno Division*).

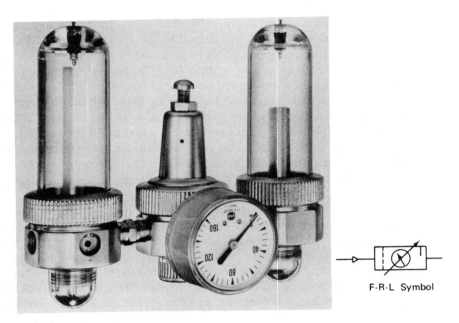

F-R-L Symbol

Figure 15-10 Filter-regulator-lubricator unit (*courtesy of Bastian-Blessing Company*).

Filter-regulator-lubricator units, such as that in Fig. 15-10, are commonly located at each workstation to condition and regulate air from the main supply line at pressures up to 150 lbf/in.^{2} and to branch circuits operating at pressures from 60 lbf/in.^{2} to 125 lbf/in.^{2}. Porous sintered metal filter elements combined with deflectors use centrifugal force to separate out free moisture which is then periodically removed by means of a drain cock located in the bottom of the filter bowl. An indicating button becomes visible when the element requires cleaning. Air from the filter unit is then passed through the regulator at a pressure set by operating or maintenance personnel and fixed by tightening the jamb nut on the T-handle adjustment screw to the lubricator where oil is added to the air stream to lubricate power tools. The amount of oil added is visible through a sight indicator and is set with a key to restrict unauthorized adjustment and tampering.

Mufflers are mounted to the exhaust ports of air valves, cylinders, and air motors to reduce noise and eliminate the possibility that the airstream or airborne particles will cause injury to personnel. OSHA requirements for permissible noise exposures are summarized in Table 15-2. To prevent the necessity of requiring employees to wear eardrum protective devices or operation of pneumatic systems from acoustical conditioned chambers, noise should be engineered out at the source.[1] Mufflers are designed to reduce noise without creating back pressure sufficient to reduce the operating efficiency of the system. The air path through a typical muffler is shown in Fig. 15-11, and indicates the cancelling effect of noise-

[1] A lengthy discussion of noise and OSHA standards is published in OSHA Bulletin No 2067.

TABLE 15-2 PERMISSIBLE NOISE
EXPOSURES

Duration per day, hours	Sound level dBA slow response
8	90
6	92
4	95
3	97
2	100
1½	102
1	105
½	110
¼ or less	115

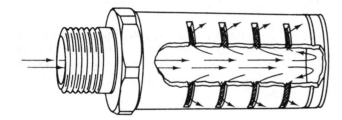

Figure 15-11 Muffler operation (*courtesy of Scovill, Fluid Power Division*).

causing vibrations.[2] Appropriate sizing for each application is necessary to assure maximum performance.

15-7 SUMMARY

Physically, air behaves approximately according to the laws for a perfect gas. These laws describe the behavior of a gas with respect to changes in volume, temperature, and pressure. They can be used to solve for one of two variables, or combined into what is known as the general gas law to solve for one of three variables if the other two are known or cancel when one variable is held constant.

Standard air has a barometric pressure of 29.92 in. Hg (760 mm Hg) at sea level and 32°F (0°C). The moisture content of standard air depends upon the relative humidity which is measured as a percent of the moisture required for saturation at a given temperature. The temperature at which air becomes saturated and moisture begins to condense out is called the dew point. Normally, air contains about 1.1 lbf of moisture per 1000 ft³ air at room temperature.

When air is pressurized, its volume is decreased, and the moisture content increased in direct proportion to the pressure. For example, if 1000 ft³ of air is compressed from atmospheric pressure (14.7 lbf/in.² absolute) to 130 lbf/in.² gauge, both the pressure and the moisture content would increase ten times, so

[2] An engineering project which elaborates on the testing procedure for mufflers titled M-100 Muffler Engineering Project Report is available from Scovill Fluid Power Division, Wake Forest, N.C. 27587.

that at 130 lbf/in.^{2} the moisture content would be about 11 lbf, or 1.32 gallons of water.

The presence of more than about 10% of the moisture at atmospheric conditions not only damages pneumatic components, but interferes with sensitive controls. Normally, compressing and then cooling the air near the compressor outlet will condense out most of the moisture, which is drained from the receiver, or from taps placed downstream. The air is then conditioned for use by placing filter-regulator-lubricator units before air tools and machine circuits to further reduce the water content, remove airborne particles, set the pressure for the circuit, and add lubricant to reduce component friction and wear. If completely dry air is needed, without any lubrication whatsoever, oilless compressors are used to increase the pressure of the air supply. The pressurized air then is passed through after-coolers, moisture separators, and air dryers to remove nearly all of the unwanted moisture, after which the clean, dry, oil-free air supply is reheated to prevent pipe sweating.

Finally, mufflers are installed on cylinders, valves, and other components to reduce noise levels and prevent airborne particles in the exhausting air from injuring operating personnel. Noise exposure levels for operating personnel are set by the Occupational Safety and Health Administration.

REVIEW QUESTIONS AND PROBLEMS

1. Convert a temperature of 180°F to °R and °K.

2. A gauge calibrated to read °F temperature in 50° increments from 0°F to 350°F is to be rescaled with a second set of numbers at the same increments but in °C. List those seven increments with their °F counterparts in two columns.

3. If the gauge pressure in a 50-gal tank half-filled with liquid is 100 lbf/in.^{2}, what would be the gauge pressure if the tank were empty?

4. An air over oil loading dock lift with a 90-gal reservoir is raised by 125 lbf/in.^{2} shop air piped through a check valve. The reservoir is 2/3 full when the lift is lowered and 1/3 full when the lift is raised. If the raised lift is overloaded sufficiently to lower it, what pressure could be expected in the reservoir?

5. A pump feeds coolant into a 25-liter tank until the pressure rises to 3000 kPa gauge. What is the remaining volume in the tank?

6. The charge pressure in a 5-gal accumulator drops from 3000 lbf/in.^{2} to 1000 lbf/in.^{2}. If the gas volume at 3000 lbf/in.^{2} is 200 in.^{3}, how much oil was expelled from the accumulator?

7. A ballast-type accumulator with a 3,000-in.^{3} capacity is heated from 70°F to 225°F. What will be the new volume occupied by the gas?

8. When a machine is switched to stand-by, 200 in.^{3} of shock absorbing charge gas in an accumulator cools from 180°F to 75°F with no change in pressure. What was the original volume occupied by the gas in the accumulator?

9. It is noticed that a dead weight accumulator with a 10-in. diameter bore, and holding against a constant load, raises 6 in. when the machine warms up. If the initial volume was 2000 in.^{3} and room temperature is 80°F, estimate the temperature of the gas.

10. If 300 in.³ of a gas held at 250°F decreases in volume by 25% as it cools with no change in pressure, how much will the temperature be expected to change?

11. A constant volume pressure vessel is heated from 75°F at 500 lbf/in.² to 400°F. Compute the new pressure reading.

12. If a constant volume of gas under a pressure of 1.725 MPa is heated from 30°C to 150°C, what will be the new pressure reading?

13. How much would the temperature of a constant volume pressure vessel operating at 450°F and 1000 lbf/in.² have to decrease to lower the pressure to 800 lbf/in.²?

14. To demonstrate the principle of the Savery steam engine (circa 1702), a pressure vessel is heated to 300°F, isolated by closing a valve, and then cooled with 70°F water. What final pressure might be expected on a gauge monitoring the experiment?

15. A ballast-type accumulator with a 3,000-in.³ capacity under 1,500 lbf/in.² pressure is reduced in volume to 2,000 in.³ while the temperature changes from 100°F to 400°F. What is the final gauge pressure reading?

16. An accumulator under a pressure of 10.3 MPa is reduced in volume from 0.03 m³ to 0.02 m³ while the temperature increases from 37°C to 175°C. What is the final pressure reading?

17. Compute the free air consumed at 68°F by a 3-in. bore spring returned cylinder with a 12-in. stroke operating at 250 cycles/min, 150 lbf/in.² pressure and 200°F.

18. If a double-acting air cylinder with a 40-mm diameter bore and 20-cm stroke operates at 550 kPa and cycles at 100 cycles/min, how long will it take to consume 100 m³ of free air? (Assume the rod diameter is negligible and there is no change in the temperature of the air.)

19. Upon impact, 100 in.³ of gas in a shock absorber is reduced to 50 in³. If the initial charge pressure of 500 lbf/in.² at 80°F is increased three times, what would be the expected temperature rise in the gas?

20. When an oscillating load operating against a hydraulic system accumulator comes to a stop in mid-cycle, the charge gas expands from 175 in.³ to 250 in.³ and cools from 210°F to 80°F. If the pressure in the accumulator is 1000 lbf/in.², what was the charge gas pressure when the system was operating?

21. What is meant by the dew point, and how is it determined?

22. How do after-coolers and dryers remove moisture from the pneumatic system?

CHAPTER 16

SIZING PNEUMATIC SYSTEMS

16-1 INTRODUCTION

Pneumatic systems are sized to meet output power requirements. Pipe and flexible lines carry compressed air from the receiver to air tools powered from flexible drops and to production equipment. System air pressure is regulated at the compressor, while the pressure for tools and equipment is regulated at each pressure drop. The air distribution system is sized to carry the required air flow with minimum friction losses through sections of pipe and various fittings.

The compressor pumps air from the atmosphere into the receiver, reducing its volume and raising its pressure level. The receiver stores the air supply. When the demand is low, this permits the compressor to be switched off; and when the demand is greater than the output of the compressor, additional air is available. The compressor motor is controlled by the high-low pressure setting at the receiver. When the pressure drops below a predetermined level the motor is switched on, and when the pressure has been raised to the high-pressure setting, the motor is switched off. In practice, the compressor and receiver are sized as one unit, with the control selected to provide the most efficient use of the compressor.

Pneumatic valves regulate the pressure, direction, and flow rate in response to system demand and safety requirements. The receiver is equipped with a pressure relief valve as a safeguard against rupture and personnel injury. Each pressure drop is equipped with a pressure regulator and sometimes a combination filter-regulator-lubricator unit to condition the air. Directional control valves are used on machinery to direct movement, while flow control valves are used to control actuator velocity and stiffen the movement. Directional control valves are sized with C_v factors that combine the variables of pressure temperature and flow in one formula. The use of C_v factors across the industry also allows comparisons between competitive products.

Rotary actuators are sized to meet the force and velocity requirements at the output. The flow is determined at the required pressure, and then converted to free air, either at prevailing or standard conditions. It is also common practice to oversize components to provide additional cylinder force or motor torque at start-up.

16-2 PNEUMATIC DISTRIBUTION SYSTEMS

Air systems are plumbed to minimize losses between the receiver located near the compressor and the point of use. A loss of 10% pressure is allowable under normal operating conditions with less than half the loss being attributed to the main transmission line. Loop systems such as that in Fig. 16-1 provide air through more than one path to the air drops, thereby reducing run length restrictions. Where loops are particularly long, installing more than one compressor at convenient locations reduces the run length and resulting pressure drop. Provision also is made to pitch air lines slightly from 0.1–.25 in./ft, to collect moisture accumulation in lines at drain points where water can be removed periodically.

Distribution pipe is selected for its mechanical and noncorrosive properties rather than for its ability to withstand pressure, since 250 lbf/in.^{2} is usually maximum, although shocks can increase this value substantially. Schedule 40 pipe or tubing with flare of hard soldered fittings is appropriate with primary consideration being given to provide the mechanical support necessary to prevent sagging

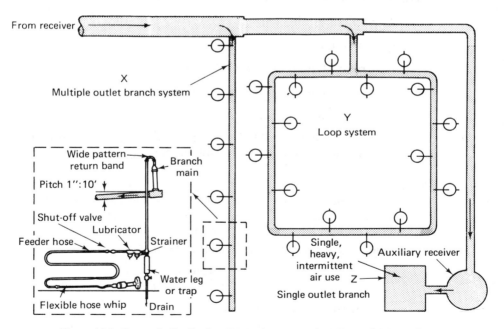

Figure 16-1 Loop air distribution system (*courtesy of Air Power Division, Ingersoll-Rand Corporation*).

and vibration. Sound transmission from vibration through the main transmission line can be reduced further by using a flexible connection between the compressor and the main trunk line. The flexible coupling also compensates for thermal expansion and relieves pipe strains attributable to minor misalignment.

Harris Formula

Pressure losses in transmission lines can be calculated using the Harris formula, or convenient tables so derived. From the Harris formula, pressure loss due to friction (p_f) is computed from

$$p_f = \frac{C \times L \times Q^2}{CR \times d^5} \tag{16-1}$$

where the friction coefficient (C) for schedule 40 commercial pipe has been found experimentally to be

$$p_f = \frac{0.1025}{d^{0.31}} \tag{16-2}$$

making the computed loss due to friction equal to

$$p_f = \frac{0.1025 \times L \times Q^2}{CR \times d^{5.31}} \tag{16-3}$$

where p_f = pressure drop due to friction (lbf/in.^{2})

L = length of pipe (ft)

Q = ft^3/sec of free air

CR = ratio of compression at the pipe entrance

d = actual internal diameter of the pipe (in.)

C = experimental coefficient

Calculated values for d$^{5.31}$ for several pipe sizes are listed in Table 16-1 to reduce the difficulty associated with solving this equation.

Example 1

A compressor delivers 650 ft^3/min of free air through a 2-in. steel pipe at a receiver pressure of 250 lbf/in.^{2}. For a section of 500 ft, use the Harris formula to compute the pressure drop.

Solution   Using the Harris formula (Eq. 16-3),

$$p_f = \frac{0.1025 \times L \times Q^2}{CR \times d^{5.31}}$$

The compression ratio equals

$$CR = \frac{250 \text{ lbf/in.}^2 + 14.7 \text{ lbf/in.}^2}{14.7 \text{ lbf/in.}^2} = 18.01 : 1$$

TABLE 16-1 CALCULATED VALUES FOR $d^{5.31}$ FOR SCHEDULE 40 PIPE SIZES

Pipe size	Inside diameter d (in.)	$d^{5.31}$ (in.)
$\frac{3}{8}$	0.493	0.0234
$\frac{1}{2}$	0.622	0.0804
$\frac{3}{4}$	0.824	0.3577
1	1.049	1.2892
$1\frac{1}{4}$	1.380	5.5304
$1\frac{1}{2}$	1.610	12.5384
2	2.067	47.2561
$2\frac{1}{2}$	2.469	121.4191
3	3.068	384.7707
$3\frac{1}{2}$	3.548	832.5501
4	4.026	1628.8448
5	5.047	5048.8738
6	6.065	14349.3284

From Table 16-1, the value for $d^{5.31}$ for a 2-in. pipe is 47.2561. Substituting these values in the Harris formula,

$$p_f = \frac{0.1025 \times 500 \text{ ft} \times (650 \text{ ft}^3/\text{min} \times 1 \text{ min}/60 \text{ sec})^2}{18.01 \times 47.2561}$$

and

$$p_f = 7 \text{ lbf/in.}^2$$

The friction loss through fittings is determined for each size and type of fitting experimentally, and expressed as an equivalent length of straight pipe of the same size. The equivalent lengths of all fittings in the system then are added together and combined with the length term in the Harris formula to determine the pressure drop for the system.

The friction loss for a number of common fittings of various sizes is given in Table 16-2. It should be noticed that for any particular fitting the equivalent length increases dramatically with pipe size. For example, air passing from a straight section of 2-in. pipe through the side outlet of a standard tee generates a friction loss equivalent to 10.3 ft of straight pipe, and this is approximately twice the loss for a 1-in. side outlet tee.

Example 2

An air run contains eight 2-in. standard elbows, two 2-in. globe valves, and one 2-in. side outlet tee. Calculate the equivalent length and pressure drop associated with these 11 fittings if the run transmits 500 ft^3/min of free air at a receiver pressure of 50 lbf/in.^{2}.

TABLE 16-2 FRICTION LOSS THROUGH FITTINGS AND VALVES

Valves and fittings

Nominal pipe size inches	Schedule number	Globe valve	Angle valve	Gate valve	Swing check valve	Plug cock	45° std. elbow	90° std. elbow	90° long radius elbow	Standard tee		Close return bend	90° welding elbow	
										Run of tee	Side outlet		Short radius	Long radius
$\frac{1}{2}$	40	17.6	7.5	.67	7.0	.93	.83	1.55	1.04	1.04	3.11	2.59		
$\frac{3}{4}$	40	23.3	9.9	.89	9.2	1.23	1.10	2.06	1.37	1.37	4.11	3.43		
1	40	29.7	13.6	1.14	11.8	1.57	1.40	2.62	1.74	1.74	5.2	4.36	1.4	1.1
$1\frac{1}{2}$	40	45.5	19.4	1.74	18.1	2.41	2.14	4.02	2.68	2.68	8.1	6.7	2.1	1.6
2	40	59	25.0	2.24	23.2	3.10	2.75	5.2	3.44	3.44	10.3	8.6	2.8	2.1
$2\frac{1}{2}$	40	70	29.9	2.68	27.8	3.70	3.30	6.2	4.12	4.12	12.4	10.3	3.3	2.5
3	40	87	37.1	3.32	34.6	4.60	4.10	7.7	5.1	5.1	15.4	12.8	4.1	3.1
4	40	114	48.5	4.35	45.2	6.0	5.4	10.1	6.7	6.7	20.1	16.8	5.4	4.0
5	40	143	61	5.5	57	7.6	6.7	12.6	8.4	8.4	25.2	21.0	6.7	5.1
6	40	172	73	6.6	68	9.1	8.1	15.1	10.1	10.1	30.3	25.3	8.1	6.1

All valves and cocks fully open

Check valves require 0.50 lbf/in.² pressure loss to open fully

Solution The equivalent length values for each of the fittings is taken from Table 16-2 and added together.

$$L = 8(5.2) + 2(59) + 1(10.3) = 169.9 \text{ ft}$$

This value is then substituted directly in the Harris formula:

$$p_f = \frac{0.1025 \times 169.9 \times (500 \text{ ft}^3/\text{min} \times 1/60 \text{ min/sec})^2}{14.6 \times 47.256} = 5.8 \text{ lbf/in.}^2$$

16-3 COMPRESSORS

Air compressors are pumps. Their purpose is to supply clean, dry air for power tools and industrial processes. They draw air in through the intake, compress it which raises its pressure and temperature, and then store it in a receiver. Air is conditioned by filtering it at the intake, by cooling it, and then by drying it to remove the water. Lubricating oil which migrates downstream is removed to prevent problems in some applications.

There are a number of ways to classify compressors:

1. Positive displacement.
2. Nonpositive displacement.
3. Rated output.
4. Single or multiple stage.
5. Intended use and duty cycle.

Positive displacement compressors use pistons, vanes, diaphragms, and lobes to reduce the volume of the air in the compression cycle. A cutaway view of a typical industrial compressor of the nonlubricated piston type is illustrated in Fig. 16-2. This unit keeps oil segregated from the piston area by wiper rings that confine wet lubrication to the crankcase portion of the compressor, virtually assuring oil-free air downstream of the compressor. The piston uses Teflon compression rings which do not require lubrication. Standard single-stage units of this type provide plant air at 100 lbf/in.^{2}. Piston-type compressors are available in vertical and horizontal models, as well as vee configurations in sizes to 200 hp and above. Reciprocating piston compressors are known to consume less power than equivalent capacity rotary screw compressors at all operating loads, although their initial cost is usually higher. Most large stationary industrial compressors are of the reciprocating piston type that use the piston to compress the air with single- or double-acting strokes—that is, compress air on one or both sides of the piston.

There is a practical limit to the pressure that can be developed in single-stage piston type air compressors. This is reached at about 150 lbf/in.^{2} where the compression ratio ($164.7/14.7 - 11.2 : 1$), compressing chamber size, and heat of compression act to reduce efficient pumping action.

Staging has the effect of increasing pumping efficiency by dividing the total pressure between two or more stages and reducing the heat associated with ineffi-

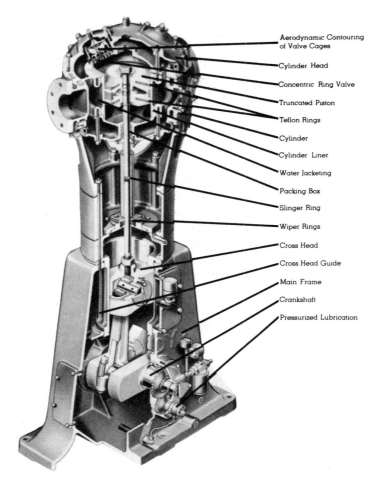

Labels on figure:
- Aerodynamic Contouring of Valve Cages
- Cylinder Head
- Concentric Ring Valve
- Truncated Piston
- Teflon Rings
- Cylinder
- Cylinder Liner
- Water Jacketing
- Packing Box
- Slinger Ring
- Wiper Rings
- Cross Head
- Cross Head Guide
- Main Frame
- Crankshaft
- Pressurized Lubrication

Figure 16-2 Stationary piston compressor (*courtesy of Joy Manufacturing Company*).

ciency by intercooling between stages. Multistage piston-type compressors pipe the outlet of each succeeding cylinder into the next, thereby stepping up the pressure. Cylinder size decreases in the progression and intercooling removes a substantial portion of the heat of compression, thereby increasing air density and volumetric efficiency of the compressor. Typical pressure ranges with the degrees of staging are listed in Table 16-3.

Rotary lobe screw compressors raise the pressure of the air by the action of two intermeshing helical lobe rotors housed in a casing having suction and discharge ports. The rotor assembly shown as an insert in Fig. 16-3 has a male rotor with four convex lobes formed helically along its length, and the female rotor has six lobes with corresponding concave interlobe spaces or flutes. The casing is machined with two precise bores with very little clearance between the rotors and

TABLE 16-3 PRESSURE RANGES
WITH STAGING

Piston type compressors	Pressure range (psi)
Single-stage	80–150
Two-stage	100–250
	(500 intermittant)
Three-stage	400–2500
Four-stage	2000–5000

the casing. Essentially, there are three phases to the compression cycle in a rotary lobe compressor. First, atmospheric air enters the suction port of the compressor at one end and fills the cavity between the rotor and the compressor housing. In the second phase, the rotors continue to turn past the intake port,

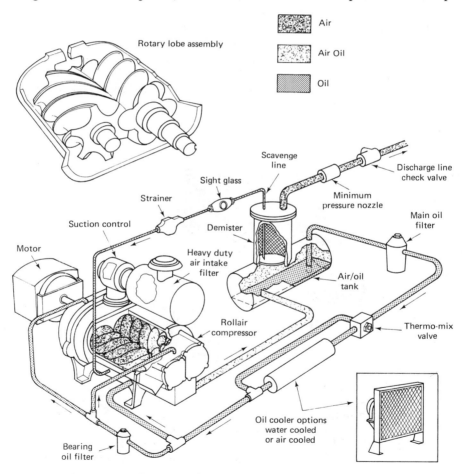

Figure 16-3 Asymmetrical screw rotors of rotary lobe compressor (*courtesy of Worthington Compressors, Inc.*).

trapping the air between the lobe and the interlobe housing. As rotation continues, the male rotor rolls into the female rotor interlobe space and the point of intermeshing moves progressively along the rotor length with a continuous reduction in volume and a corresponding increase in pressure. The third phase of the cycle is completed as compressed air is discharged through the discharge port. The discharge of one lobe overlaps the next with the result that delivery is continuous and pulsation-free. Lubrication for bearings which support the lobes, and provides a seal between the lobes and the housing, is delivered with the compressed air to the receiver where it is separated, filtered, cooled, and reinjected into the compressor.

Compressor units and accessories used in plants are usually located in a compressor room or installed outside the plant in a designated area to conserve floor space and reduce in-plant noise. Large packaged units are also available for transport to job sites above ground, or below ground level in mines.

Unit air compressors, sometimes called automotive air compressors because of their widespread use in service stations and garages, are available in a variety of types and sizes, from less than 1 ft^3/min for the hobbyist to more than 20 ft^3/min for commercial users. A unit air compressor is shown in Fig. 16-4. They are characterized by their completeness, and usually include the compressor, the receiver if one is used, pressure controls, and all instrumentation for immediate hook-up and use.

Figure 16-4 Unit air compressor (*courtesy of Ingersoll-Rand Corporation*).

The smaller positive displacement compressors are shown in Fig. 16-5. Shown in the figure are a rotary vane compressor, a diaphragm compressor, and an opposed two-cylinder single-stage compressor. These particular compressors are of the oilless type, which eliminates the requirement for an oil separation system, and assures an oil-free air supply to power air tools and paint guns.

Centrifugal air compressors raise the pressure level of the fluid by using impellers to increase the velocity of the air. Centrifugal air compressors are avail-

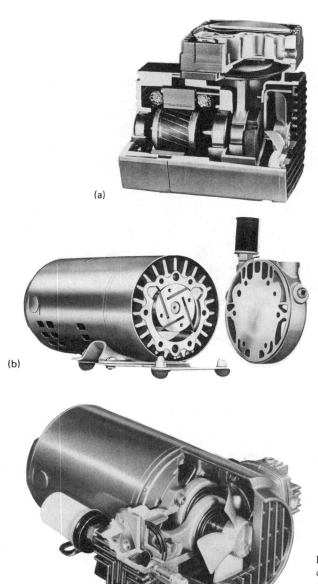

(a)

(b)

(c)

Figure 16-5 Positive displacement compressors; (a) diaphragm compressor, (b) vane compressor, (c) opposed piston compressor (*courtesy of Gast Manufacturing Corporation*).

able in sizes to 3000 hp. Figure 16-6 illustrates a three-stage impeller-type air compressor that supplies plant air at 100–125 lbf/in². Two-stage compressors are available to about 65 lbf/in², four-stage models will supply compressed air at 150 lbf/in²., and five-stage compressors can boost pressure to 350–400 lbf/in². Two-stage air compressors are used in the manufacturing of glass, textile fibers, and

Figure 16-6 Centrifugal compressor (*courtesy of Joy Manufacturing Company, Air Power Group*).

pharmaceuticals. Three-stage compressors are used for standard plant air applications. Four- and five-stage compressors are used for chemical processing and
high pressure industrial manufacturing processes. By the nature of the action of
the impeller, the delivery is inversely proportional to the pressure. This means
that as the pressure drops, the delivery increases. This is unlike the action of a
reciprocating piston compressor that sweeps a constant volume at a given speed.
Thus, centrifugal compressors will supply more air at reduced working pressures
than will their positive displacement counterparts.

Centrifugal compressors use intercoolers between the impeller stages to
reduce temperature and increase volumetric efficiency. The flow path of air
through three stages and two intercoolers of a centrifugal compressor is illustrated
in Fig. 16-7. Air enters the first stage of the compressor at atmospheric pressure
and temperature, where its pressure and temperature are raised by the velocity
imparted by the impeller. The first stage intercooler reduces the velocity of the
air, and by circulating it around the water cooled tubes which are bundled in
finned plates, reduces its temperature. The second stage of the compressor raises
the pressure to 60–70 lbf/in.^{2}, and again the velocity imparted by the impeller
increases its temperature. The second-stage intercooler acts the same as the first
and directs cooled air to the third stage which brings it to 100–125 lbf/in.^{2} for use
throughout the plant. Centrifugal compressors virtually assure 100 percent oil-
free air because the impeller pressurizes the space surrounding the impeller on the
air chamber side of the lubricated bearings.

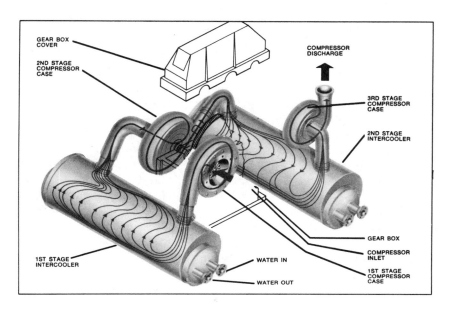

Figure 16-7 Centrifugal compressor staging and intercooling (*courtesy of Joy Manufacturing Company, Air Power Group*).

16-4 SIZING COMPRESSORS

Air compressors are sized to supply present equipment, with a 25%–50% capacity built in for future expansion. First the pressure range is selected, usually 80 to 140 lbf/in². Then the free air demands of all tools and equipment using the air are totaled—both the *continuous* demand and the *average* air demand. The continuous demand is the amount of air required if all the air tools and equipment were operated continuously; whereas the average demand is the free air consumption rate multiplied by the percentage of time the equipment will be in use. The compressor cannot be sized below the average value of the air consumption because, even with a large receiver supplying air during peak demands, the compressor could not recover.

Air compressors are selected by their capacity to supply free air, even though air tools are commonly rated at air consumption values for a given pressure range. Free air means air at standard atmospheric conditions: 29.92 in. of Hg, 36 percent humidity, and 68°F. Corrections to this standard are made using the General Gas Law.

For normal use, the compressor can be sized from the average demand, which is 30% to 60% of the continuous demand, depending upon the range of operating pressure. Table 16-4 matches the compressor horsepower with the average and continuous demand of pneumatic equipment in three pressure ranges. In the 90–125 lbf/in.² pressure range, for example, a 2-hp compressor will supply free air at an average of 22.5 to 30.4 ft³/min. Notice that the value for continuous operation is just under 30% of the average use, but this holds true only for a 2-hp compressor operating in the range of 90 to 125 lbf/in². As the horsepower rating and pressure increase, there is a noticeable drop in delivery.

A number of reciprocating positive displacement compressors are rated from their piston displacement, but this information is misleading. In single-stage units, the piston displacement equals the product of the swept volume of the cylinder and the rpm at which the compressor operates. In multiple-stage compressors, only the swept volume of the first stage is used in the calculation because it determines the air intake. Piston displacement does provide information about the size of the unit, but because volumetric efficiency decreases both with pressure and speed, it does not give an accurate estimate of the available delivery from the compressor. This can be learned only when delivery is given as free air at one or more of the pressure ranges. It is noticed in Table 16-4, for example, that the free air delivery from a 2-hp compressor decreases as the pressure range increases, showing that for a given displacement, the volumetric efficiency drops as pressure increases. It is also apparent from this example that the pressure range selected should be no higher than necessary to power the air equipment because power can be needlessly wasted, generating unused high pressures.

TABLE 16-4 COMPRESSOR HORSEPOWER BASED UPON FREE AIR CONSUMPTION OF TOOLS AND EQUIPMENT AT A GIVEN PRESSURE

Horsepower of compressor required (two stage)	Pressure required (lbf/in.²) Cut in Cut out	Free air consumption of all equipment used ft³/min (continuous operation)	Free air consumption of all equipment used ft³/min (average use)
1½	90–125	4.3– 6.4	14.8– 22.4
2	90–125	6.5– 8.7	22.5– 30.4
3	90–125	8.8– 13.2	30.5– 46.2
5	90–125	13.3– 20.0	46.3– 60.0
7½	90–125	20.1– 29.2	60.1– 73.0
10	90–125	29.3– 40.0	73.1–100.0
15	90–125	40.0– 60.2	100.1–127.0
20	90–125	60.3– 85.0	127.1–154.0
25	90–125	85.1–100.0	154.1–192.0
1½	120–150	3.7– 5.7	12.7– 20.0
2	120–150	5.8– 7.4	20.1– 25.9
3	120–150	7.5– 11.2	26.0– 39.2
5	120–150	11.3– 17.3	39.3– 51.9
7½	120–150	17.4– 27.0	52.0– 67.5
10	120–150	27.1– 37.0	67.6– 92.5

Example 3

Using Table 16-4 and a pressure range of 90–125 lbf/in.², size an air compressor to handle the air tools listed in Table 16-5. Assume that demand will be average, and the system will be oversized 25 percent for future expansion.

TABLE 16-5 CONSUMPTION OF
AIR TOOLS AND EQUIPMENT
FOR EXAMPLE 3

	ft³/min
Air hoist	4
¼-in. Rotary in drill	7
Horizontal grinder	20
Paint gun	5
Air ratchets (2 at 6)	12
Impact wrench (2 at 6)	12
Total	60

Solution Totaling the air consumption of the tools in Table 16-5 and adding 25 percent capacity for future expansion.

$$\text{Total Air Consumption} = 60 \text{ ft}^3/\text{min} + (25\% \times 60 \text{ ft}^3/\text{min}) = 75 \text{ ft}^3/\text{min}$$

Referring to Table 16-4 in the pressure range 90–125 lbf/in.², a 10-hp compressor would supply 73.1 to 100.0 ft³/min of free air at standard conditions. Also notice that it would supply only about 29.3 to 40.0 ft³/min on a continuous basis, or about 50% of the air consumed if all the tools and equipment were run continuously. The catalog ratings for a 10-hp compressor using piston displacements commonly show outputs in the 50 ft³/min range, but this information is calculated using higher volumetric efficiency estimates than Table 16-4, which is based upon air tool consumption.

16-5 RECEIVERS

Receivers are sized to consider system output pressure and flow rate requirements, air input capability from the compressor, and duty of operation. An overload factor for future expansion of production facilities also should be incorporated in the design to justify the initial investment of capital in this unit, which is normally a one-time-buy item.

To properly size a receiver, its functions must be understood. The receiver is an air reservoir. It dampens pressure pulses from the compressor at the inlet and supplies air at substantially constant and steady pressure at the outlet. Pulsation generated in the outlet line from valve shifting and component operation is transmitted back to the receiver and damped. During times of excessive demand, the receiver supplies an output flow in excess of the compressor input delivery capability. Receivers with large volumes also reduce the frequency of compressor operation, thus reducing operating costs and wear associated with excessive starting and stopping of the unit. Moisture that accumulates in the receiver is condensed and drained off through the petcock or automatic drain provided at the bottom of the receiver.

The time that a receiver can supply air between a maximum operating pressure (p_1) and minimum acceptable pressure (p_2) is computed from

$$t = \frac{V_r(p_1 - p_2)}{14.7 \times Q_r}$$ (16-4)

where t = the time (min)

V_r = receiver displacement (ft³)

p_1 = maximum pressure (lbf/in.²)

p_2 = minimum pressure (lbf/in.²)

Q_r = consumption rate from the receiver (ft³/min)

If the compressor is running and delivering an input to the receiver (Q_c), the formula becomes

$$t = \frac{V_r(p_1 - p_2)}{14.7(Q_r - Q_c)}$$ (16-5)

The formula can also be solved for the receiver volume (V_r) and converted either to gal, or to diameter and length dimensions. Table 16-6 lists the diameter and length of standard receivers of a given capacity.

Example 4

Compute the size of a receiver that must supply air to a pneumatic circuit consuming 10 ft³/min for 5 min between 125 lbf/in.² and 100 lbf/in.² before the compressor resumes operation. Select an appropriate size reservoir from the data in Table 16-6.

Solution Rearranging Eq. 16-4 to solve for the displacement of the receiver,

$$V_r = \frac{14.7 \times Q_r \times t}{p_1 - p_2}$$

TABLE 16-6 DIAMETER AND LENGTH OF STANDARD SIZE RECEIVERS*

Diameter, in.	Length ft.	Actual compressor capacity*	Volume cu. ft.
14	4	45	4½
18	6	110	11
24	6	190	19
30	7	340	34
36	8	570	57
42	10	960	96
48	12	2,115	151
54	14	3,120	223
60	16	4,400	314
66	18	6,000	428

* Reprinted from *The Compressed Air and Gas Handbook*

and

$$V_r = \frac{14.7 \times 10 \times 5}{125 - 100} = 29.4 \text{ ft}^3 \ (219.9 \text{ gal})$$

In SI units

$$V_r = \frac{(101 \text{ kN/m}^2)(10 \text{ ft}^3/\text{min} \times 0.028 \text{ m}^3/\text{ft}^3)(5 \text{ min})}{862 \text{ kN/m}^2 - 690 \text{ kN/m}^2)} = 0.8309 \text{ m}^3 \ (831 \text{ l})$$

From Table 16-6 we know that an appropriate size receiver (34 ft^3) would have the dimensions 7 ft in length by 30 in. in diameter. Increasing the size of the reservoir by 25% for unexpected overload and 25% for future expansion of production capacity is common.

16-6 COMPRESSOR CONTROLS

Even with proper sizing of the compressor and receiver, additional controls are needed to regulate the compressor to meet the fluctuating demand of components running off the system. Here there are a number of options depending upon how often the compressor must cycle to keep up with demand.

On piston type compressors of less than 25 hp, operating less than 50% of the time, the *automatic start-stop* control is the most common. The control consists of a pressure sensitive switch connected to the motor. When air is used from the receiver and the pressure drops below a predetermined lower limit—for example to 90 lbf/in.^{2}—the motor control switch closes and the compressor kicks on. When the pressure reaches the predetermined high limit, for example 125 lbf/in.^{2}, the motor control switch opens and the compressor stops. Automatic start-stop control is appropriate when the compressor cycles no more than five to seven times per hour, which is once about every ten minutes.

A compressor with an automatic start-stop control must be unloaded when it starts to prevent the electric motor from being overloaded. The best way to unload the compressor is with a centrifugal governor valve that vents the line to the receiver until the compressor reaches a predetermined speed. Modern compressors use a flyweight governor mounted inside the case to actuate a ball-check release valve (Fig. 16-8), although historically the governor was mounted externally on the flywheel to directly operate a poppet valve in the head of the compressor. Making the governor mechanism part of the compressor assures that the system is self-protecting.

When the compressor starts, the line between the compressor and receiver is vented. At the speed where the electric motor has full power the flyweights move out allowing the pin-operated check valve to seal and close the vent. Air pressure continues to build as the receiver fills with air, and the check valve at the receiver prevents the return flow when the start-stop switch opens and the compressor stops.

A second method used to unload piston compressors is a combination pressure-switch/unloading valve, but this is less reliable than an integral unloading valve on the compressor with a pressure switch mounted on the receiver.

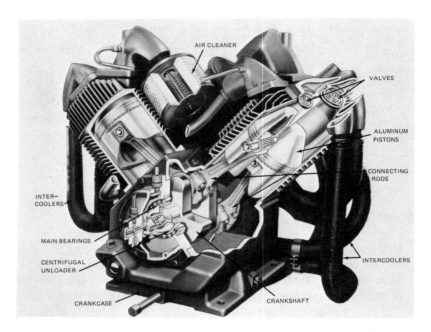

Figure 16-8   Flywheel unloading valve.

If the compressor operates more than 50% of the time, and the pressure control cycles more than five to seven times per hour, a *constant speed control* should be used. This allows the motor to run continuously, but the compressor is loaded and unloaded by a pressure switch that closes off the inlet or lifts the inlet valves with suction unloaders. Both of these mechanisms prevent the compressor from pumping air and lower the stand-by power required to turn the compressor.

Dual controls combine the advantages of the automatic start-stop control with the constant speed control. Having the compressor equipped with both allows the operator to select the mode that best matches the operating conditions simply by moving a selector switch.

Screw compressors use a *modulating control*. A receiver is not required because the air flow from the outlet does not pulsate like that from a piston compressor. The control mounts on the inlet side of the compressor, and air capacity is matched with demand by a valve that modulates the inlet open and closed to control air flow. In a screw compressor system with an air receiver, a cut-in, cut-out control switch is used to unload the compressor. When the preset pressure limit has been reached, the electropneumatic switch closes the inlet and the system goes to stand-by. A check valve at the receiver prevents backflow, and the compressor blows down, consuming only about 25% of full power. Air usage from the receiver causes the pressure to drop and when the lower pressure limit is reached, the electropneumatic valve snaps the inlet open, and the compressor comes back on line. The air receiver is required to provide air while the compressor is on stand-by.

16-7 PNEUMATIC VALVES

Pneumatic valves are used to regulate the pressure, air velocity, and flow rate, and to manage the direction of flow. Pressure relief valves protect the compressor and components by setting the maximum system pressure within safe working limits. Pressure control valves, usually called air line regulators, are installed at or near the pressure drop from the main transmission line and establish the working pressure of the branch circuit. They are also installed in circuits to provide two or more pressures for different component functions within one circuit. Flow or speed control valves regulate the volume flow of air at a specified working pressure to the component and thus regulate the speed of operation. Directional control valves are used to stop, start, divert, shuttle, and otherwise manage the direction of the fluid (air) in two, three, four, or more ways to effect the desired direction of component movement.

A direct-acting pressure relief valve is shown in Fig. 16-9. Valves of this type are available in sizes to $1\frac{1}{4}$ in. for pressures from 25 to 400 lbf/in². Valve action can be checked by pulling the ring to open the valve. The pressure setting of the valve is dependent upon the force applied by the coil spring against the plug or poppet valve element.

An in-line pressure regulator is illustrated in Fig. 16-10. In operation, air at the inlet is directed to the pressure control valve for pressure reduction and

RA SERIES
(Atmospheric Discharge)

RP SERIES
(Piped Discharge)

Figure 16-9 Direct-acting pressure relief valves (*courtesy of Fairchild, Industrial Products Division*).

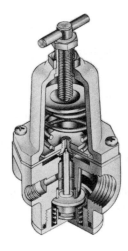

Symbol

Figure 16-10 Air-line pressure regulator (*courtesy of Aro Corporation*).

regulation. The desired pressure setting is adjusted by the T-handle which exerts a force on the compression spring. This spring force is transferred to the diaphragm regulating the opening and closing of the control valve, which in turn regulates the air flow at a precise pressure in the downstream line.

Figure 16-11 illustrates a typical poppet-type valve for flow regulation. Valves of this type are available in flow capacities to 420 ft³/min. In the regulated direction, speed control is adjusted with the fine metering screw thread adjustment secured by a lock nut. The poppet design allows maximum flow with minimum restriction and pressure drop. Reverse flow causes the integral check valve to open, allowing unrestricted flow in the reverse direction. Reverse flow also cleans the cone-shaped stainless steel metering valve making the unit tolerant to the effects of foreign matter which might become momentarily lodged in the valve.

Directional control valves are available in a variety of designs to provide actuator control. Because most production applications do not require a null or neutral position, two-position valves are usually sufficient. Typical applications require pneumatic cylinders to extend and retract in sequence with progressive and varying rates and forces and air motors to rotate continuously or intermit-

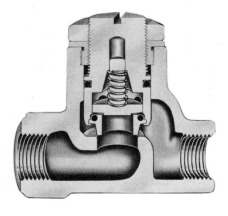

Figure 16-11 Flow control valve operation (*courtesy of Scovill, Fluid Power Division*).

tently in one direction. In most cases, the transmission medium, air, is exhausted to the atmosphere and requires no return line.

Figure 16-12 illustrates a two-way or three-way valve for full flow production control. Valves of this type are available with port sizes to 2 in. Accompanying symbols in Fig. 16-12 show a few of the several configurations available from the basic design shown. Response times of 200 milliseconds are reported by manufacturers of these and similar valves.

Symbols

Figure 16-12 Two- or three-way valve for production (*courtesy of Rex-Hanna Fluid Power Products*).

Four-way and five-way, two- and three-position valves are used typically to pressurize and exhaust both ends of a double-acting cylinder. Both may have center or null positions, if necessary, that allow the piston to coast to a stop when the valve is open center, or positively lock the piston when the center is closed or connected through check valves to pressurized ports. Valves may be operated manually or solenoid controlled. Because substantial force is required to shift large volume flow valve spools under pressure, the solenoid is usually used to direct pilot air against the spool ends to shift position. Pilots are either internally directed through a passage integral with the housing (in-pilots) or plumbed from other components with separate tubing to apply the force to the spool ends for shifting. A manual-operated and two solenoid-operated valves are shown in Fig. 16-13. The solenoid valve at the top of the figure uses an in-pilot to shift the spool.

There are a number of other mechanisms used to operate pneumatic control valves including palm buttons, foot pedals, foot treadles, ball cams, and roller cams. Several of these are shown in Fig. 16-14. The palm button provides hand operation that directly moves the spool. It is available for two-position valves. The foot pedal actuator provides tiptoe operation on two-position valves. The pedal is depressed to move the spool, and when the toe is removed, a spring returns the spool and pedal to its normal position. The foot treadle is used with two- or three-position valves. When used with a two-position valve, actuation is provided by depressing the treadle with the heel. The spool is returned to its

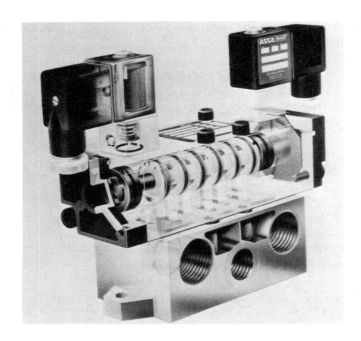

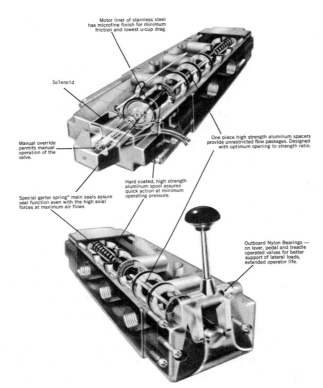

Motor liner of stainless steel
has microfine finish for minimum
friction and lowest u-cup drag.

Solenoid

Manual override
permits manual
operation of the
valve.

One piece high strength aluminum spacers
provide unrestricted flow passages. Designed
with optimum opening to strength ratio.

Hard coated, high strength
aluminum spool assures
quick action at minimum
operating pressure.

Special garter spring* main seals assure
seal function even with the high axial
forces at maximum air flows

Outboard Nylon Bearings —
on lever, pedal and treadle
operated valves for better
support of lateral loads,
extended operator life.

Figure 16-13 Solenoid and manual operated four-way valves (*courtesy of Automatic Switch Company, and Skinner Valve Division of Honeywell, Inc.*).

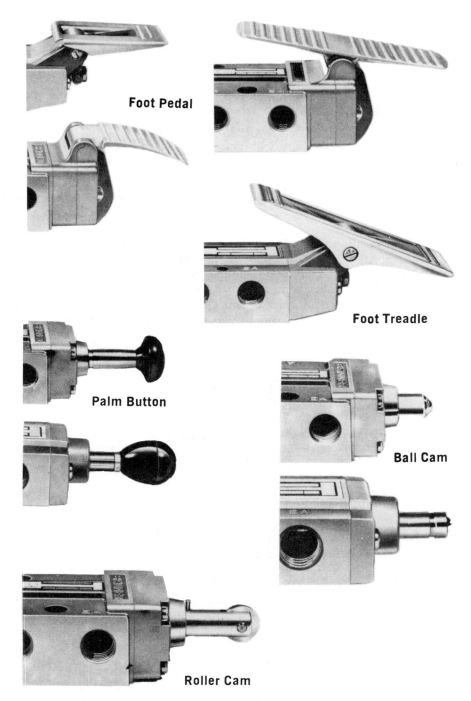

Figure 16-14 Pneumatic control vavle operators (*courtesy of Skinner Valve Division of Honeywell, Inc.*).

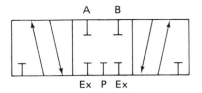

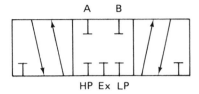

Ex P Ex

HP Ex LP

Figure 16-15 Four- and five-way valves.

normal position through a spring or by depressing the treadle with the toe. In three-position valves, the spool is moved by toe and heel motions. Detents are used to give the operator a definite feel when the valve is in a specific position. Ball cam and roller cam actuators are operated by in-line push devices or by cams. They are spring returned. This allows the pressure angle to be off the center line of the valve spool by as much as 15°. Because bottoming the spool

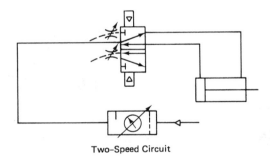

Two–Speed Circuit

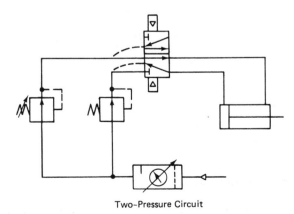

Two–Pressure Circuit

Figure 16-16 Four- and five-way valve applications.

could damage the valve, provision is made for overtravel of the spool within the valve body.

Where four-way hydraulic valves typically direct fluid from one pressurized port P to cylinder ports A and B, and provide one return port T to the reservoir, a fifth port on pneumatic valves allows additional flexibility to the design of air circuits. For example, look at the valve schematic symbols in Fig. 16-15. In the first one, because air is exhausted through two ports, the valve has two such connections. The valve is still considered a four-way valve, and one connection or port could have been used if there were a connecting passage within the valve. Notice in the center position that all ports are blocked and the cylinder or other actuator comes to rest. If, however, the two exhaust ports were changed to pressure ports, high pressure (HP) and low pressure (LP), the valve becomes a true five-way valve. By having two inlet pressurized ports as shown in the second schematic, the valve can supply air at one pressure to extend a cylinder during the work portion of the cycle, and at another pressure to retract it, thus varying the force applied in each direction. This is shown in Fig. 16-16. Another option shown is to use one pressurized port and two exhaust ports with flow control valves to meter out exhaust air, thus providing different speeds in each direction.

16-8 SIZING PNEUMATIC VALVES

Flow rates through valves and other devices imposing restrictions on flowing air are computed from formulas derived for orifices. But unlike an orifice, the flow through each valve is different, from size to size, and from manufacturer to manufacturer. Thus there is a problem with standardization when comparing valves from different manufacturers because two valves of the same type and size may not pass the same flow of air. In the past, using the port size to designate the capacity of the valve was common, but in recent years improvements in valve design have increased the flow rates within the same size valve body making this method unreliable.

One means used to standardize the flow capacity of valves is to use a flow capacity coefficient C_v. The value of C_v is an indication of how much air the valve will flow in ft^3/min. Higher values for C_v indicate higher flow rates. And valves from different manufacturers with the same C_v have the same flow rate. Valve applications can be made from flow charts that solve for the C_v when the inlet pressure and air flow are known, or from orifice formulas that have been modified to include C_v. Computed methods for determining valve size are slower but more accurate because they allow the designer to select a valve based upon an ideal pressure drop across the valve, rather than from a chart developed for a given pressure drop of say 15 lbf/in^2.

One such formula to calculate C_v is

$$C_v = \frac{Q}{22.67} \sqrt{\frac{T}{(p_1 - p_2)K}} \qquad (16\text{-}6)$$

where C_v = flow capacity coefficient

$\qquad Q$ = flow rate (scfm)

$\qquad T$ = absolute temperature (°F + 460)

$\qquad p_1$ = inlet pressure (lbf/in.² abs)

$\qquad p_2$ = outlet pressure (lbf/in.² abs)

$\qquad K$ = constant

The valve of the constant K depends upon the pressure drop across the valve. For a pressure drop of 10% or less of supply pressure, $K = p_2$. If the pressure drop is between 10% and 25% of supply pressure, $K =$ the average of p_1 and p_2—that is, $(p_1 + p_2)/2$. For pressure drops greater than 25% of supply pressure, $K = p_1$, but in no case can the pressure drop be greater than $0.53 p_1$ because at this point the critical velocity is reached and the formula is no longer valid. In practical applications this means that 0.53 times the upstream pressure is the limiting factor for passing air through a valve to an actuator. Increasing the pressure above this value will result in a greater pressure drop across the valve, but will have little constructive effect on increasing production rates from cylinders and air motors receiving the air.

Example 5

An air motor that operates at 90 lbf/in.² and at a temperature of 80°F consumes approximately 35 scfm of air. What size air valve would be required to keep the pressure drop to 10 lbf/in²?

Solution The constant K to be used in Eq. 16-6 is determined for an application where the pressure drop is approximately 11.1%, and must be computed from the average of p_1 and p_2. With a pressure drop of 10 lbf/in.², $p_2 = 80$ lbf/in.²

$$K = \frac{p_1 + p_2}{2} = \frac{(90 + 14.7) + (80 + 14.7)}{2} = 99.7$$

substituting in Eq. 16-5

$$C_v = \frac{35}{22.67} \sqrt{\frac{(80 + 460)}{(104.7 - 94.7)(99.7)}} = 1.14$$

Thus a control valve with a $C_v = 1.14$ would be required to pass 35 scfm at 80°F with a pressure drop of 10 lbf/in².

Valve selection charts commonly incorporate a 10–15 lbf/in.² pressure drop. One such chart for a family of air valves is given in Fig. 16-17. To read the chart, the air inlet pressure to deliver the required force is established first and read across the bottom. The flow rate for the particular application is then determined from the size and speed of the actuator and read from the left ordinate of the chart. Finally, the flow capacity coefficient C_v is found by following the curve band above to the place where the inlet pressure and flow rate line intersects the right margin.

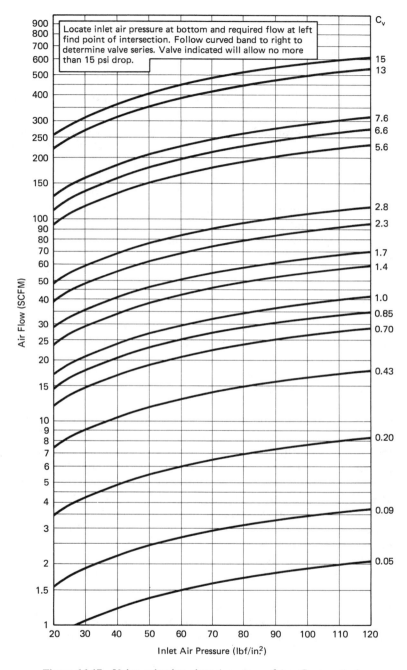

Figure 16-17 Valve selection chart (*courtesy of Aro Corporation*).

Sec. 16-8   Sizing Pneumatic Valves

Example 6

What flow capacity valve would be required to deliver 30 scfm at 70°F, if the inlet pressure is 90 lbf/in.²?

Solution Referring to Fig. 16-17, the pressure is found across the bottom, and then the air flow is located along the left ordinate. Following the C_v curve above this intersection to the right margin shows a value just above $C_v = 0.85$.

While generalized C_v values are helpful as a rough guide to make a valve selection, they have obvious limitations. The chart given is accurate only at 70°F, and for a pressure drop across the valve of 15 lbf/in². If a pressure drop of less than 15 lbf/in² is required, the next larger value for C_v would be selected. After determining the required C_v, flow charts for specific valves are examined to see that the valve being selected has the flow capacity at the required inlet pressure, temperature, and pressure drop across the valve. This is necessary because as pressure increases, so too does the flow capacity within the established limits of pressure drop.

16-9 CYLINDERS

Pneumatic cylinder construction makes extensive use of aluminum and other nonferrous alloys to reduce weight and the corrosive effects of air. Aluminum cylinders also transfer heat efficiently. Figure 16-18 illustrates three representative cylinder designs. The double-acting cylinder shown in Fig. 16-18a is constructed with an aluminum body and a chrome-plated cylinder rod with an NFPA standard threaded end. Air cushioning at both ends of the stroke is accomplished by tapered ends on the piston that seal off the exhaust port near the end of the stroke, causing air to be metered out through the adjustable needle valves in both end caps. Ball check valves permit free reverse flow to the power side of the piston. Figure 16-18b illustrates a double-wall cylinder. The inner body is stainless steel for long wear, and the outer body is aluminum, which gives the cylinder strength and dissipates heat. The hard chrome-plated steel rod transfers force through one of four NFPA threaded rod ends. Figure 16-18c illustrates what is called a *flat* air cylinder for use where space is limited. The bore and stroke dimensions are approximately equal, with sizes ranging from 3/4 to 4 in. The stainless steel body is held between the anodized aluminum end caps with tie rods, and the stainless steel piston rod is guided by an oil impregnated bronze bushing. Buna-N O-rings are used at the cylinder ends, piston to cylinder bore, and for the rod seal, simplifying the design.

Air cylinders are also cushioned with polyurethane shock pads placed between the piston and cylinder end caps. The design results in reduced noise levels near the working cylinder. Shock pads increase the length of the cylinder approximately 1/4 in. per pad. The pneumatic cylinder shown in Fig. 16-19 illustrates the use of shock pads on a pneumatic cylinder with a splined antirotation device. This allows the rod to extend and retract with the cylinder rod end held in a fixed position.

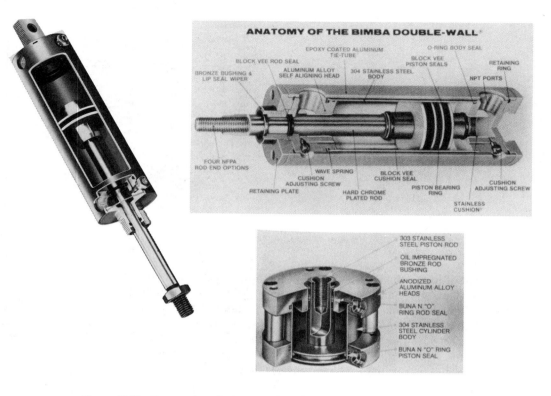

Figure 16-18 Pneumatic cylinder equipped with shock pads (*courtesy of Pneumatic and Hydraulic Development Company*).

Figure 16-19 Pneumatic cylinder equipped with shock pads (*courtesy of Pneumatic and Hydraulic Development Company*).

Figure 16-20 illustrates a cylinder design that incorporates a meter out device in the port fitting (see inset). The fitting also swivels to position the threaded outlet in the direction of connecting lines without overtightening the tapered thread where it enters the end cap. Fittings of this type are called banjo fittings. On the power stroke side of the piston, air flows freely through the swivel end, around the lip seal, and into the cylinder. When the power stroke is reversed, the lip seal checks to prevent air from exhausting through this flow path and directs the air past the tapered needle valve in the center of the fitting. Essentially, then,

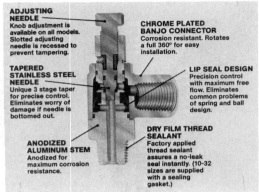

Figure 16-20 Air cylinder equipped with swivel banjo fittings that check and meter air flow (*courtesy of Bimba Manufacturing Company*).

in addition to the swivel function, the fitting acts as a reverse free-flow check with an adjustable meter-out valve to control the stroke velocity.

Cylinders are sized to meet the demand of the load resistance. The bore is determined from the force requirement and the stroke from the length of travel. The air flow rate is determined from the cylinder displacement and cycle rate in ft³/min of free air. Unless the rod size is large compared to the bore diameter, double-acting cylinders are considered to have the same displacement in both directions. Oversizing the bore slightly is a common practice to compensate for the pressure drop through the control valve and cylinder ports.

Example 7

A single-acting, spring-return, 3-in. bore air cylinder with a 4-in. stroke is used to close a hold-down clamp. If the cylinder runs on 90 lbf/in.² air and cycles 20 times per minute, compute free air consumption in ft³/min.

Solution   A 3-in. bore cylinder has an area of

$$A = \frac{3.14 \times 3^2}{4} = 7.065 \text{ in.}^2$$

and at a cycle rate of 20 times/min the air consumption would be

$$Q_v = \frac{7.065 \text{ in.}^2 \times 4 \text{ in.}}{1728 \text{ in.}^3/\text{ft}^3} \times 20 \text{ c/min} = 0.33 \text{ ft}^3/\text{min at } 90 \text{ lbf/in.}^2$$

converting to free air at the same temperature using Boyle's law

$$V_2 = \frac{p_1 V_1}{p_2} \frac{(90 \text{ lbf/in.}^2 + 14.7 \text{ lbf/in.}^2)(0.327 \text{ ft}^3/\text{min})}{(14.7 \text{ lbf/in.}^2)} = 2.33 \text{ ft}^3/\text{min free air}$$

Approximate values for air consumption of cylinders can be determined from charts such as that shown in Fig. 16-21. Notice that for any given bore size, the air consumption increases as the pressure increases. This means that to determine the flow rate, the lowest pressure at which the cylinder will move the load

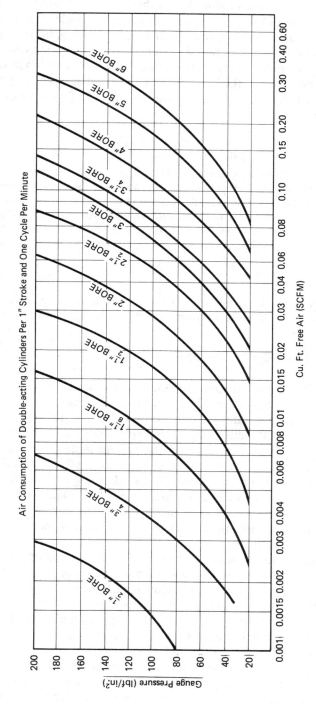

Figure 16-21 Air consumption chart for various size cylinders at given pressures (*courtesy of Aro Corporation*).

resistance must be determined first, and that value is located on the left margin. The horizontal line of the graph is then traced to the right until it intersects with bore diameter, and the ft³/min of free air is read at the bottom. The chart determines air consumption values for each inch of stroke and one cycle per minute for a double-acting cylinder. This requires that the value determined for the air consumption be multiplied by the stroke length and number of cycles per minute. For single-acting cylinders that value is divided by 2. The following example demonstrates the use of Fig. 16-21.

Example 8

Solve Example 7 using the air consumption chart given in Fig. 16-21.

Solution   First, the air pressure of 90 lbf/in.² is located at the left margin, and then moving across to the right, the place is found where this pressure line intersects with the line labeled 3-in. bore. Looking down at the bottom of the graph, the ft³ of free air is found to be between 0.06 and 0.07 ft³, or approximately 0.065 ft³ of free air per inch of stroke for one cycle. For a stroke of 4 in. and a cycle rate of 20 cycles/min for a single-acting cylinder, air consumption would be

$$Q_v = \frac{0.065 \text{ ft}^3/\text{in.} \times 4 \text{ in./c} \times 20 \text{ c/min}}{2} = 2.6 \text{ ft}^3/\text{min}$$

Notice that the value obtained by using the air consumption chart is approximately 15% larger than the value determined from the physical dimensions of the cylinder.

16-10 ROTARY PNEUMATIC ACTUATORS

Air motors offer several positive advantages. Motors are available in sizes from 1/20th to 10 hp in speeds from 300 to 10,000 rpm. Because air motors are explosion-proof, they offer a safe source of power in flammable areas. Speed is infinitely variable within the operating range and instantly reversible when used with four-way valves without damage to the motor. Air motors also are burnout-proof and may be stalled repeatedly for indefinite periods without damage. So long as adequate lubrication is provided, usually by admitting oil into the air stream, air motors can be expected to operate in any position. The expanding gas has a cooling effect reducing the heat normally generated during operation. Finally, pressurized air driving the motor acts as a cushion to allow smooth engagement against the load resistance.

Air motors differ in several ways from hydraulic motors, and their unique operating characteristics must be considered when selecting an air motor for a particular job. For example, the horsepower and speed of an air motor may be changed simply by throttling the air inlet. Because of this, it is a common rule of thumb to select a motor that will provide the required horsepower using only two-thirds of the available line pressure. This will leave full line pressure for starting and overloads.

The horsepower of an air motor is related to the speed (rpm) and air line pressure. Typically, an air motor slows down when the load increases, and at the same time, its torque increases to a point where it matches the load. The motor will continue to provide increased torque with decreased speed until a stall condition is reached. Stall occurs without harm to the motor. Conversely, as the load is reduced, the motor will increase speed with a decrease in torque to match the requirements of the resistance. When the load on an air motor is either increased or decreased, the speed can be controlled by increasing or decreasing air line pressure. This makes it necessary to have higher air line pressure available for starting under heavy loads. Air consumption increases as either speed or air pressure is increased.

Continuous rotation air motors are available in a number of designs. The construction details of a typical rotary vane motor are illustrated in Fig. 16-22. Eight- as well as four-vane designs are available where low-speed accuracy of operation under full load is necessary. If the motor rotates only in one direction, the exhaust port is fitted with a muffler to reduce noise and airborne particles. Overrunning loads require flow control of the exhaust port, or external braking devices to control rotation. The speed of rotation is controlled with a meter-in circuit. Continuous rotation air motors are available as radial piston motors, axial piston motors, and more recently as orbital motors which feature speed reduction. Speed reduction as well as rigidity of control is accomplished by right angle worm drives which stiffen the drive and prevent overrunning and reversals of the output shaft when the air service is discontinued.

In addition to continuous rotation motors, there are a number of limited rotation torque motors that convert the linear motion of cylinders to limited rotation of the output shaft from 90° to several turns. These are used for rotating, bending, oscillating, transfers and flipovers, positioning, dumping, and indexing operations. The three most common types are chain cylinders, single and double

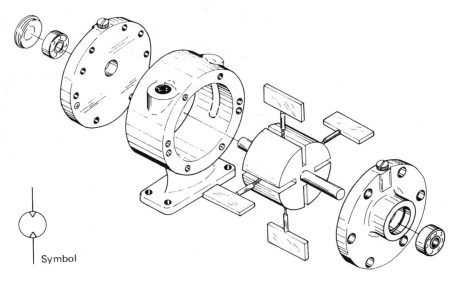

Symbol

Figure 16-22 Rotary vane motor construction detail (*courtesy of Gast Manufacturing Corporation*).

rack and pinion cylinders, and helical rod cylinders. These are illustrated in Fig. 16-23. The chain cylinder (see p. 242) uses two pistons to seal the endless chain as it travels around the idler and power gear. The larger piston is the power piston, while the smaller piston or secondary is used for a seal. The net piston area against which the air pressure acts equals the difference between the primary and secondary piston areas. The torque available from the output shaft equals net force on the primary piston multiplied by the pitch diameter of the power gear. Rack and pinion and double rack and pinion torque motors apply pressure alternately to both ends of the tube chamber, causing the shaft to rotate. These can be high-output devices with torque to 10,000 lbf-in. Helical torque actuators employ a helical rod, which is free to rotate, over which a mating piston is fitted. The piston slides in the bore over two guide rods that prevent it from turning. Applying air to one port causes the piston to be moved down the bore turning the helical rod clockwise, while applying air to the other port returns the piston turning to rod counterclockwise. Air cushioning with a meter-out control is used to prevent shock bottoming at each end of the stroke. Speed is controlled by metering the air supply.

Air motors are sized by their horsepower at a rated input, usually 90 lbf/in².

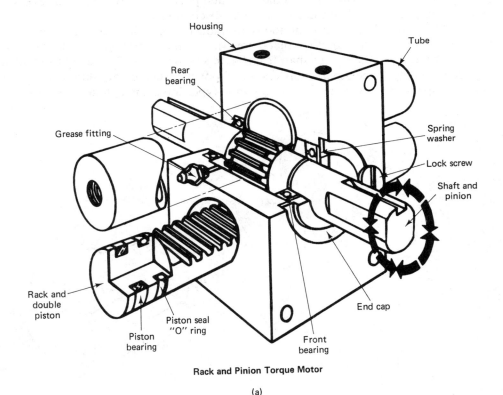

Rack and Pinion Torque Motor

(a)

Figure 16-23 Limited rotation torque motors. (*See also* pp. 441–42.)

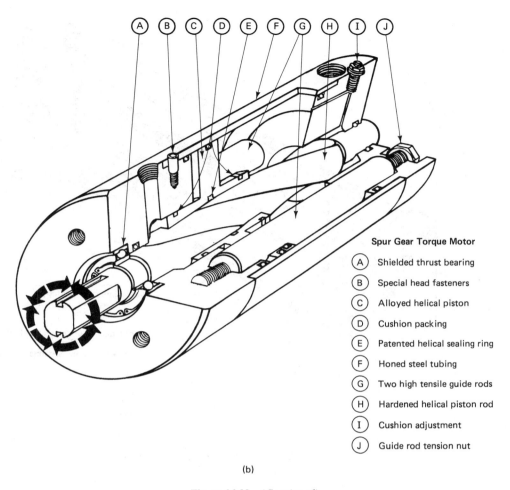

Spur Gear Torque Motor

(A) Shielded thrust bearing

(B) Special head fasteners

(C) Alloyed helical piston

(D) Cushion packing

(E) Patented helical sealing ring

(F) Honed steel tubing

(G) Two high tensile guide rods

(H) Hardened helical piston rod

(I) Cushion adjustment

(J) Guide rod tension nut

(b)

Figure 16-23 (*Continued*)

This is converted to air consumption in scfm to determine the size compressor needed. Table 16-7 lists the approximate air consumption for air motors from 1/4 to 2½ hp. The horsepower and speed of a given motor can be changed by throttling the inlet, and so the common rule is to size the motor to provide the torque needed at 2/3 of the line pressure available. Full line pressure would then be used for breakaway starting and overloads. Each manufacturer supplies power curves for their products. They are computed for a range of pressures. Figure 16-24 illustrates a family of curves with the power output range given for several models. If, for example, a 2-hp motor was required with a speed of 2000 rev/min, the motor would be selected with a range that provided this power at 2/3 of the line pressure available. Looking at the graph in Fig. 16-24, an "x" has been placed at the intersection of 2.0 hp and 2000 rpm. Both the Model 6-AM and 8-AM will deliver the necessary power because the "x" falls within the range of

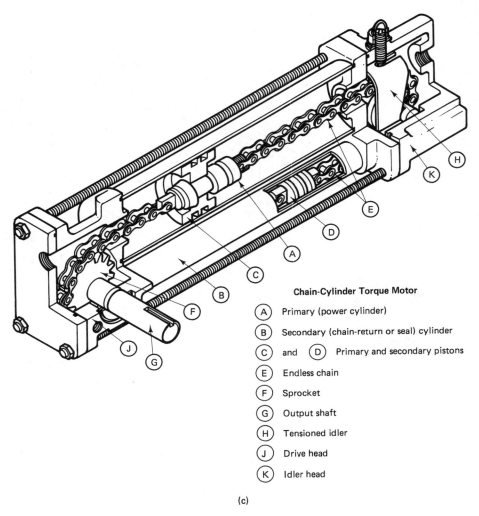

Chain-Cylinder Torque Motor

(A) Primary (power cylinder)

(B) Secondary (chain-return or seal) cylinder

(C) and (D) Primary and secondary pistons

(E) Endless chain

(F) Sprocket

(G) Output shaft

(H) Tensioned idler

(J) Drive head

(K) Idler head

(c)

Figure 16-23 *(Continued)*

TABLE 16-7 AIR CONSUMPTION
OF AIR MOTORS

Power motor air consumption	
Horsepower	Approximate CFM
$\frac{1}{4}$	20
$\frac{1}{2}$	40
$\frac{3}{4}$	45
1	50
2	102
$2\frac{1}{2}$	107

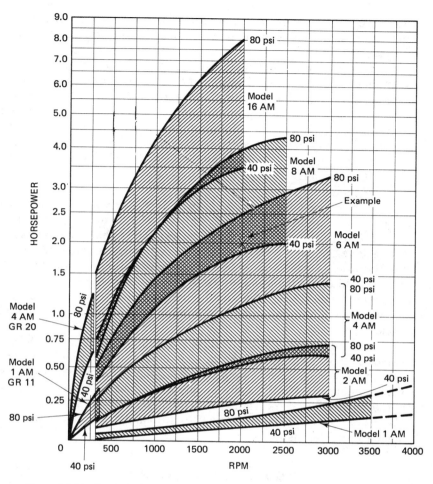

Figure 16-24 Typical air motor power curves (*courtesy of Gast Manufacturing Corporation*).

power using air pressures between 40 lbf/in.² and 80 lbf/in². Notice that the power requirement falls in the upper pressure range of the Model 6-AM, but in the lower pressure range of the Model 8-AM. Which motor is chosen would depend upon the magnitude of the starting load and upon how much overload capacity was needed.

16-11 SUMMARY

Pneumatic components are sized to meet the force and flow requirements of the system. The pressure is determined from the bore size of cylinders or the displacement per revolution of rotary motors. Both cylinders and motors are commonly oversized 10% to 15% to provide start-up and overload capacity. The pressure at the compressor is then set at the lowest level necessary to power actuators with some margin added to account for flow losses in the distribution system.

The distribution system is sized to carry the flow requirements of pneumatic components and machines connected to drops at various places. Structurally, the piping system must withstand the mechanical stress and vibration caused by pressure surges, and the main runs of the system are oversized to places where expansion is planned. Flow losses cause a drop in pressure at the outlet. They result as air flows through straight pipe sections as well as fittings. The pressure drop in straight pipe runs is computed using the Harris formula. To this must be added the pressure drop through fittings, valves, and other restrictions computed using C factors.

Compressors and receivers are sized to meet air flow requirements during periods of continuous and peak demand. The compressor control consists of a pressure sensitive electric switch that responds to high and low pressure levels at the receiver. Which type of control is used depends upon the percentage of time the compressor will be in operation. Low-use systems switch the compressor on and off, unloading the compressor to prevent overloading the drive motor. High-use compressor systems run continuously, but unload the compressor by lifting the inlet valve to reduce the power requirements during the slack portion of the cycle.

Air valves are similar to their hydraulic counterparts. Directional control valves exhaust directly into the atmosphere through a combination filter muffler to reduce noise and prevent injury to personnel. Control valves are sized using a velocity constant C_v. This allows valves from different manufacturers to be compared against the same standard. The larger the C_v, the greater the flow rate. Values of C_v range from less than 0.10 to 15 and flow rates of one to 250 ft^3/min at standard conditions.

STUDY QUESTIONS AND PROBLEMS

1. A compressor delivers 200 ft^3/min of free air through a 1-in steel pipe line at a receiver pressure of 125 lbf/in^2. Use the Harris formula to compute the pressure drop for a 100-ft section of pipe.

2. If the air pressure at the receiver is 160 lbf/in.^{2}, what flow of free air could be expected through a 2-in. schedule 40 pipe 250 ft away with a pressure drop of 5 lbf/in.^{2}?

3. A compressor supplies 20 ft^3/min of free air through a 1/2-in. schedule 40 pipe to an air tool 100 ft away. If the pressure drop is limited to 2 lbf/in.^{2}, what would be the pressure setting at the receiver?

4. A compressor set at 150 lbf/in.^{2} delivers 40 ft^3/min of free air to an air hammer 300 ft away. What size pipe would be required if the pressure drop is limited to 4 lbf/in.^{2}?

5. A compressor delivering 50 ft^3/min of free air through a 1-in. schedule 40 pipe is set at 125 lbf/in^2. How long could the line be before the pressure drops to 120 lbf/in.^{2}?

6. List the steps to size an air compressor.

7. Why is the pressure range selected no higher than necessary to power the air tools and equipment?

8. What four functions does the air receiver serve?

9. How long will a 250-gal receiver supply air to a pneumatic circuit consuming 6 ft³/min of air between 150 and 110 lbf/in.² before the compressor resumes operation?

10. What volume receiver must be used to supply air between 175 lbf/in.² and 150 lbf/in.² at 10 ft³/min for 12 min. without the compressor operating?

11. What volume receiver must be used to supply air for 1 min. between 850 kPa and 650 kPa at the rate of 0.20 m³/min without the compressor operating? Compute the answer in liters.

12. A 60-gallon receiver must supply air at 6 ft³/min for 5 min before the compressor resumes operation. If the low-pressure setting is 100 lbf/in.², what would be the high-pressure setting?

13. A 200-gal reservoir supplies air at 10 ft³/min for 8 min. If the high-pressure setting is 140 lbf/in.², what would be the low-pressure setting?

14. What continuous air flow rate would a 100 gal receiver supply between 160 lbf/in.² and 140 lbf/in.² for 7.5 min?

15. What is the difference between an *automatic stop-start* control and a *constant speed control,* and when would each be used?

16. A control valve must direct 75°F air at 20 scfm to an actuator at 100 lbf/in². If the pressure drop can account for no more than 10% of the supply pressure, what C_v valve should be selected?

17. A directional control valve with a C_v of 15.0 shows a pressure drop across the ports of 15 lbf/in.². If the operating pressure and temperature are 115 lbf/in.² and 75°F, what flow rate will the valve have?

18. A directional control valve with a C_v of 1.7 has a flow rate of 55 ft³/min with a pressure drop of 10 lbf/in.² across the valve. If the inlet pressure is 100 lbf/in.², what is the upstream temperature in °F?

19. How much pressure drop could be expected from a directional control valve with an advertised C_v of 3.5 and a K of 80 if the manufacturer rates the valve at 140 ft³/min at an inlet pressure and temperature of 90 lbf/in.² and 80°F?

20. Using the valve selection chart in Fig. 16-17, determine what C_v valve would be required to deliver 30 ft³/min at an inlet pressure of 90 lbf/in².

21. Compute the free air consumed at 68°F by a 3-in. bore, spring-returned cylinder with a 12-in. stroke operating at 250 cycles/min., 150 lbf/in.² and 200°F.

22. If a double-acting cylinder of 40-mm diameter bore and 20-mm stroke operates at 550 kPa and cycles at 100 cycles/min, how long will it take to consume 100 m³ of free air? (Assume no cylinder rod displacement and no change in temperature.)

23. A double-acting pneumatic cylinder with a 1.5-in. cylinder bore and 15-in. stroke operates at 300 cycles/min on 90 lbf/in.² air. If the cross section area of the cylinder rod is not considered and the temperature remains constant, what is the flow rate of free air through the cylinder?

24. A double-acting pneumatic cylinder with a 4-in. diameter bore and 10-in. stroke cycles 8 times per minute. If the cylinder runs on 75 lbf/in.² air and the area of the cross section area of the rod is ignored, use Fig. 16-21 to compute the free air consumed.

25. A rotary vane motor with a displacement of 3 in.³/cycle operates at 1500 rev/min on 95 lbf/in.² air. Compute the free air consumption and estimate the horsepower of the motor. (Assume no temperature change.)

CHAPTER 17

PNEUMATIC CIRCUIT APPLICATIONS AND CONTROLS

17-1 INTRODUCTION

Pneumatic circuits are used for both power and control. Power circuits direct air to pneumatic cylinders and air motors and exhaust the return. Pneumatic control circuits use air and logic functions to manage the operation of either hydraulic or pneumatic systems. Control rather than direct actuation is the output objective, but even in control circuits substantial force is commonly needed to shift the main spool of high flow directional control valves.

The force exerted by a cylinder and the torque exerted by an air motor are determined by the air pressure. The speed is determined by the air flow. The air pressure is controlled by an air regulator placed at the drop near the machine, while the speed is controlled by flow control valves and metering devices at various places in the circuit.

Because air is a spongy fluid, controlling the speed of actuators is a special problem. A pressurized cylinder that is not under load will extend or return suddenly if the exhaust air is not controlled. This could damage the cylinder. To solve this problem, cushioning devices are incorporated in the ends of cylinders, and meter-out flow-control valves are installed in the return lines. Air motors with gear drivers are also used to stiffen the movement of air circuits.

In the pneumatic hoist circuit shown in Fig. 17-1, for example, the load resistance is lifted by releasing back pressure on the blank end of the piston through the three-way valve to the atmosphere. Lowering the load resistance is accomplished by piping air to both sides of the piston, thereby providing a low-force regenerative circuit. The lifting force exerted by the hoist is determined by the area on the rod side of the piston; whereas adjustable flow control valves located in the spring-centered, three-position, three-way flow control valve and at the cylinder regulate the speed.

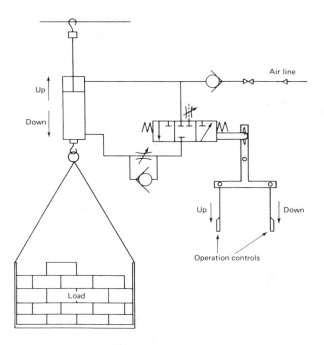

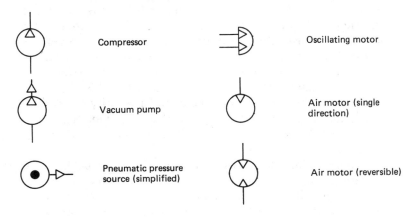

Figure 17-1 Pneumatic hoist circuit.

17-2 BASIC SYMBOLS

The symbols used to construct the pneumatic circuits are taken from ANSI Y-32.10. While many of the symbols in the standard can be used interchangeably in hydraulics and pneumatics, there are differences in a number of the symbols used for pneumatic devices and these will be reviewed here.

Energy conversion devices, such as compressors and vacuum pumps, oscillating motors and air motors, incorporate an open diamond to show that the device operates on air rather than oil (Fig. 17-2). The same is true for fluid

Figure 17-2 Pneumatic energy conversion devices.

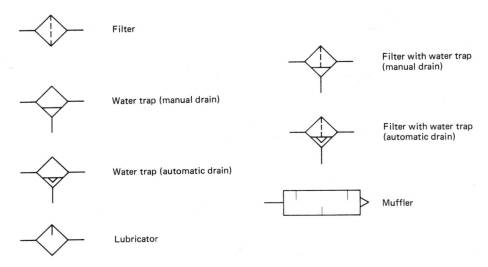

Figure 17-3 Conditioning devices.

conditioning devices which incorporate a diamond shaped symbol in filters, water traps, lubricators, and mufflers (Fig. 17-3).

Single-acting cylinders require a return force. In hydraulics this is accomplished most often by the return of the load resistance. Dump cylinders, for example, use the weight of the dump bed to retract the cylinder. The same is true for pneumatic cylinders, but springs also receive widespread use, either to extend or retract the cylinder. A number of these cylinder symbols are shown in Fig. 17-4. Double-acting cylinders are the most like their hydraulic counterparts. Also notice that cushioning devices are incorporated in the cylinder to slow down the drive at the ends of the stroke and reduce shock loading.

The symbols for pressure and flow control valves are shown in Fig. 17-5. They are similar to those used for hydraulics, but pneumatic devices relieve the over-pressure to the atmosphere when they reset from a higher to a lower pressure. Sequence valves are used in pneumatics to control two cylinders in sequence by requiring the pressure to step up to a higher level before the second one will operate—for example, a clamp cylinder would have to stall, causing the pressure to step up before the work cylinder could extend. Flow control valves are used to regulate the velocity of actuators on bleed-in and bleed-out circuits. They can also be used to give one cylinder priority over another by causing one to lag. And they are used as timing devices in pilot circuits to time-in and time-out pilot signals to power valves. The reverse check permits free flow in one direction.

Combination filter-regulator-lubricator units are used only in pneumatics. The detailed and simplified symbols for these units are shown in Fig. 17-6.

Two- and three-way directional control valve symbols are much the same as those for hydraulics, but four-way valves commonly have a fifth port so the flow of return air from the opposite sides of the piston can be controlled and exhausted

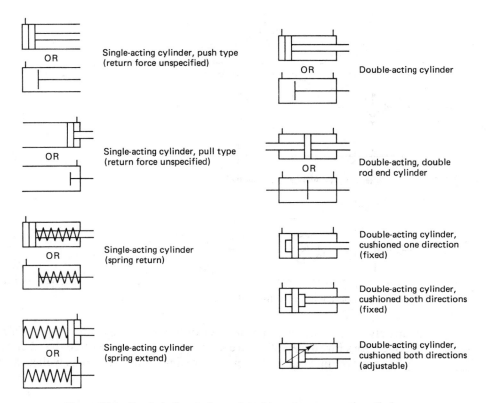

Figure 17-4 Symbols for single- and double-acting pneumatic cylinders.

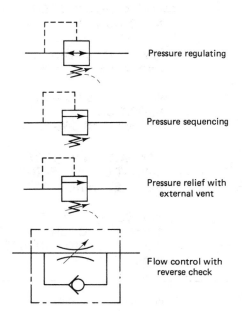

Figure 17-5 Pressures and flow control valves.

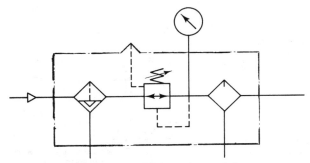

Filter regulator lubricator (composite)

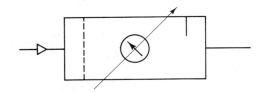

Filter regulator lubricator (simplified)

Figure 17-6 Combination filter-regulator-lubricators.

separately. The symbols for a number of four-way valves are shown in Fig. 17-7. The operators for these valves are shown in Fig. 17-8. Notice that a control valve may have more than one operator—for example, a solenoid as well as a remote air pilot.

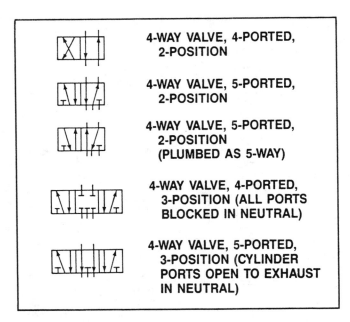

4-WAY VALVE, 4-PORTED, 2-POSITION

4-WAY VALVE, 5-PORTED, 2-POSITION

4-WAY VALVE, 5-PORTED, 2-POSITION (PLUMBED AS 5-WAY)

4-WAY VALVE, 4-PORTED, 3-POSITION (ALL PORTS BLOCKED IN NEUTRAL)

4-WAY VALVE, 5-PORTED, 3-POSITION (CYLINDER PORTS OPEN TO EXHAUST IN NEUTRAL)

Figure 17-7 Pneumatic directional control valve symbols.

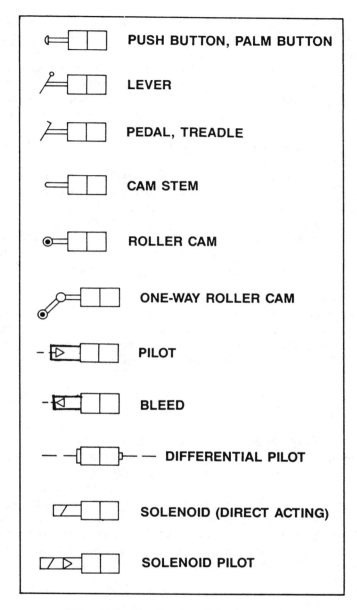

	PUSH BUTTON, PALM BUTTON
	LEVER
	PEDAL, TREADLE
	CAM STEM
	ROLLER CAM
	ONE-WAY ROLLER CAM
	PILOT
	BLEED
	DIFFERENTIAL PILOT
	SOLENOID (DIRECT ACTING)
	SOLENOID PILOT

Figure 17-8 Directional control valve operators.

17-3 BASIC CIRCUIT DESIGN

The force requirement in a circuit is used to size the actuator for a given operating pressure. The flow requirement to achieve the velocity at a given pressure is then used to determine the free air capacity of the compressor. The operating cycle and frequency of operation are then used to size the receiver and select the appropriate controls based upon percent of time the compressor will be in operation.

The function of the circuit is used to determine the initial layout. Starting with the actuator, component symbols are drawn and connected by lines back to the control valve and air source. The compressor and receiver are sized separately and are not usually shown. After the initial circuit is drawn, it is checked through each operation in sequence.

While the output function of the circuit is used to create the first layout, it must be remembered that the safe operation of the circuit is equally important. Not only must the operator be protected from unintended operation or failure, but safety interlocks must be built in to prevent personal injury. This means the circuit would include emergency stop, retract, and shut-down provisions. Consideration for safety also must include maintenance personnel as well. Remember that compressed air is silent but acts suddenly. This could cause injury to the mechanic who accidentally starts an actuator while checking or removing components. Installing isolation valves and pressure gauges is a must to monitor the system.

The initial cost of the circuit is important, but so too are the reliability and repairability because they affect the long-term cost. Reliability describes the tolerance of the circuit to adverse conditions, including heat, and dirt and water in the air stream. The machine also must be easy to repair. Unreliable machinery, or difficult to repair machinery, increases operating expenses throughout the life of the machine.

17-4 SINGLE-ACTING CIRCUITS

The most basic circuit uses a three-way valve to power and exhaust a single-acting cylinder. Figure 17-9 illustrates the single-acting direct control circuit. The power valve in this case is palm operated and spring returned. In the spring-returned position, air from the cylinder is exhausted. The single-acting cylinder extends under power and returns under the force of the spring. Pressing the palm button extends the cylinder, while releasing the palm button returns the cylinder. For the cylinder to extend or hold the force, the palm button must be held depressed because the valve return spring has a bias or memory to the return position which retracts the cylinder.

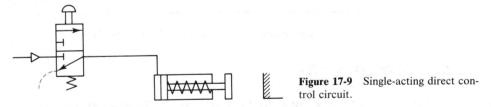

Figure 17-9 Single-acting direct control circuit.

Figure 17-10 illustrates the same circuit but includes provisions for safety, air conditioning, and speed control. Air is supplied through a shut-off valve and

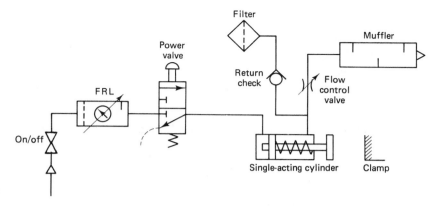

Figure 17-10 Single-acting direct control circuit with shut-off valve, FRL unit, flow control return check of filtered air and cushioning device.

an FRL unit, which supplies filtered, pressure regulated, and lubricated air to the cylinder. Air passes through the three-way directional control valve to extend the cylinder at a rate determined by the variable flow control valve metering the exhaust on the rod side of the cylinder. A muffler is installed to illustrate the importance of noise reduction. The filter and check valve located on the rod end of the cylinder allows clean air to enter the cylinder on the return stroke. Since the return line has no variable flow control device, the cylinder can be expected to retract quickly. To prevent bottoming, a cushioning device is located at the head end of the piston. Notice again that the cylinder returns when the palm button is released. If the requirements of the circuit had required the clamping cylinder to stop or hold in the clamped position when the operator released the palm button, then another type valve would be required.

The sequence of operation of a circuit is checked by listing what happens during each operation. For the circuit in Fig. 17-10 this is as follows:

1. When the air is turned on, the pressure is read on the FRL unit, but the circuit remains inactive.
2. Pressing the palm button shifts the three-way directional control power valve to extend the cylinder and close the clamp.
3. When the palm button is released, the spring in the cylinder retracts the rod and opens the clamp.
4. If the operator releases the palm button in any intermediate position, the cylinder will retract.

Step four in the circuit check requires the operator to hold the palm button depressed for the cylinder to remain extended, preventing the operator from attending other work. This could be remedied by installing a detent power valve like that shown in Fig. 17-11 which would eliminate step four.

Sec. 17-4   Single-Acting Circuits

Figure 17-11 Single-acting direct control valve with detent.

Shifting the power valve by hand has the advantage of simplifying the circuit, but considerable operator force could be required for larger valves. And unless the power valve is close to the cylinder the response time will be increased. One way to solve this problem is to install the power valve close to the cylinder and operate it with a small detent valve that controls only the pilot signal. Figure 17-12 includes a remote operated power valve in the circuit. Operator fatigue is now reduced since the remote valve controls only a pilot signal to the end of the spool to shift the power valve. A pilot signal of 100 lbf/in.^{2} on the end of a 1-in. diameter spool, for example, would supply a shifting force of 78.5 lbf. And because the power valve and cylinder are mounted near each other, the speed is not restricted.

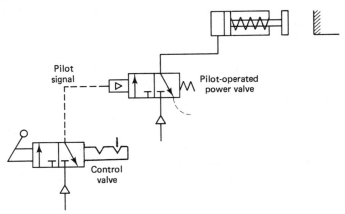

Figure 17-12 Single-acting remote control circuit with detent pilot valve.

With only minor modifications, the control portion of the circuit in Fig. 17-12 can be used to pilot operate a two-way valve to power an air motor. Such a circuit could power a fan, for example. Shifting the hand-operated pilot valve in Fig. 17-13 delivers a pilot signal to shift the two-way spring-returned directional control power valve. This supplies air from the FRL unit to the pneumatic motor which is equipped with a muffler to reduce noise and eliminate airborne particles from leaving the exhaust of the air motor. When the detent pilot valve is shifted to vent the pilot signal, the air supply to the motor is discontinued, the spring in the power valve returns the spool or poppet to the closed position, and the motor coasts to a stop. Notice that the circuit provides no braking action to the motor and air enters the circuit through the check valve to protect the motor against

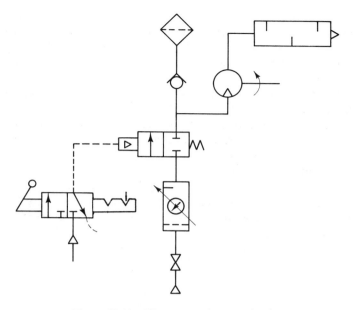

Figure 17-13 Pilot operated motor circuit.

overrunning loads. If braking were required, either metering the exhaust or providing an external brake would be necessary.

17-5 DOUBLE-ACTING CIRCUITS

Two- and three-way directional control power valves are simple in their operation and relatively inexpensive, but they have some limitations. Two-way valves are *on* or *off* and supply air in one direction, for example, to an air motor. Three-way valves are used to power and exhaust air cylinders, but the return stroke must be driven by a spring in the cylinder or the force of the load resistance. To power a cylinder in both directions, four-way valves with five ports are used. Here, some explanation about the use of a fifth port on pneumatic valves should be given.

In a typical application of a four-way two-position valve, (Fig. 17-14), air enters one port on the supply side and exhausts through the other. This allows the

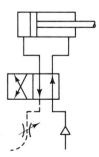

Figure 17-14 Two position, four port, four-way valve.

use of a common port for the supply, but it also reduces the flexibility of the valve because the same metering requirements would be used when the cylinder rod extends as when it retracts. It may be desirable to have the cylinder extend at one speed, and retract at another. Metering a common exhaust port does not allow this provision.

Metering reverse check valves could be added to each leg of the circuit at the cylinder, but this would mean additional hardware.

Adding a third port on the supply side of the control valve makes the valve a two-position, four-way, five-ported control valve (Fig. 17-15). Supply air still enters a common port on the supply side of the valve, but the exhaust air returns through one of the remaining two ports when the cylinder extends, and through the other when the cylinder retracts. This allows metering the air at two different rates, for example, providing a slow advance by metering air from the rod end of the piston, and a rapid return by metering the air from the blank end of the piston.

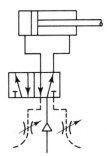

Figure 17-15 Two position, five port, four-way valve.

Another variation of the five-ported valve converts it to a two-position, five-way, five-ported valve. This configuration is commonly referred to as a four-way, five-ported valve, plumbed as a five-way valve (Fig. 17-16). It uses two air sources and a common exhaust port which may be metered, but it is more common to meter the return air from each cylinder port. The advantage of such an arrangement is that high-pressure air can be supplied to the port labeled 1 in Fig. 17-16 to extend the cylinder, and low-pressure air supplied to port 2 to retract the

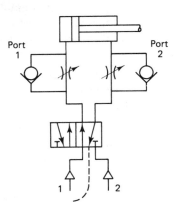

Figure 17-16 Two position, five port, four-way valve (plumbed as a five-way).

cylinder. Under load, the high-pressure air supplies the force necessary to complete operation, while on the return stroke, low-pressure air is used. This not only provides more control, but reduces air consumption since the return stroke volume is supplied at a lower pressure. Also notice in Fig. 17-16 that metering reverse check valves are used at the cylinder to control the velocity in both directions. If the load were constant, this control could be provided by adjusting the inlet pressure for the extension stroke, but because the cylinder is unloaded when it retracts, this additional control is used to prevent bottoming at high velocity.

Remote control is added by mounting the pilot valve at the operator station and the power valve near the cylinder. Figure 17-17 shows a double-acting remote control circuit with a two-position spring return, five-port, four-way valve. The sequence of operation is as follows:

1. The spring return power valve will return the cylinder when the detent pilot control valve is in the off return position.
2. When the detent pilot valve is manually shifted to clamp, pilot air shifts the power valve against spring pressure directing air to the blank end of the piston, causing the cylinder to clamp.
3. With the detent pilot control valve in the clamp position, the clamping cylinder will hold with a force determined by the cylinder bore and supply pressure.
4. Shifting the detent pilot valve to the open position, vents the pilot pressure on to the power valve. The power valve will then shift to the return position, venting the air to the blank end of the piston, returning the cylinder and opening the clamp.

The circuit in Fig. 17-17 allows the operator to hold the clamp open or closed with the detent valve, but has the undesirable feature of moving when the shop air is turned on. This could be unsafe if supply air were turned off and then turned on

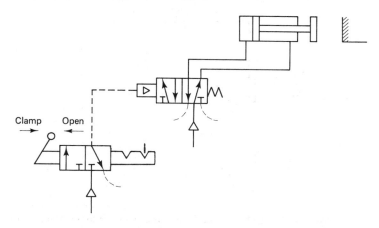

Figure 17-17 Double-acting remote control.

while the operator were loading the machine. For example, if the cylinder were clamped when the supply air was turned off, and then the detent pilot valve, shifted to the open position, when the supply air was resumed, the cylinder would return and the clamp would open. This occurs because the power valve has a spring bias which gives it the quality of *memory*. That is, it always shifts to the return position when pilot pressure from the control valve is removed.

One way to prevent the unwanted movement of the circuit in Fig. 17-17 when the air supply is interrupted is to install a power valve that does not have a bias spring or memory to one position, and pilot control valves that are vented until operated (Fig. 17-18). The sequence of operation is as follows:

1. When the supply air is turned on, the unbiased power valve will retain whatever position it had when the supply air was turned off.
2. Pressing the palm button momentarily on the *clamp* start valve sends a pilot signal to the power valve shifting it to the right and directing air to the blank end of the piston. This causes the cylinder to extend and close the clamp.
3. Pressing the pilot valve palm button on the *open* pilot valve sends a pilot signal to the power valve, shifting it to the left and directing air to the rod end of the piston. This causes the cylinder to retract and open the clamp.

This is an improvement over the circuit in Fig. 17-17 in that pressing the *clamp* or *open* palm buttons after the supply air has been discontinued will not cause the circuit to change the position of the power valve, and when the supply air is resumed, the circuit will remain in the same state as it was when the service was discontinued. If, for example, a piece were clamped when the air service

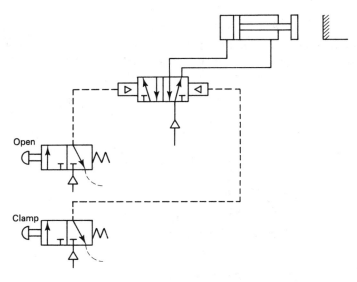

Figure 17-18 Unbiased four-way power valve; pilot operated by two three-way control valves.

were discontinued, it would remain clamped when the service were resumed, and if it were open when the air service was discontinued, it would remain open. Holding clamped or open may be desirable, but the cylinder still remains under pressure and so could move when the air service is turned on; for example, if the clamped piece were removed leaving an open space while the air service were off, the clamp would move to close the gap when the air service were turned back on. If there is to be no movement when the air service is discontinued and resumed, the power valve would have to have a center neutral position with all ports closed.

17-6 SEQUENCE CIRCUITS

Until now, the circuits discussed respond to signals given by the operator. Sometimes these are called *combinational* circuits because they react to a combination of signals without regard to order. More often than not, pneumatic circuits are started and then complete a series of events in a prescribed order using limit valves. These are called *sequence* circuits.

Figure 17-19 illustrates four operations of a sequence circuit that extends and retracts a cylinder in response to one momentary signal when the operator depresses the palm button on the *start* control valve. In Fig. 17-19a, the start valve is actuated sending a signal to the left side of the power valve. Figure 17-19b shows the power valve shifted to send supply air to the blank end of the piston extending the cylinder rod. In Fig. 17-19c, the power valve memorizes the signal from the start valve and continues to direct air to the blank end of the piston until the rod reaches the limit valve, sending a pilot signal to shift the power valve to the return position. Notice that the palm button has been released, and the pilot signal has been vented from the left side of the power valve. In Fig. 17-19d, the cylinder rod is returning and will continue to do so until it is fully retracted, at which time it will remain idle until the palm button is reactuated.

Figure 17-20 illustrates a pilot-controlled reciprocating circuit. When the manually operated two-way control valve is opened, the circuit will reciprocate at a rate determined by the size of the load resistance and the setting of the flow control devices located at the power valve that meters the return air from the cylinder. Automatic cycling is achieved by using pilot air directed alternately through cam-actuated limit valves located in the piston rod travel path to shift the main control valve. The control valve holds one position until an alternate signal shifts the spool in the opposite direction. Each time the cylinder rod reaches the end of its stroke, it depresses the three-way pilot sequence valve directing pilot air to the left side of the power valve. This shifts the unbiased power valve to direct air to the blank end of the cylinder. The events are the same when the cylinder rod extends to the right. The sequence valve on the left directs pilot air to the right end of the power valve. The circuit will continue to cycle until the two-way valve is closed, with the length of stroke and shifting points determined by the location of the adjustable cams on the cylinder rod.

When the duration of the pilot signal is controlled by the movement of the limit valve, some circuits become unreliable, particularly as the speed is in-

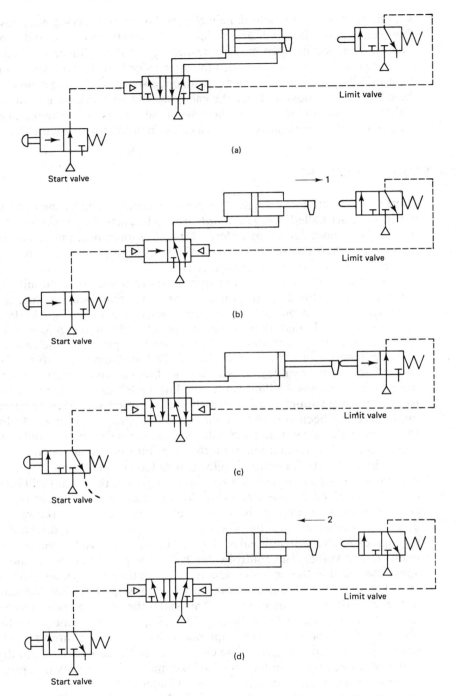

Figure 17-19 Extending and retracting a double-acting cylinder with a limit valve (*courtesy of Aro Corporation*).

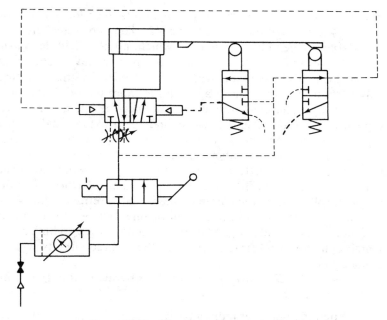

Figure 17-20 Pilot-control automatic circuit.

creased. This is because there is no means to time how long the pilot signal is sent to the power valve before it is vented. One solution to this problem is the "one-shot" circuit shown in Fig. 17-21. Here the one-shot valve controls the return of a single cylinder. Another common application is the clamp-work circuit that uses two cylinders. In Fig. 17-21, the power valve V2 shifts to extend the cylinder when the start valve V1 is depressed and released. The cylinder rod extends until

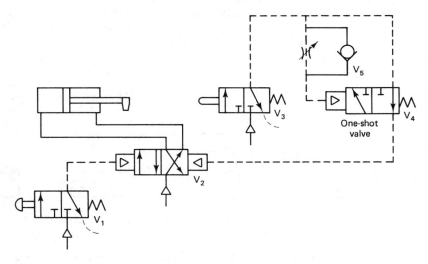

Figure 17-21 Basic "one shot" control circuit.

it depresses the limit valve V3, which directs the pilot signal through the one shot valve V4 to reverse the power valve V2. The pilot signal also bleeds through the adjustable orifice in bleed valve V5 which, after a slight delay, shifts the one shot valve V4 to vent the pilot signal to the power valve. The duration of the pilot signal through the one shot valve is controlled by the adjustment on the bleed valve V5, from a fraction of a second to more than one second.

It is also common practice to sequence the operation of two cylinders. One typical application requires two cylinders to extend and retract together. This type of circuit could be used to move a heavy metal sheet across a table into a shear, for example. Another typical application is the clamp-drill circuit which requires one cylinder to advance and clamp the work, followed by the second cylinder moving a drill head toward the clamped work. The drill cylinder then retracts, followed by the clamp cylinder which retracts. Still a third two-cylinder circuit would require the operation of one cylinder to follow another. That is, cylinder one extends, cylinder two extends, then cylinder one retracts, after which cylinder two retracts. Each of these three circuits will be shown and explained here.

Figure 17-22 shows a foot pedal operated circuit that has the following requirements:

1. When the operator momentarily steps on the pedal valve, two double-acting cylinders will extend to their limits.
2. When both cylinders are fully extended, they should automatically retract.
3. If either cylinder reaches full extension before the other, it should remain in that position until both are fully extended.

The main power valve in Fig. 17-22 is a two-position, four-way, five-port, unbiased pilot-operated valve. The foot pedal valve is a two-position, three-way, spring-returned valve. And the sequence valves are two-position, three-way, spring-returned valves plumbed in series. That is, both valves have to be depressed for the return pilot signal to be transmitted to the power valve. With both

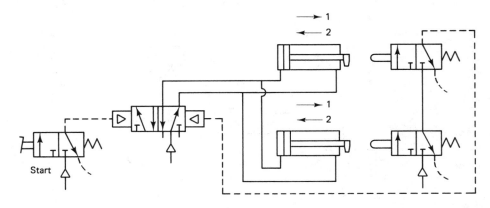

Figure 17-22 Two double-acting cylinders extending and retracting together (*courtesy of Aro Corporation*).

cylinders retracted, momentarily depressing the pedal valve sends a signal that shifts the power valve to direct air to the blank ends of both cylinders, and they begin to extend. When either cylinder rod reaches the end of its travel, it depresses a limit valve. And when both cylinders have fully extended, both limit valves will be depressed allowing the pilot signal to shift the power valve to the return position. This will direct air to the rod ends of the cylinders and they will return and stay at rest under pressure. Notice that the circuit will not return in mid-stroke. It must advance to the end before returning. Also notice that there is no provision to synchronize the cylinders, except at the ends of their strokes. If the cylinders are of the same bore, it can be assumed they would advance together if the loads were the same, but if the loads were different they would not. Flow control valves with reverse checks could be added to the rod sides of both cylinders to meter out return air when the cylinders advance. This would help keep one cylinder from advancing faster than the other unless one were stuck. Finally, if the operator does not release the foot pedal, the cylinder will remain extended until the pedal is released. This is probably not objectionable since it could be used as an additional control.

Figure 17-23 illustrates an air circuit that extends and retracts two double-acting cylinders in sequence. The circuit has the following requirements:

1. Momentary actuation of the palm button will cause cylinder 1 to extend to its limit.

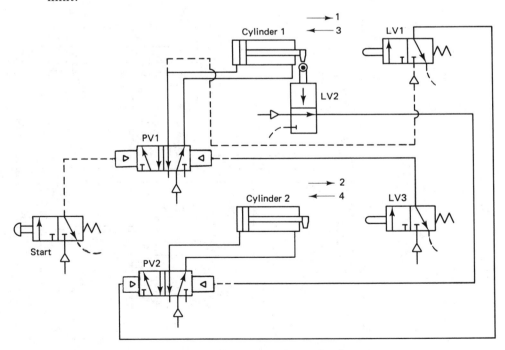

Figure 17-23 Extending and retracting two double-acting cylinders in sequence (*courtesy of Aro Corporation*).

2. When cylinder 1 is fully extended, a second double-acting cylinder 2 will extend to its limit.

3. When cylinder 2 is fully extended, cylinder 1 should retract.

4. When cylinder 1 is fully retracted, cylinder 2 should retract.

Notice that the pilot pressure for limit valve 1 is taken from the line that connects to the blank side of the piston in cylinder 1, allowing limit valve 1 to shift power valve 2 only when cylinder 1 is under pressure to extend. This prevents power valve 2 from shifting accidentally during set up, which would extend cylinder 2 if limit valve 1 were connected directly to an air supply. This should keep both cylinders in the retracted position until the start valve is depressed, even if the limit valves are shifted accidentally.

Figure 17-24 is a variation of the two-cylinder sequence circuit that extends one cylinder while it retracts the other. This is accomplished by adding a second cam-operated valve to cylinder 2 and eliminating the second power valve. Also notice that the start valve receives control air after it passes through limit valve 3 and limit valve 2 in series. The second power valve is eliminated because both cylinders operate at the same time, even though one is extending and the other is retracting. To keep the cylinders in sequence, limit valves 1, 2, 3, and 4 are located at the end of each stroke and connected in series. And because the cylinders operate in opposite directions, the limit valve at the end of the extension

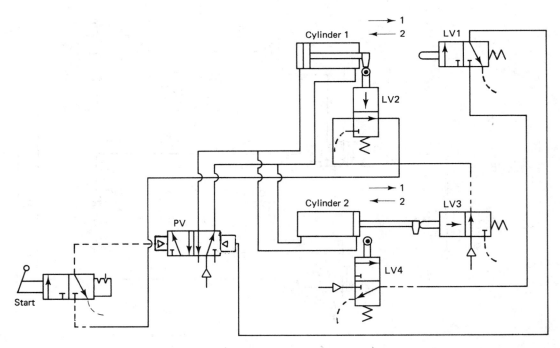

Figure 17-24 Extending one cylinder while retracting the other in sequence (*courtesy of Aro Corporation*).

stroke on cylinder 1 is connected in series with the limit valve that signals cylinder 2 has fully retracted. It also follows that the limit valve at the end of the extension stroke on cylinder 2 is connected in series with the limit valve that signals cylinder 1 has fully retracted.

The circuit operates as follows. When the hand operated three-way start is shifted open, pilot air flows through limit valve 3 and limit valve 2, and then through the start valve to shift the power valve to the right. This permits air to flow through the power valve to the blank side of the piston in cylinder 1 and to the rod side of the piston in cylinder 2. This causes cylinder 1 to extend while cylinder 2 retracts. When cylinder 1 has fully extended, it shifts limit valve 1, and when cylinder 2 has fully retracted, it shifts limit valve 4. This connects pilot air entering limit valve 4 in series through limit valve 1 which, in turn, shifts the power valve to the left, reversing the cycle. The power air is now connected to the rod side of the piston in cylinder 1, and the blank end of the piston in cylinder 2. The cylinders will continue to cycle, with one cylinder extending while the other retracts until the three-way start valve is shifted to the nonpassing condition. With the start valve in the nonpassing condition, the signal from limit valves 2 and 3 cannot reach the four-way power valve. The circuit will continue to operate until cylinder 1 is fully retracted and cylinder 2 is fully extended, at which point it will stop.

Pressure sequence circuits allow one directional control valve to control two cylinders. The sequence valve gives priority flow to one cylinder until it bottoms and the pressure steps up, shifting the valve to direct flow to the second cylinder. Pressure sequence circuits are less common in pneumatics because air is spongy,

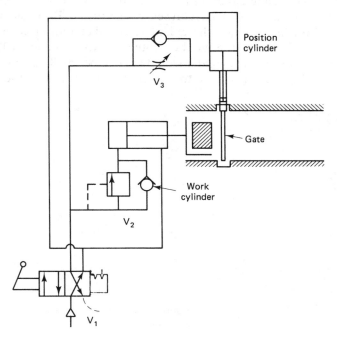

Figure 17-25 Pressure sequence circuit.

making precision control more difficult. Their use is restricted to less critical operations. The basic pressure sequence circuit is used in a clamp-work operation. One cylinder advances to clamp the work, followed by the advance of the second cylinder which performs a work operation. The cylinders then retract in the opposite sequence, first the work cylinder, followed by the clamp cylinder.

In the basic pressure sequence circuit in Fig. 17-25, when the control valve is shifted, the position cylinder lifts the gate. When it has fully retracted, the pressure steps up, say from 80% to 100% of line pressure, and sequence valve V2 opens, allowing the work cylinder to extend and move the load through the open gate. When the control valve is shifted back, the cylinders operate in the reverse order. First, the work cylinder retracts and air is exhausted through the one-way check valve in sequence valve V2, and then the position cylinder extends, closing the gate, forcing air through adjustable flow control valve V3 which has the check closed. The flow control valve is required, or a second sequence valve is required, so that the position cylinder will lag the work cylinder. The check valve is used in V3 to permit free reverse flow when the gate lifts.

17-7 INTERLOCK CIRCUITS

Interlock circuits provide for the safety of the operator by requiring two-handed operation of the controls. This keeps the operator's hands occupied, preventing them from becoming trapped or pinched by the machine. Most production presses are so equipped.

Interlock circuits commonly connect the control valves in series, but this is not always the case. The single-acting circuit in Fig. 17-26 uses two palm button valves in series to actuate a cylinder that is returned by a spring. Both palm buttons must be depressed for air to reach the cylinder. If either button is released, air from the blank end of the piston will be vented and the cylinder will retract. For such a circuit to operate, the cylinder would have to be small because the control valves are handling the air supply to power the cylinder. Another

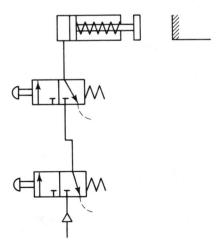

Figure 17-26 Basic single-acting interlock circuit.

problem is that the circuit can be operated with either one of the palm buttons tied down. This means they would have to be mounted on a vertical surface and recessed to prevent the operator from working around the circuit. A number of ingenious mechanisms to circumvent interlock circuits has been discovered after an operator was injured.

The pilot valves could be used to operate a four-way spring-returned power valve as shown in Fig. 17-27. This would allow sizing the power valve to carry the flow required by the air cylinder and permit the use of miniature three-way, spring-return palm button valves. The cylinder rod would retract if either button were released, but it still does not prevent the operator from tying down one of the palm buttons and working around the circuit.

Figure 17-28 shows a circuit that would solve the operator tie-down problem by controlling both the air entering the cylinder and the air leaving the cylinder.

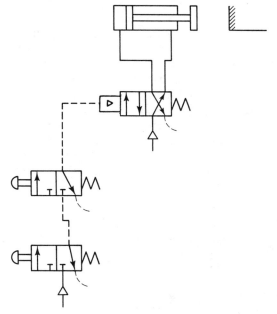

Figure 17-27 Double-acting pilot operated interlock circuit.

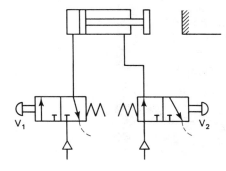

Figure 17-28 Direct operated no-tie down interlock circuit.

Depressing valve 1 directs air from the supply to the blank end of the cylinder, but the return air is blocked by the supply pressure through valve 2. Before the air can be vented from the rod side of the cylinder, valve 2 must be depressed. With both valves depressed, the cylinder will advance. If valve 2 is released and valve 1 depressed, the cylinder will stop with air pressure applied to both sides of the cylinder. If valve 1 is released while valve 2 is held depressed, the cylinder will stop with both sides of the piston vented. Releasing both valves vents valve 1, directs air through valve 2 to the rod side of the piston, and retracts the cylinder. The air pressure to valves 1 and 2 would have to be regulated separately, with the pressure at valve 2 being slightly higher to compensate for the difference in piston areas. This would prevent the tendency for the cylinder to extend if valve 1 were depressed by itself. Also notice that the control valves must handle the power air supply which limits the size of the cylinder.

Air-oil boosters like that shown in Fig. 17-29 use interlock circuits. They are operated from 100–125 lbf/in.^{2} shop air to power a reciprocating hydraulic piston pump. They receive widespread use as portable power supplies for bench pressure, test stands, and other applications where shop air is readily available and where costs and inconvenience do not warrant a more expensive hydraulic system. Fluid pressure to 100,000 lbf/in.^{2} are available. The operation of an air-oil booster can be seen in Fig. 17-30. Notice that the air piston and hydraulic plunger are driven on the downstroke, and returned by the action of the return spring when the air control valve releases the pressure. Cycling is automatic. The hydraulic portion of the pump booster is a single-acting reciprocating plunger which uses two check valves to control the flow of hydraulic fluid. Fluid is drawn into the lower chamber at low pressure through the liquid inlet on the upward stroke of the plunger. The multiplication of pressure is determined by the relative sizes of

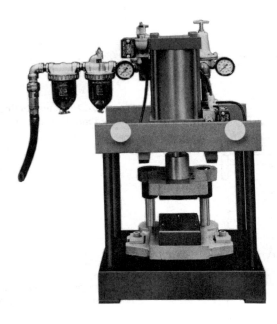

Figure 17-29 Air bench press (*courtesy of Miller Fluid Power Division*).

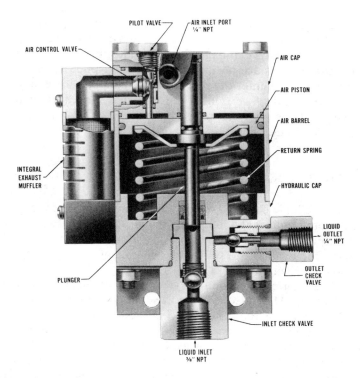

Figure 17-30 Air-oil booster (*courtesy of Haskel Engineering and Supply*).

the plunger piston and air piston. For example, a ratio of air piston diameter to plunger piston diameter of 30:1 would theoretically deliver a hydraulic pressure of 3000 lbf/in.^{2} from shop air supplied at 100 lbf/in^2. In actual practice there are some losses, and the supply pressure would have to be increased slightly.

The interlock circuit for the bench press is shown in Fig. 17-31. The palm

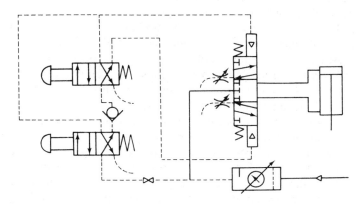

Figure 17-31 Pilot operated, no-tie-down interlock circuit with center STOP position.

actuated four-way pilot valves direct pilot air to the spring-centered three-way, five-port power valve. Both buttons must be depressed to actuate the power valve to extend the cylinder. Both also must be released to retract the cylinder. If the palm buttons are not operated together, pilot air is vented, the power valve control springs center the power valve spool and the cylinder stops. The same is true if the control portion of the circuit is closed off with the manual shut-off valve. And one pilot control valve in the interlock circuit cannot be tied down to make the circuit operate.

17-8 PNEUMATIC LOGIC CONTROL

Pneumatic logic control systems use air and logic functions to manage the operation of other adjacent fluid power systems. Control rather than direct actuation is the output objective, but even in control systems, substantial force is commonly needed and available to shift such components as high-pressure control valves for the main circuit.

Pneumatic logic control systems are available with varying degrees of sophistication. They also operate at several levels of output force. At the highest level, shop air at 100–150 lbf/in.^{2} is used to power miniature pneumatic devices that perform logic functions and deliver substantial force outputs, typically through cylinders. At the lowest level, conditioned air at 10–15 lbf/in.^{2} is used to power fluidic devices. The output force of these systems is small and interface elements are used to raise the output energy to the level necessary to shift electric switches and pilot operate main control valves. At some intermediate level and pressure, moving part logic control system devices combine logic functions with low force outputs to actuate such components as pilot valves.

Logic control systems are not new. They have been used for some time in pneumatics. Electrical relay logic and switching, for example, have been used for more than 50 years to control systems of all types. Pneumatic logic control systems are novel, however, in that they offer several positive advantages to specialized applications. For example, air devices seldom self-destruct as many electrical devices do when subjected to undue resistance, heat, or seizing. Pneumatic logic control is particularly applicable in explosive environments. The life expectancy of pneumatic logic control systems is longer than their electrical counterparts, on the order of 50 to 100 times for such components as high-maintenance switches. Service is usually simplified because problems can be quickly isolated from telltale signs or indicators on the components themselves. Systems are simple in design and easily fabricated by available personnel. Reliability is typically high compared with other methods of logic control.

Air logic control systems frequently respond faster than their electrical control counterparts. While electrical transmission of signals from point to point is nearly instantaneous and faster than an air control signal that travels on the order of 1 ft per millisecond (ms), the response time of switching components is typically longer. For example, the response time of air logic elements is on the order of 10 to 12 ms compared with 50 to 60 ms for most industrial switches and relays and 75 ms for valve solenoids. For transmission line lengths that typically are less than 50 ft, then, component switching rather than signal transmission consumes the major portion of time. In real industrial applications, air logic control systems often respond twice as fast as electrical controls.

17-9 LOGIC CONTROL CONCEPTS

Pneumatic logic control systems use five basic control functions to manage air flow through control valves and actuators. These are the AND, OR, NOT, MEMORY, and TIME.

In an AND circuit, two or more input signals must be present to obtain an output signal. Conversely, the output signal is off when so much as one input signal is off. Figure 17-32 shows an air logic circuit using standard ANSI fluid power symbols. The circuit illustrates that to obtain an output at D, valves A, B, and C must be manually operated. The valves are in series. Releasing any one of the valves results in the loss of the output at D.

In an OR circuit, if a signal is present at A or B or C, an output signal will be present at D. Conversely, all outputs must be absent to discontinue the output at D. In Fig. 17-33, depressing any one of the manually operated valves results in an

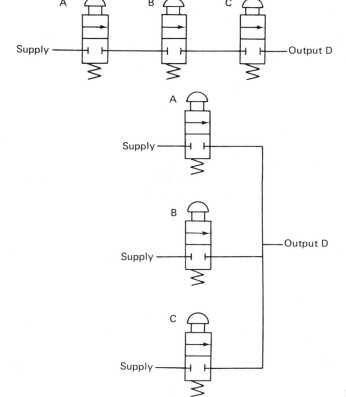

Figure 17-32 AND circuit.

Figure 17-33 OR circuit.

output at D. Because all three of the two-way valves block air flow in the off position, no other valving is necessary. A problem occurs, however, if it is desirable to permit air to return from the output D to vent after momentary application of the pilot signal. The circuit in Fig. 17-34 accomplishes this task by using three-way control valves that provide the return vents, and two shuttle valves that prevent the signal from going to vent. When the signal is applied by operating manual control to valve A, shuttle valves 1 and 2 move to block the return air to vent through valves B and C. Thus, the signal is sent to the output D. Releasing valve A vents the pilot circuit. The same results occur if valves B or C are used to connect the supply to output D. In each case, when all valves are released, the pilot circuit is vented and free to return air from the output D.

In the NOT, the output is on when the input is off. Figure 17-35 shows a

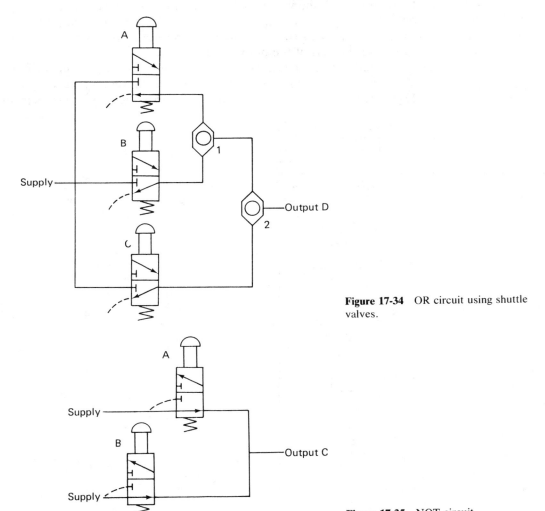

Figure 17-34 OR circuit using shuttle valves.

Figure 17-35 NOT circuit.

NOT circuit in which two manually operated three-way valves, A and B, connect the supply inputs to the output C. The supply is on when both manually operated valves A and B are not operated. If either valve A or B is depressed, the supply is vented and the output at C is lost.

Memory is the ability of the circuit to retain a bias. That is, the circuit will remain in one position until a pilot signal is given to cause it to revert to a former state. In unlimited memory, the command pilot signal causes the circuit to shift to one state and remain in that state until an opposite pilot signal is given manually by an operator. In limited memory, the command signal reverts automatically after a given time, causing the circuit to shift back to its former position.

Figure 17-36 illustrates an unlimited MEMORY circuit. Three-way manual valves A and B pilot the main control valve causing the supply to be connected to output D. In the position shown, a momentary signal from manually operated valve B has piloted the main control valve causing the supply to be connected to output D. The valve will retain this bias indefinitely. If both valves are depressed simultaneously, nothing will occur since pilot pressure applied in the opposite direction to the main control valve will be balanced.

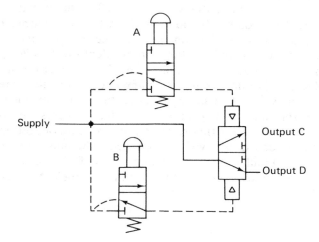

Figure 17-36 Unlimited MEMORY circuit.

Figure 17-37 illustrates a limited MEMORY circuit. Momentarily depressing manually operated pilot valve A sends a signal through the one-way check valve to the main control valve and at the same time charges the accumulator acting as a capacitance for the pilot signal. The manual control valve shifts against the force of the valve spool return spring and connects the supply to output B. Releasing pilot valve A causes the pilot signal to return through the variable flow restrictor and then to vent through the pilot valve. Since the accumulator is charged, the pilot circuit will remain under pressure for a time, depending on the rate at which the flow restrictor valve vents the signal. When the force from the pilot pressure against the main control valve spool becomes less than the return force of the valve spool spring, the main control valve will revert to its original position connecting the supply to output C.

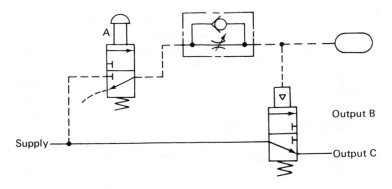

Figure 17-37 Limited MEMORY circuit.

TIME functions are introduced to delay operation of the circuit, usually by placing a resistance in the circuit. Timing-in and timing-out circuits commonly use flow restrictors to accomplish this objective. The limited memory circuit in Fig. 17-37 is also a timing-out circuit. After valve A actuates the circuit and is released, the supply is directed to output B and then to C when pilot air has been exhausted through the flow restrictor. The amount of time elapsed before the main control valve reverts to its original position is dependent on the pilot pressure, the capacity of the accumulator, the setting of the flow restrictor, and the force exerted by the valve spool return spring. A timing-in circuit is illustrated in Fig. 17-38. Depressing pilot valve A sends a signal to the main control valve through the flow restrictor shifting the main control valve that connects the supply to B. Pilot air flow is limited by the restrictor, while pressure cannot built in the pilot circuit until air has charged the reservoir capacitance. The pilot valve must be held depressed until the accumulator becomes charged to shift the main control valve. Releasing pilot valve A allows pilot air to be vented through the one-way check valve with no delay.

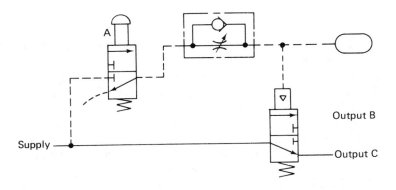

Figure 17-38 Timing-in circuit.

17-10 BASIC AIR LOGIC CIRCUIT DESIGN

When designing air logic control circuits, it is necessary to have clearly established objectives. Usually this requires that the desired sequence of operations that accomplishes the output objective be written out. Manual controls should then be listed together with their function. Input signals other than those from the manual input should be identified from such sources as limit valves, sensors, and other controls. Output devices, such as cylinders, piloted valves, electric motors, and other components that are to be controlled by the circuit should then be listed. Before beginning the design problem, mechanical and safety interlocks that must exist are established. Finally, the circuit problem is solved using logic control concepts and the required components to achieve the desired sequence. After the diagram is drawn, the circuit is checked for proper actuation during start up, shut down, loss of air, panic stops in the middle of the cycle, restarts in the middle of the cycle, and control at other events that are likely to occur.

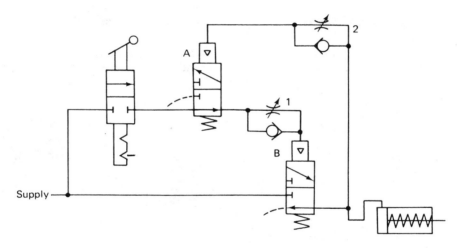

Figure 17-39 Automatic cycling circuit.

Figure 17-39 illustrates an automatic cycling circuit. When the manual control valve is shifted on, air is directed through normally open valve A, flow restrictor 1, and after a delay, pilots normally closed valve B. This directs supply air to the blank end of the cylinder and extends it. Air to the blank end of the cylinder is also directed through flow restrictor 2, and after a delay, pilots valve A. This vents the pilot signal to valve B, which is spring returned and exhausts the blank end of the cylinder and the pilot circuit to valve A. The cylinder returns and the cycle restarts. Elapsed time-in before the circuit is shifted to extend the cylinder is controlled by flow restrictor 1. Elapsed time-out before the circuit is shifted to return the cylinder is controlled by flow restrictor 2.

17-11 MINIATURE PNEUMATICS

Miniature Pneumatics is a recent development adopted from the instrument industry that uses reduced size components. Small actuators are directed by components usually assembled in modules. Actuator output may be used to transfer power directly to the load resistance or to act as a transducer to control larger electrical, mechanical, hydraulic, and pneumatic equipment.

While logic control circuits have been used for some time in the compressed air industry, a complete and uniform line of miniature components similar to full-scale components operating on standard shop air has not been available. Today, a complete line of these components with similar functions and configuration to full-scale components are stock items, including cylinders, valves, fittings, and modular components.

Figure 17-40 illustrates a precision miniature pneumatic cylinder. They are available in sub-miniature sizes from 5/32-in. (4 mm) bore and 1/4-in. (6 mm) stroke, to 1⅛-in. (29 mm) bore and 20-in. stroke (508 mm). Shop air pressures to 250 lbf/in.^{2} (1725 kPa) are used with these cylinders, which are available in double-acting and single-acting designs. Cushions are standard to reduce cylinder shocks and reduce noise during cycling. Cylinders with 1⅛-in. bore deliver a force at the cylinder rod equal to the air pressure since the area equals 1 in^2. Thus a pressure gauge inserted in the system which reads in lbf/in.^{2} also reads the force directly.

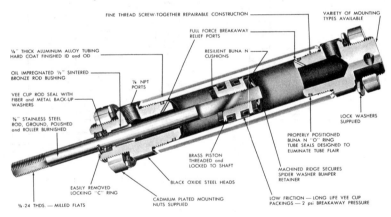

17-12 MOVING PART LOGIC AND MODULAR AIR COMPONENTS

Moving part logic systems control air volumes to 1 ft^3/min to pilot air circuits for larger air power valves. Air to electric and air to hydraulic interfaces are also standard. Moving part logic systems have the capacity to process and qualify information, as well as to accomplish the action sequence. The input comes from the operator at the control station using selector valves and push buttons to designate which operation should be accomplished and how the sequence should begin. The control station typically includes the selectors, starting devices, visual indicators, and most of the logic components mounted in a standard electrical enclosure. A number of these components are shown in Fig. 17-41.

Logic devices are connected by tubing to process information and send momentary or maintained signals. They come as standard plug-in components. Port locations commonly identify standard functions. Logic components include three-way valves, four-way valves, shuttle valves, flow control valves, sensing elements, and a number of interface components. Logic control valves are shifted with pilot air pressure in the 20–40 $lbf/in.^2$ range to control main air at pressures to 150 lbf/in^2. Main air pressure, in turn, acts to pilot larger power valves which control cylinders, motors, and other fluid power components.

Moving part logic circuits are constructed using logic device symbols which conform to ANSI and NFPA standards to build ladder diagrams of machine functions and sequences. Symbols are similar to their electrical counterparts in that detached symbols (nonpictorial representations) are used.

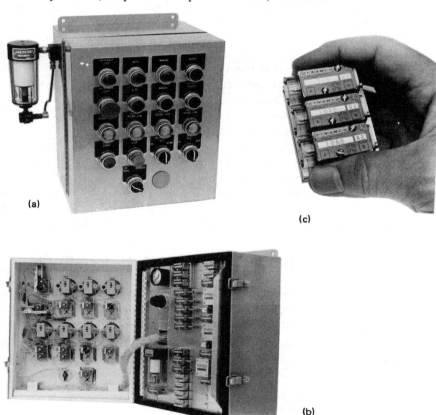

(a)

(c)

(b)

Figure 17-41 Moving part logic control system (*courtesy of Dyamco*); (a) outside operator panel, (b) inside operator control panel, (c) logic elements.

17-13 SUMMARY

Pneumatic circuits are designed from output objectives, the same as hydraulic circuits, but because air is spongy, special provisions must be made to control the speed and stiffen the movement of actuators. The force of a given bore cylinder or torque from a given displacement motor is determined by the air pressure. The speed of an actuator under constant load can be controlled by regulating the flow, and a meter-out circuit will stiffen the drive. The pneumatic hoist circuit shown in Fig. 17-1, for example, lifts the load resistance by releasing the back pressure on the blank end of the piston through a three-way directional control valve and meters it through the built-in flow control valve at the exhaust port. Pressure on both sides of the piston stiffen the drive, and the flow control valve regulates the speed.

The procedure to construct a pneumatic circuit is determined from the sequence of operations the circuit is to perform, followed by drawing the layout using standard symbols. The circuit must be safe, not only during operation, but when it is shut off to prevent injury to mechanics and set-up personnel. This means the circuit must be equipped with two-hand, no-tie-down operators, as well as safety interlocks, emergency stop buttons, and retract and shut-down provisions. Pressure gauges mounted at the control panel and near actuators will let personnel know whether the system is under pressure. And installing isolation valves will permit maintenance personnel to shut off the system and bleed off pressure at cylinders that could actuate with accidental operation of directional control and limit valves.

Pneumatic logic control circuits operate the air power system by switching valves and other components. Logic components include three-way valves, four-way valves, shuttle valves, flow control valves, sensing elements, accumulators, and a number of interface elements. In a typical logic control system, logic control valves in a panel box enclosure are shifted with pilot air pressures in the 20–40 lbf/in.^{2} range to control main air at 150 lbf/in.^{2} to pilot the power valves mounted near cylinders, motors, and other fluid power components.

STUDY QUESTIONS AND PROBLEMS

1. From the discussion of circuit design, list the steps to arrive at a circuit application.
2. What are the differences between a *power* circuit and a *control* circuit?
3. What safety hazards are associated with compressed air circuits?
4. Basically, how do you position and stiffen the drive in pneumatic circuits?
5. Complete the double-acting circuit in Fig. 17-42 to fulfill the following conditions:
   (a) Velocity control—both directions.

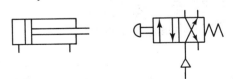

Figure 17-42 Figure for Problem 5.

(b) Velocity control—extension only.

(c) Velocity control—cylinder rod can extend two times faster than it retracts.

6. Figure 17-43 shows a double-end-rod cylinder controlled by two three-way directional control valves. List the four possible operating conditions for the circuit.

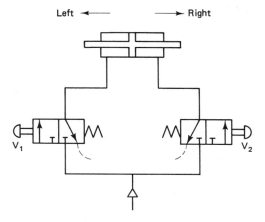

Figure 17-43 Figure for Problem 6.

7. In Problem 6, Fig. 17-43, the cylinder rod runs free in the unoperated position. Redraw the circuit using the same components to fulfill the following conditions:

(a) When valve V1 is operated, the cylinder rod moves left.

(b) When valve V2 is operated, the cylinder moves right.

(c) Releasing either valve in mid-travel stops the cylinder rod, even under an overrunning load.

(d) When both valves are operated, the cylinder rod runs free.

8. How could velocity control in both directions be added to the circuit in Problem 7?

9. The circuit in Fig. 17-44 operates an air motor. How might the circuit be modified with a pressure regulator to provide two-speed operation?

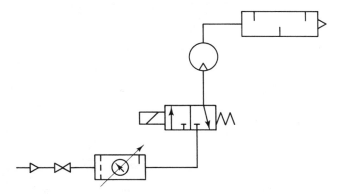

Figure 17-44 Figure for Problem 9.

10. Modify the circuit in Fig. 17-44 to include a palm-operated "Stop" button, and substitute a pilot-operated power valve for the solenoid-operated power valve.

11. The circuit in Fig. 17-45 is supposed to lift and return two single-acting cylinders in parallel. The load on the left cylinder is twice as large as the one on the right. What might be expected to happen when the control valve is shifted to lift?

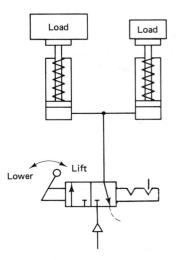

Figure 17-45 Figure for Problem 11.

12. How could the circuit in Fig. 17-45 be modified to make the cylinders lift and retract in parallel?

13. The cylinder rod in Fig. 17-46 will extend and return to remain at rest when the "Start" button is depressed and released. Redesign the circuit so that depressing a "Stop" button will stop the circuit cylinder rod travel in either direction until the "Start" button again is depressed, after which the cylinder will extend.

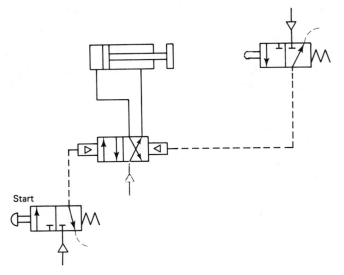

Figure 17-46 Figure for Problem 13.

14. Modify the circuit in Problem 13 by adding a "Retract" button. The functions then would be "Start", "Stop," and "Retract". Remember that the circuit still must advance when the "Start" button is depressed.

15. Figure 17-47 shows a circuit that clamps and then flares a tube at one end. What happens when the power valve V1 is actuated? What happens when the power valve is returned?

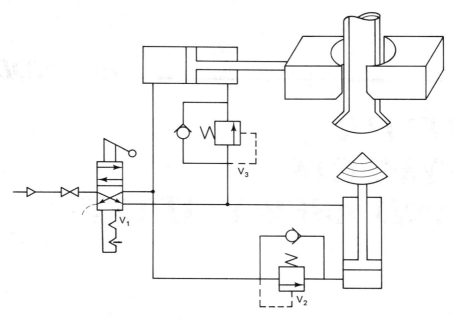

Figure 17-47 Figure for Problem 15.

16. List the conditions that the circuit in Problem 15 can accomplish.

17. When the power valve is reversed in Fig. 17-47, the blank end of the clamp cylinder is vented and may lose its grip on the work piece. How could the circuit be modified to prevent this?

18. Use three-way and shuttle valve symbols to construct a logic circuit that will accomplish the following statement?

A and B or C will result in an output at D.

19. How would the circuit in Problem 18 be changed if the logic statement were: A and not B or C will result in an output at D?

20. What is the difference between *limited* and *unlimited* memory, and how is each of these accomplished?

21. Basically, what is the difference between a *timing-in* and a *timing-out* circuit, and how is one changed to the other?

22. What is *capacitance* in a timing-in or timing-out circuit and how is it increased? How is the *time delay* increased in a timing-in or timing-out circuit?

LETTER SYMBOLS AND ABBREVIATIONS

Symbol	Definition	Units
A	Area	square inches (in²)
		square meters (m²)
		square centimeters (cm²)
C	Various constants	
D,d	Diameter	inches (in.)
		meters (m)
		centimeters (cm)
E	Electric pressure	volts
F	Force	pounds (lbf)
		Newton (N)
		dyne
G	Work	foot-pounds (ft-lbf)
		meter-Newtons (m-N)
		centimeter-dyne (cm-dyn)
		joule (J)
H	Head energy	feet (ft)
		metres (m)
I	Electric energy flow	amperes (amp)
J	Joule's constant	778 (ft-lbf/Btu)
		(4.26×10^4 gm-cm/calorie)
K	Various constants	

Symbol	Definition	Units
L,l	Distance	inches (in.) feet (ft) meters (m) centimeters (cm)
M,m	Mass	pound mass (lbm) slugs (lbf-sec²/ft) kilograms (N-sec²/m) gram (dyne-sec²/cm)
N	Speed Newton	revolutions per minute (rpm) kilogram-meter per second per second (kg-m/sec²)
p	Pressure	pounds per square inch (lbf/ in²) (psi) Newtons per square meter (N/m²) Pascals (Pa) (1 Pa = 1 N/m²) kilograms force per square centimeter (kgf/cm²) bars (1 bar = approx. 1 atm. = 10⁵ N/m²)
Q,q	Liquid volume flow rate	gallons per minute (gal/min) liters per minute (lpm)
R	Radius	inches (in.) meters (m) centimeters (cm)
S	Stroke	feet (ft) inches (in.) meters (m) centimeters (cm)
T	Torque Temperature	pound-feet (lbf-ft) pound-inches (lbf-in) ounce-inches (oz-in.) Newton-meter (N-m) kilogram-meters (kgf-m) gram-centimeters (gmf-cm) degrees Fahrenheit (°F) degrees Celsius (°C)
V	Volume	cubic feet (ft³) cubic inches (in³) cubic meters (m³) cubic centimeters (cm³)
W	Electrical power	watts

Symbol	Definition	Units
a	Acceleration	feet per second per second (ft/sec²)
		meters per second per second (m/s²)
e	Efficiency	percent (%)
f, CF, C_f	Coefficient of friction	without dimension
g	Acceleration due to gravity	32.2 feet per second per second (32.2 ft/sec²)
		9.8 meters per second per second (9.81 m/s²)
g_a	Acceleration factor	without dimension
h	Height or head	feet (ft)
		meters (m)
n	Number	
p	Hardness	
r	Radius	inches (in.)
		centimeters (cm)
	Ratio	without dimension
s	Percent slip	percent (%)
t	Time	hours
		minutes (min)
		seconds (sec)
u	Specific volume	cubic feet per pound (ft³/lbf)
		cubic meters per Newton (m³/N)
v	Velocity	feet per minute (ft/min, fpm)
		feet per second (ft/sec, fps)
		meters per second (m/s)
		centimeters per second (cm/s)
w	Weight	pounds (lbf)
		kilogram force (kgf)
		Newtons
		dynes

Greek Symbols	Definition	Units
γ (gamma)	Specific weight	pounds per cubic foot (lbf/ ft³)
		Newtons per cubic meter (N/ m³)
		dynes per cubic centimeter (dynes/cm³)
ρ (rho)	Density	slugs per cubic foot (slugs/ ft³)
		kilograms per cubic meter (kg/m³)
		grams per cubic centimeter (g/cm³)
μ (mu)	Absolute or dynamic viscosity	reyns (lb-sec/in.²)
		Poise (dyne-sec/cm²)
		centipoise (.01 poise) or cP
υ (nu)	Kinematic viscosity	Newts (in.²/sec)
		Stokes (cm²/sec) or St
		centistokes (.01 Stokes) or cSt

Other Abbreviations	Definition
SSU	Saybolt seconds universal viscosity
RMS	Root mean square
CR	Compression ratio

CONVERSION FACTORS

Units

bar $\times$ 10⁵ = Pa (N/m²)

bar = 14.5 lbf/in² (psi) (approximately 1 atmosphere)

dyne $\times$ 10⁻⁵ = N

gal $\times$ (3.785 $\times$ 10⁻³) = m³

gal $\times$ 231 = in³

hp $\times$ 746 = P (watts)

in $\times$ (2.54 $\times$ 10⁻²) = m

lbf $\times$ 4.448 = N

metric ton $\times$ 10³ = kgf

short ton (2000 lbf) $\times$ (9.072 $\times$ 10²) = kgf

statute mile $\times$ (1.609 $\times$ 10³) = m

α_{water} = 9802 N/m³ = 62.4 lbf/ft³ @ 10°C (50°F)

ρ_{water} = 1000 kg/m³ = 1.94 slugs/ft³ @ 4°C (39.4°F)

Absolute Viscosity

lbf·sec/ft² (no special name) $\times$ 47.88 = Pa·s (N·s/m²)

lbf·sec/in² (reyns) $\times$ (6.895 $\times$ 10³) = Pa·s (N·s/m²)

Poise (dyne·sec/cm²) $\times$ 10⁻¹ = Pa·s(N·s/m²)

Poise $\times$ 10² = cP (Centipoise)

cP $\times$ 10⁻³ = Pa·s (N·/m²)

Kinematic Viscosity

ft²/sec (no special name) $\times$ (9.29 $\times$ 10⁻²) = m²/s

in²/sec (Newts) $\times$ (6.45 $\times$ 10⁻⁴) = m²/s

Stoke (cm²/s) $\times$ 10⁻⁴ = m²/s

Stoke $\times$ 10² = cSt (Centistoke)

Pressure
(See Table 2-1)

APPROXIMATE VISCOSITY CONVERSIONS

SSU	cSt	ft^2/sec	SI (m^2/s)
31.0	1.00	10.76×10^{-6}	1.0×10^{-6}
31.5	1.13	12.16×10^{-6}	1.13×10^{-6}
32.0	1.81	19.48×10^{-6}	1.81×10^{-6}
32.6	2.00	21.53×10^{-6}	2.0×10^{-6}
33.0	2.11	22.71×10^{-6}	2.11×10^{-6}
34.0	2.40	25.83×10^{-6}	2.40×10^{-6}
35	2.71	29.17×10^{-6}	2.71×10^{-6}
36	3.00	32.29×10^{-6}	3.0×10^{-6}
38	3.64	39.18×10^{-6}	3.64×10^{-6}
39.2	4.00	43.06×10^{-6}	4.00×10^{-6}
40	4.25	45.75×10^{-6}	4.25×10^{-6}
42	4.88	52.53×10^{-6}	4.88×10^{-6}
42.4	5.00	53.82×10^{-6}	5.00×10^{-6}
44	5.50	59.20×10^{-6}	5.50×10^{-6}
45.6	6.00	64.58×10^{-6}	6.00×10^{-6}
46	6.13	65.98×10^{-6}	6.13×10^{-6}
46.9	7.00	75.35×10^{-6}	7.00×10^{-6}
50	7.36	79.22×10^{-6}	7.36×10^{-6}

SSU	cSt	ft²/sec	SI (m²/s)
52.1	8.00	86.11×10^{-6}	8.00×10^{-6}
55	8.88	95.58×10^{-6}	8.88×10^{-6}
55.4	9.00	96.88×10^{-6}	9.00×10^{-6}
58.8	10.00	10.76×10^{-5}	1.00×10^{-5}
60	10.32	11.11×10^{-5}	1.03×10^{-5}
65	11.72	12.62×10^{-5}	1.17×10^{-5}
70	13.08	14.08×10^{-5}	1.31×10^{-5}
75	14.38	15.48×10^{-5}	1.44×10^{-5}
80	15.66	16.86×10^{-5}	1.57×10^{-5}
85	16.90	18.19×10^{-5}	1.69×10^{-5}
90	18.12	19.50×10^{-5}	1.81×10^{-5}
95	19.32	20.80×10^{-5}	1.93×10^{-5}
100	20.52	22.09×10^{-5}	2.05×10^{-5}
120	25.15	27.07×10^{-5}	2.52×10^{-5}
140	29.65	31.91×10^{-5}	2.97×10^{-5}
160	34.10	36.70×10^{-5}	3.41×10^{-5}
180	38.52	41.46×10^{-5}	3.85×10^{-5}
200	42.95	46.23×10^{-5}	4.30×10^{-5}
300	64.60	69.53×10^{-5}	6.46×10^{-5}
400	86.20	92.78×10^{-5}	8.62×10^{-5}
500	108.00	11.63×10^{-4}	1.80×10^{-4}
600	129.40	13.93×10^{-4}	1.29×10^{-4}
700	151.00	16.25×10^{-4}	1.51×10^{-4}
800	172.6	18.58×10^{-4}	1.73×10^{-4}
900	194.2	20.90×10^{-4}	1.94×10^{-4}
1000	215.8	23.23×10^{-4}	2.16×10^{-4}
2000	431.7	46.47×10^{-4}	4.32×10^{-4}
5000	1078.8	11.61×10^{-3}	10.78×10^{-4}
7000	1510.3	16.26×10^{-3}	15.10×10^{-4}
9000	1941.1	20.92×10^{-3}	19.41×10^{-4}
10000	2157.6	23.22×10^{-3}	21.57×10^{-4}
15000	3236.5	34.83×10^{-3}	32.36×10^{-4}
20000	4315.3	46.45×10^{-3}	43.15×10^{-4}

Data generated in part from *Cameron Hydraulic Data,* Appendix 4-24, Ingersoll-Rand Corporation, 1977.

Approximate Viscosity Conversions

STANDARD GRAPHIC SYMBOLS FOR FLUID POWER DIAGRAMS*

1. Introduction

1.1 General

Fluid power systems are those that transmit and control power through use of a pressurized fluid (liquid or gas) within an enclosed circuit.

Types of symbols commonly used in drawing circuit diagrams for fluid power systems are Pictorial, Cutaway, and Graphic. These symbols are fully explained in the USA Standard Drafting Manual (Ref. 2).

1.1.1 *Pictorial symbols* are very useful for showing the interconnection of components. They are difficult to standardize from a functional basis.

1.1.2 *Cutaway symbols* emphasize construction. These symbols are complex to draw and the functions are not readily apparent.

1.1.3 *Graphic symbols* emphasize the function and methods of operation of components. These symbols are simple to draw. Component functions and methods of operation are obvious. Graphic symbols are capable of crossing language barriers, and can promote a universal understanding of fluid power systems.

Graphic symbols for fluid power systems should be used in conjunction with the graphic symbols for other systems published by the USA Standards Institute (Ref. 3–7 inclusive).

1.1.3.1 Complete graphic symbols are those which give symbolic representation of the component and all of its features pertinent to the circuit diagram.

1.1.3.2 Simplified graphic symbols are stylized versions of the complete symbols.

1.1.3.3 Composite graphic symbols are an organization of simplified or complete symbols. Composite symbols usually represent a complex component.

1.2 Scope and Purpose

1.2.1 *Scope*

This standard presents a system of graphic symbols for fluid power diagrams.

1.2.1.1 Elementary forms of symbols are:

Circles	Triangles	Lines
Squares	Arcs	Dots
Rectangles	Arrows	Crosses

1.2.1.2 Symbols using words or their abbreviations are avoided. Symbols capable of crossing language barriers are presented herein.

1.2.1.3 Component function rather than construction is emphasized by the symbol.

1.2.1.4 The means of operating fluid power components are shown as part of the symbol (where applicable).

1.2.1.5 This standard shows the basic symbols, describes the principles on which the symbols are based, and illustrates some representative composite symbols. Composite symbols can be devised for any fluid power component by combining basic symbols.

Simplified symbols are shown for commonly used components.

---

* Extracted from *USA Standard Graphic Symbols for Fluid Power Diagrams* [USAS Y32.10-1967 (Reaffirmed 1979)], American Society of Mechanical Engineers, with the permission of the publisher.

1.2.1.6 This standard provides basic symbols which differentiate between hydraulic and pneumatic fluid power media.

1.2.2 *Purpose*

1.2.2.1 The purpose of this standard is to provide a system of fluid power graphic symbols for industrial and educational purposes.

1.2.2.2 The purpose of this standard is to simplify design, fabrication, analysis, and service of fluid power circuits.

1.2.2.3 The purpose of this standard is to provide fluid power graphic symbols which are internationally recognized.

1.2.2.4 The purpose of this standard is to promote universal understanding of fluid power systems.

1.3 Terms and Definitions

Terms and corresponding definitions found in this standard are listed in Ref. 8.

2. Symbol Rules (See Section 10)

2.1 Symbols show connections, flow paths, and functions of components represented. They can indicate conditions occurring during transition from one flow path arrangement to another. Symbols do not indicate construction, nor do they indicate values, such as pressure, flow rate, and other component settings.

2.2 Symbols do not indicate locations of ports, direction of shifting of spools, or positions of actuators on actual component.

2.3 Symbols may be rotated or reversed without altering their meaning except in the cases of: a.) Lines to Reservoir, 4.1.1; b.) Vented Manifold, 4.1.2.3; c.) Accumulator, 4.2.

2.4 Line Technique (See Ref. 1)

Keep line widths approximately equal. Line width does not alter meaning of symbols.

2.4.1 Solid Line

(Main line conductor, outline, and shaft)

2.4.2 Dash Line

(Pilot line for control)

2.4.3 Dotted Line

(Exhaust or Drain Line)

2.4.4 Center Line

(Enclosure outline)

2.4.5 Lines Crossing
(The intersection is not necessarily at a 90 deg angle.)

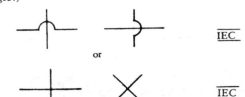

2.4.6 Lines Joining

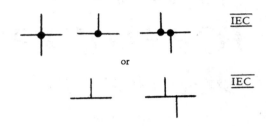

2.5 Basic symbols may be shown any suitable size. Size may be varied for emphasis or clarity. Relative sizes should be maintained. (As in the following example.)

2.5.1 Circle and Semi-Circle

2.5.1.1 Large and small circles may be used to signify that one component is the "main" and the other the auxiliary.

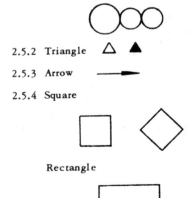

2.5.2 Triangle

2.5.3 Arrow

2.5.4 Square

Rectangle

Standard Graphic Symbols for Fluid Power Diagrams

2.6 Letter combinations used as parts of graphic symbols are not necessarily abbreviations.

2.7 In multiple envelope symbols, the flow condition shown nearest an actuator symbol takes place when that control is caused or permitted to actuate.

2.8 Each symbol is drawn to show normal, at-rest, or neutral condition of component unless multiple diagrams are furnished showing various phases of circuit operation. Show an actuator symbol for each flow path condition possessed by the component.

2.9 An arrow through a symbol at approximately 45 degrees indicates that the component can be adjusted or varied.

2.10 An arrow parallel to the short side of a symbol, within the symbol, indicates that the component is pressure compensated.

2.11 A line terminating in a dot to represent a thermometer is the symbol for temperature cause or effect.

See Temperature Controls 7.9, Temperature Indicators and Recorders 9.1.2, and Temperature Compensation 10.16.3 and 4.

2.12 External ports are located where flow lines connect to basic symbol, except where component enclosure symbol is used.

External ports are located at intersections of flow lines and component enclosure symbol when enclosure is used, see Section 11.

2.13 Rotating shafts are symbolized by an arrow which indicates direction of rotation (assume arrow on near side of shaft).

3. Conductor, Fluid

3.1 Line, Working (main)

3.2 Line, Pilot (for control)

3.3 Line, Exhaust and Liquid Drain

3.4 Line, sensing, etc. such as gage lines shall be drawn the same as the line to which it connects.

3.5 Flow, Direction of

   3.5.1 Pneumatic

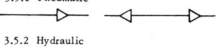

   3.5.2 Hydraulic

3.6 Line, Pneumatic
    Outlet to Atmosphere

   3.6.1 Plain orifice, unconnectable

   3.6.2 Connectable orifice (e. g. Thread)

3.7 Line with Fixed Restriction

3.8 Line, Flexible

3.9 Station, Testing, measurement, or power take-off

   3.9.1 Plugged port

3.10 Quick Disconnect

3.10.1 Without Checks

Connected

Disconnected

3.10.2 With Two Checks

Connected

Disconnected
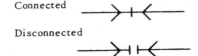

3.10.3 With One Check

Connected

Disconnected
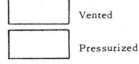

3.11 Rotating Coupling

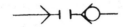

4. Energy Storage and Fluid Storage

4.1 Reservoir

Vented

Pressurized

Note: Reservoirs are conventionally drawn in the horizontal plane. All lines enter and leave from above. Examples:

4.1.1 Reservoir with Connecting Lines

Above Fluid Level

Below Fluid Level

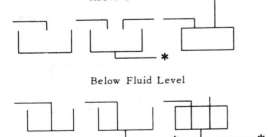

* Show line entering or leaving below reservoir only when such bottom connection is essential to circuit function.

4.1.2 Simplified symbol

The symbols are used as part of a complete circuit. They are analogous to the ground symbol of electrical diagrams. ——|ı $\overline{\text{IEC}}$. Several such symbols ⊔⊔ may be used in one diagram to represent the same reservoir.

4.1.2.1 Below Fluid Level

4.1.2.2 Above Fluid Level

(The return line is drawn to terminate at the upright legs of the tank symbol.)

4.1.2.3 Vented Manifold

4.2 Accumulator

4.2.1 Accumulator, Spring Loaded

4.2.2 Accumulator, Gas Charged

4.2.3 Accumulator, Weighted

4.3 Receiver, for Air or Other Gases

Standard Graphic Symbols for Fluid Power Diagrams

4.4 Energy Source
(Pump, Compressor, Accumulator, etc.)

This symbol may be used to represent a fluid power source which may be a pump, compressor, or another associated system.

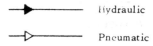 Hydraulic

Pneumatic

Simplified Symbol

Example:

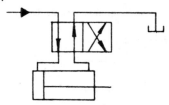

5. Fluid Conditioners

Devices which control the physical characteristics of the fluid.

5.1 Heat Exchanger

5.1.1 Heater

Inside triangles indicate the introduction of heat.

Outside triangles show the heating medium is liquid.

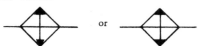

Outside triangles show the heating medium is gaseous.

5.1.2 Cooler

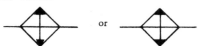 or

Inside triangles indicate heat dissipation

(Corners may be filled in to represent triangles.)

5.1.3 Temperature Controller
(The temperature is to be maintained between two predetermined limits.)

 or

5.2 Filter — Strainer

5.3 Separator

5.3.1 With Manual Drain

5.3.2 With Automatic Drain

5.4 Filter — Separator

5.4.1 With Manual Drain

5.4.2 With Automatic Drain

5.5 Dessicator (Chemical Dryer)

5.6 Lubricator

5.6.1 Less Drain

5.6.2 With Manual Drain

6. Linear Devices

6.1 Cylinders, Hydraulic & Pneumatic

6.1.1 Single Acting

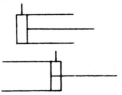

6.1.2 Double Acting

6.1.2.1 Single End Rod

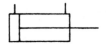

6.1.2.2 Double End Rod

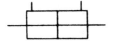

6.1.2.3 Fixed Cushion, Advance & Retract

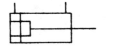

6.1.2.4 Adjustable Cushion, Advance Only

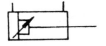

6.1.2.5 Use these symbols when diameter of rod compared to diameter of bore is significant to circuit function.

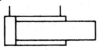

(Non-Cushion)

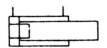

(Cushion, Advance & Retract)

6.2 Pressure Intensifier

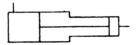

6.3 Servo Positioner (Simplified)

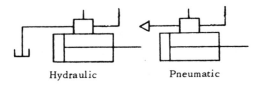

Hydraulic   Pneumatic

6.4 Discrete Positioner

Combine two or more basic cylinder symbols.

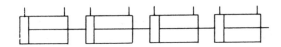

7. Actuators and Controls

7.1 Spring

7.2 Manual

(Use as general symbol without indication of specific type; i.e., foot, hand, leg, arm)

7.2.1 Push Button

7.2.2 Lever

7.2.3 Pedal or Treadle

7.3 Mechanical

7.4 Detent

(Show a notch for each detent in the actual component being symbolized. A short line indicates which detent is in use.) Detent may, for convenience, be positioned on either end of symbol.

7.5 Pressure Compensated

7.6 Electrical

7.6.1 Solenoid (Single Winding)

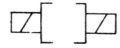

7.6.2 Reversing Motor

7.7 Pilot Pressure

7.7.1

Remote Supply

7.7.2

Internal Supply

7.7.3 Actuation by Released Pressure

by Remote Exhaust

by Internal Return

7.7.4 Pilot Controlled, Spring Centered

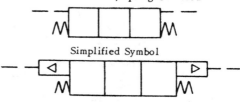

Simplified Symbol

Complete Symbol

7.7.5 Pilot Differential

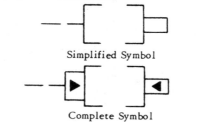

Simplified Symbol

Complete Symbol

7.8 Solenoid Pilot

7.8.1 Solenoid or Pilot

External Pilot Supply

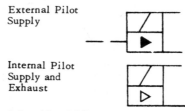

Internal Pilot Supply and Exhaust

7.8.2 Solenoid and Pilot

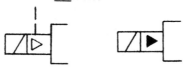

7.9 Thermal

A mechanical device responding to thermal change.

7.9.1 Local Sensing

7.9.2 With Bulb for Remote Sensing

7.10 Servo

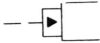

(This symbol contains representation for energy input, command input, and resultant output.)

7.11 Composite Actuators (and, or, and/or)

Basic One signal only causes the device to operate.

And One signal and a second signal both cause the device to operate.

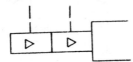

Or One signal or the other signal causes the device to operate

And/Or The solenoid and the pilot or the manual override alone causes the device to operate.

The solenoid and the pilot or the manual override and the pilot

The solenoid and the pilot or a manual override and the pilot or a manual override alone.

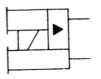

8. Rotary Devices

8.1 Basic Symbol

8.1.1 With Ports

8.1.2 With Rotating Shaft, with control, and with Drain

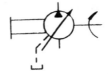

8.2 Hydraulic Pump

8.2.1 Fixed Displacement.

8.2.1.1 Unidirectional

8.2.1.2 Bidirectional

8.2.2 Variable Displacement, Non-Compensated

8.2.2.1 Unidirectional

Simplified

Standard Graphic Symbols for Fluid Power Diagrams

Complete

8.2.2.2 Bidirectional

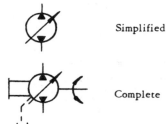

Simplified

Complete

8.2.3 Variable Displacement, Pressure Compensated

8.2.3.1 Unidirectional

Simplified

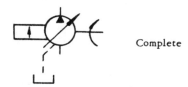

Complete

8.2.3.2 Bidirectional

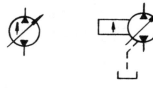

Simplified   Complete

8.3 Hydraulic Motor

8.3.1 Fixed Displacement

8.3.1.2 Bidirectional

8.3.2 Variable Displacement

8.3.2.1 Unidirectional

8.3.2.2 Bidirectional

8.4 Pump-Motor, Hydraulic

8.4.1 Operating in one direction as a pump. Operating in the other direction as a motor.

8.4.1.1 Complete Symbol

8.4.1.2 Simplified Symbol

8.4.2 Operating one direction of flow as either a pump or as a motor.

8.4.2.1 Complete Symbol

8.4.2.2 Simplified Symbol

8.4.3 Operating in both directions of flow either as a pump or as a motor.
(Variable displacement, pressure compensated shown)

8.4.3.1 Complete Symbol

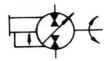

8.4.3.2 Simplified Symbol

8.5 Pump, Pneumatic

8.5.1 Compressor, Fixed Displacement

8.5.2 Vacuum Pump, Fixed Displacement

8.6 Motor, Pneumatic
8.6.1 Unidirectional

8.6.2 Bidirectional

8.7 Oscillator
8.7.1 Hydraulic

8.7.2 Pneumatic

8.8 Motors, Engines
8.8.1 Electric Motor

                                    IEC

8.8.2 Heat Engine (E. G. internal combustion engine)

9. Instruments and Accessories

9.1 Indicating and Recording
9.1.1 Pressure

9.1.2 Temperature

9.1.3 Flow Meter
9.1.3.1   Flow Rate

9.1.3.2   Totalizing

9.2 Sensing
9.2.1 Venturi

9.2.2 Orifice Plate

9.2.3 Pitot Tube

9.2.4 Nozzle

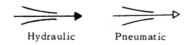

Hydraulic   Pneumatic

9.3 Accessories

9.3.1 Pressure Switch

9.3.2 Muffler

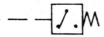

10. Valves

A basic valve symbol is composed of one or more envelopes with lines inside the envelope to represent flow paths and flow conditions between ports. Three symbol systems are used to represent valve types: single envelope, both finite and infinite position; multiple envelope, finite position; and multipe envelope, infinite position.

10.1 In infinite position single envelope valves, the envelope is imagined to move to illustrate how pressure or flow conditions are controlled as the valve is actuated.

10.2 Multiple envelopes symbolize valves providing more than one finite flow path option for the fluid. The multiple envelope moves to represent how flow paths change when the valving element within the component is shifted to its finite positions.

10.3 Multiple envelope valves capable of infinite positioning between certain limits are symbolized as in 10.2 above with the addition of horizontal bars which are drawn parallel to the envelope. The horizontal bars are the clues to the infinite positioning function possessed by the valve re-represented.

10.4 Envelopes

10.5 Ports

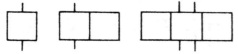

10.6 Ports, Internally Blocked

Symbol System 10.1

Symbol System 10.2

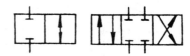

10.7 Flow Paths, Internally Open (Symbol System 10.1 and 10.2)

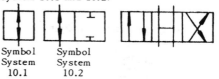

Symbol System 10.1   Symbol System 10.2

10.8 Flow Paths, Internally Open (Symbol System 10.3)

10.9 Two-Way Valves (2 Ported Valves)

10.9.1 On-Off (Manual Shut-Off)

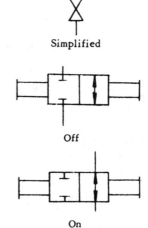

Simplified

Off

On

10.9.2 Check

 Simplified Symbol

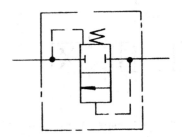

Flow to the right is blocked. Flow to the left is permitted)

(Composite Symbol)

10.9.3 Check, Pilot-Operated to Open

10.9.4 Check, Pilot-Operated to Close

10.9.5 Two-Way Valves

10.9.5.1 Two-Position

Normally Closed   Normally Open

10.9.5.2 Infinite Position

Normally Closed   Normally Open

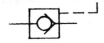

10.10 Three-Way Valves

10.10.1 Two-Position

10.10.1.1 Normally Open

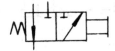

10.10.1.2 Normally Closed

10.10.1.3 Distributor (Pressure is distributed first to one port, then the other)

10.10.1.4 Two-Pressure

10.10.2 Double Check Valve

Double check valves can be built with and without "cross bleed". Such valves with two poppets do not usually allow pressure to momentarily "cross bleed" to return during transition. Valves with one poppet may allow "cross bleed" as these symbols illustrate.

10.10.2.1 Without Cross Bleed (One Way Flow)

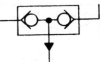

10.10.2.2 With Cross Bleed (Reverse Flow Permitted)

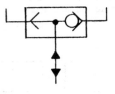

10.11 Four-Way Valves

10.11.1 Two Position

   Normal

     Standard Graphic Symbols for Fluid Power Diagrams

 Actuated

10.11.2 Three Position

(a) Normal

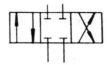

(b) Actuated Left

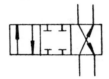

(c) Actuated Right

10.11.3 Typical Flow Paths for Center Condition of Three Position Valves

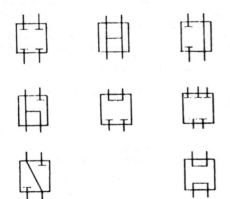

10.11.4 Two-Position, Snap Action with Transition.

As the valve element shifts from one position to the other, it passes through an intermediate position. If it is essential to circuit function to symbolize this "in transit" condition, it can be shown in the center position, enclosed by dashed lines.

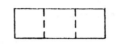

Typical Transition Symbol

10.12 Infinite Positioning (Between Open & Closed)

10.12.1 Normally Closed

10.12.2 Normally Open

10.13 Pressure Control Valves

10.13.1 Pressure Relief

Simplified Symbol
Denotes

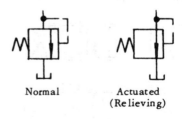

Normal   Actuated
                (Relieving)

10.13.2 Sequence

10.13.3 Pressure Reducing

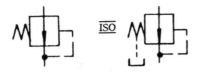

10.13.4 Pressure Reducing and Relieving

10.13.5 Airline Pressure Regulator (Adjustable, Relieving)

10.14 Infinite Positioning Three-Way Valves

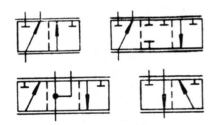

10.15 Infinite Positioning Four-Way Valves

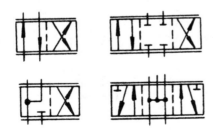

10.16 Flow Control Valves (See 3.7)

10.16.1 Adjustable, Non-Compensated (Flow control in each direction)

10.16.2 Adjustable with Bypass

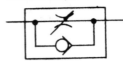

Flow is controlled to the right
Flow to the left by-passes control

10.16.3 Adjustable and Pressure Compensated With Bypass

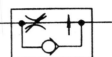

10.16.4 Adjustable, Temperature & Pressure Compensated

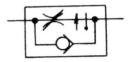

11. Representative Composite Symbols

11.1 Component Enclosure

Component enclosure may surround a complete symbol or a group of symbols to represent an assembly. It is used to convey more information about component connections and functions. Enclosure indicates extremity of component or assembly. External ports are assumed to be on enclosure line and indicate connections to component.

Flow lines shall cross enclosure line without loops or dots.

11.2 Airline Accessories (Filter, Regulator, and Lubricator)

Composite

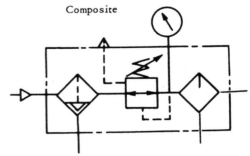

Standard Graphic Symbols for Fluid Power Diagrams

Simplified

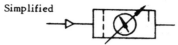

11.3 Pumps and Motors

11.3.1 Pumps

11.3.1.1 Double, Fixed Displacement, One Inlet and Two Outlets

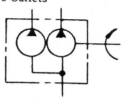

11.3.1.2 Double, with Integral Check Unloading and Two Outlets

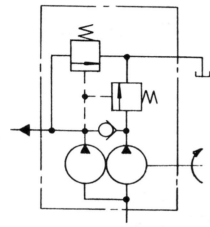

11.3.1.3 Integral Variable Flow Rate Control with Overload Relief

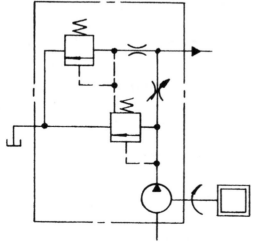

11.3.1.4 Variable Displacement with Integral Replenishing Pump and Control Valves

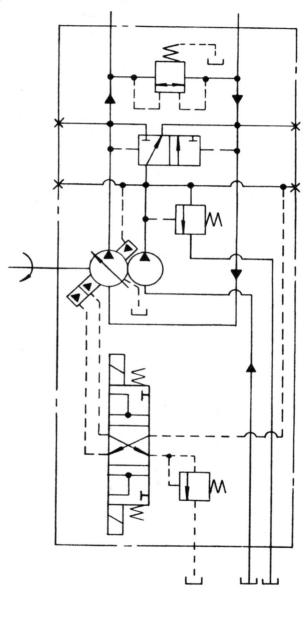

Variable displacement with manual, electric, pilot, and servo control.

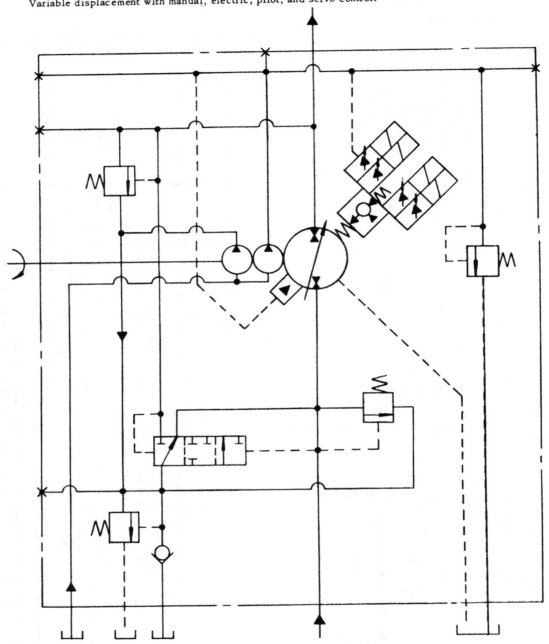

Standard Graphic Symbols for Fluid Power Diagrams

11.4 Valves

11.4.1 Relief, Balanced Type

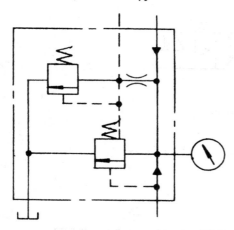

11.4.2 Remote Operated Sequence with Integral Check

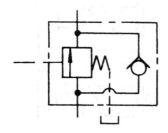

11.4.3 Remote & Direct Operated Sequence with Differential Areas and Integral Check

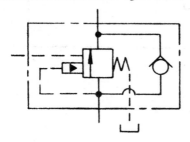

11.4.4 Pressure Reducing with Integral Check

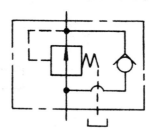

11.4.5 Pilot Operated Check

11.4.5.1 Differential Pilot Opened

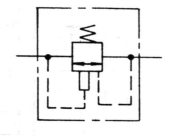

11.4.5.2 Differential Pilot Opened and Closed

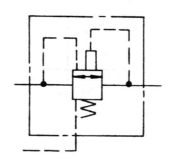

11.4.6 Two Positions, Four Connection Solenoid and Pilot Actuated, with Manual Pilot Override.

Simplified Symbol

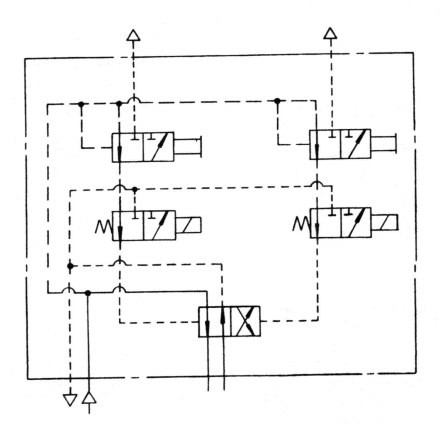

Complete Symbol

Standard Graphic Symbols for Fluid Power Diagrams

11.4.7 Two Position, Five Connection, Solenoid Control Pilot Actuated with Detents and Throttle Exhaust

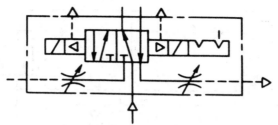

Symplified Symbol

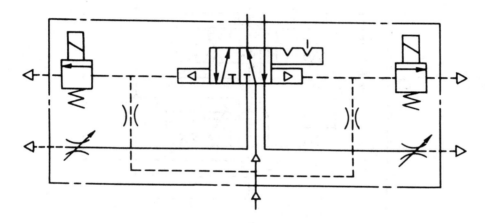

Complete Symbol

11.4.8 Variable Pressure Compensated Flow Control and Overload Relief

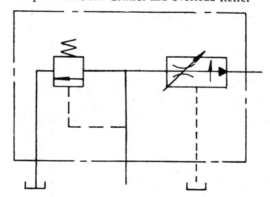

Appendix D

11.4.9 Multiple, Three Position, Manual Directional Control with Integral Check and Relief Valves

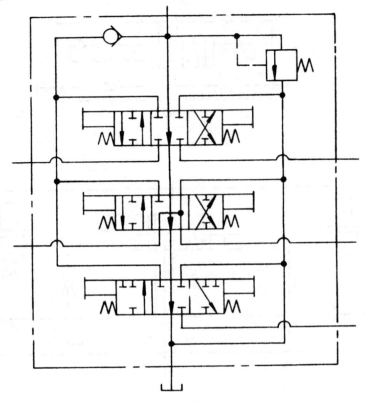

11.4.10 Cycle Control Panel, Five Position

508   Standard Graphic Symbols for Fluid Power Diagrams

11.4.11 Panel Mounted Separate Units Furnished as a Package (Relief, Two Four-Way, Two Check, and Flow Rate Valves)

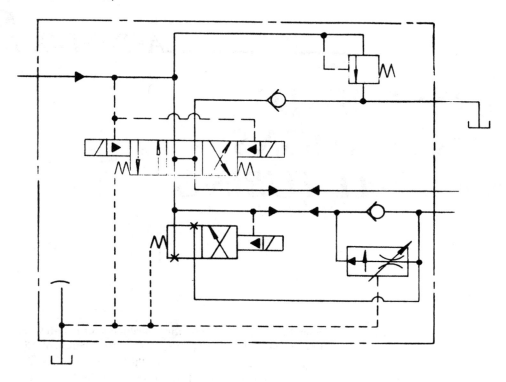

11.4.12 Single Stage Compressor with Electric Motor Drive, Pressure Switch Control of Receiver Tank Pressure

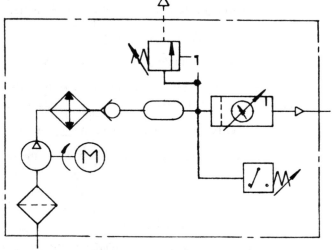

OFFSET BEND CALCULATIONS FOR STEEL TUBING

Following are the important dimensions for computing the conductor length when steel tubing is used to plumb the system (see Figure E-1). Offset bends of 15°, 30°, 45°, 60°, and 75° are given for tubing sizes from 1/8th- to 7/8-inch diameter. The dimension X equals the center to center distance between the bends, The dimension Y equals the offset dimension. Finally, the dimension Z equals the dimension between the ends of the conductor. To figure offset bends:

1) Specify the total offset desired and the offset angle.

   Offset dimension (*Y*)____in.

   Offset angle   ____degrees

2) Read the length of tubing required to make the offset bend from the tabled data in Table E-1.

   Offset bend tubing length (*X*)____in.

3) Mark the center points of each offset bend on the tubing at points *A* and *B*.

   *A*____OK

   *B*____OK

4) Align marks with the reference marks on the tubing bender.

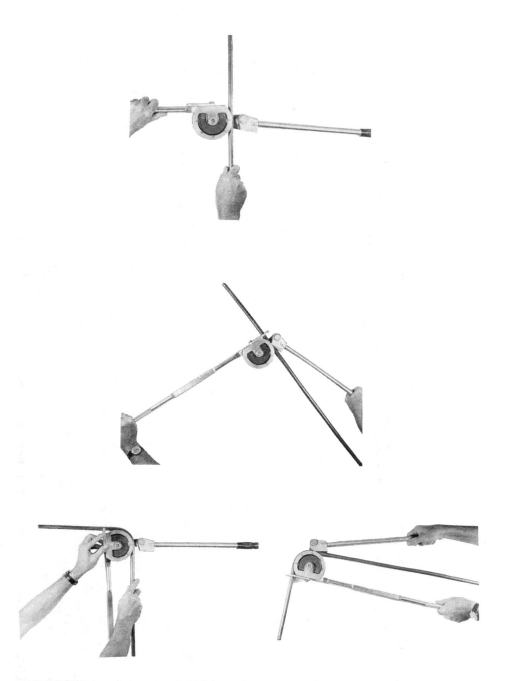

FIGURE E-1: Operation of lever-type bender *(courtesy of Imperial Eastman Corp.).*

Offset bend calculator

Amount of offset (Y dimension)	Angle of offset 15° (X dimension)	Angle of offset 30° (X dimension)	Angle of offset 45° (X dimension)	Angle of offset 60° (X dimension)	Angle of offset 75° (X dimension)
1	$3\frac{7}{8}$	2	$1\frac{13}{32}$	$1\frac{5}{32}$	$1\frac{1}{32}$
$1\frac{1}{8}$	$4\frac{11}{32}$	$2\frac{1}{4}$	$1\frac{19}{32}$	$1\frac{5}{16}$	$1\frac{5}{32}$
$1\frac{1}{4}$	$4\frac{27}{32}$	$2\frac{1}{2}$	$1\frac{25}{32}$	$1\frac{7}{16}$	$1\frac{9}{32}$
$1\frac{3}{8}$	$5\frac{5}{16}$	$2\frac{3}{4}$	$1\frac{15}{16}$	$1\frac{10}{32}$	$1\frac{7}{16}$
$1\frac{1}{2}$	$5\frac{25}{32}$	3	$2\frac{1}{8}$	$1\frac{23}{32}$	$1\frac{9}{16}$
$1\frac{5}{8}$	$6\frac{9}{32}$	$3\frac{1}{4}$	$2\frac{5}{16}$	$1\frac{7}{8}$	$1\frac{11}{16}$
$1\frac{3}{4}$	$6\frac{3}{4}$	$3\frac{1}{2}$	$2\frac{15}{32}$	$2\frac{1}{32}$	$1\frac{13}{16}$
$1\frac{7}{8}$	$7\frac{1}{4}$	$3\frac{3}{4}$	$2\frac{21}{32}$	$2\frac{5}{32}$	$1\frac{15}{16}$
2	$7\frac{23}{32}$	4	$2\frac{13}{16}$	$2\frac{5}{32}$	$2\frac{1}{16}$
$2\frac{1}{8}$	$8\frac{7}{32}$	$4\frac{1}{4}$	3	$2\frac{15}{32}$	$2\frac{3}{16}$
$2\frac{1}{4}$	$8\frac{11}{16}$	$4\frac{1}{2}$	$3\frac{3}{16}$	$2\frac{19}{32}$	$2\frac{5}{16}$
$2\frac{3}{8}$	$9\frac{3}{16}$	$4\frac{3}{4}$	$3\frac{11}{32}$	$2\frac{3}{4}$	$2\frac{15}{32}$
$2\frac{1}{2}$	$9\frac{21}{32}$	5	$3\frac{17}{32}$	$2\frac{7}{8}$	$2\frac{19}{32}$
$2\frac{5}{8}$	$10\frac{5}{32}$	$5\frac{1}{4}$	$3\frac{23}{32}$	$3\frac{1}{32}$	$2\frac{23}{32}$
$2\frac{3}{4}$	$10\frac{5}{8}$	$5\frac{1}{2}$	$3\frac{7}{8}$	$3\frac{3}{16}$	$2\frac{27}{32}$
$2\frac{7}{8}$	$11\frac{3}{32}$	$5\frac{3}{4}$	$4\frac{1}{16}$	$3\frac{5}{16}$	$2\frac{31}{32}$
3	$11\frac{19}{32}$	6	$4\frac{1}{4}$	$3\frac{15}{32}$	$3\frac{3}{32}$
$3\frac{1}{8}$	$12\frac{1}{16}$	$6\frac{1}{4}$	$4\frac{13}{32}$	$3\frac{19}{32}$	$3\frac{7}{32}$
$3\frac{1}{4}$	$12\frac{9}{16}$	$6\frac{1}{2}$	$4\frac{19}{32}$	$3\frac{3}{4}$	$3\frac{3}{8}$
$3\frac{3}{8}$	$13\frac{1}{32}$	$6\frac{3}{4}$	$4\frac{25}{32}$	$3\frac{29}{32}$	$3\frac{1}{2}$

TABLE E-1: Dimensions for computing steel tubing length.

STANDARD TEST PROCEDURES

PROPERTIES OF FLUIDS

- ASTM D 2502: Standard method of test for molecular weight of petroleum oils from viscosity measurements.
- ASTM D 941: Standard method of test for density and specific gravity of liquids by Lipkin bicapillary pycnometer.
- ASTM D 1298: Standard method of test for density, specific gravity or API gravity of crude petroleum products by hydrometer method.
- ASTM D 445: Standard method of test for viscosity of transparent and opaque liquids (kinematic and dynamic viscosities).
- ASTM D 88: Standard method of test for saybolt viscosity.
- ASTM D 2161: Standard method for conversion of kinematic viscosity to saybolt universal viscosity or to saybolt furol viscosity.
- ASTM D 97: Standard method of test for pour point.
- ASTM D 664: Standard method of test for neutralization number by potentiometric titration.
- ASTM D 974: Standard method of test for neutralization number by color-indicator titration.
- ASTM D 92: Standard method of test for flash and fire points by Cleveland open cup method.
- ASTM D 2155: Standard method of test for auto-ignition temperature of liquid petroleum products.
- ASTM D 2266: Standard method of test for wear preventive characteristics of lubricating grease (four ball method).
- ASTM D 2782: Standard method of test for measurement of extreme-pressure properties of lubricating fluids (Timken method).
- ASTM D 2711: Standard method of test for demulsibility characteristics of lubricating oils.
- ASTM D 1796: Standard method of test for water and sediment in crude oils and fuel oils by centrifuge.
- ASTM D 91: Standard method of test for the precipitation number.

- ASTM F 313: Standard method of test for insoluble contamination of hydraulic fluids by gravimetric analysis.
- ASTM D 2783: Standard method of test for measurement of extreme pressure of lubricating fluids (four ball method).
- ASTM 2271: Standard method of test for preliminary examination of hydraulic fluids (wear test).
- ASTM 2882: Tentative method for high pressure pump testing of hydraulic oils.
- ASTM D 665: Standard method of test for rust-preventing characteristics of steam-turbine oil in the presence of water.
- ASTM D 892: Standard method of test for foaming characteristics of lubricating oils.
- ASTM D 943: Standard method of test for oxidation characteristics of inhibited steam turbine oils.
- ASTM 1367: Standard method of test for lubricating qualities of graphites.
- ASTM 1401: Standard method of test for emulsion characteristics of petroleum oils and synthetic fluids.
- ASTM 313: Standard method of test for insoluble contaminants of hydraulic fluids by gravimetric analysis.
- ASTM 2882: Standard method for vane pump testing of petroleum hydraulic oils.
- ASTM 491: Standard test procedure for maintenance of cleanliness of hydraulic fluids and systems.
- MIL F-8901A: Filtration/separators, aviation and motor fuel, ground and shipboard use, performance requirements and test procedures.
- MIL TDR-64: Methods for particle counting in hydraulic fluids.
- MIL STD-1246A: Product cleanliness levels and contamination control program.
- NFPA T2.9.2: Procedure for qualifying and controlling cleaning methods for hydraulic fluid power fluid sample containers.
- NFPA T2.9.1: Method for extracting fluid samples from the lines of an operating hydraulic fluid power for particulate contamination analysis.
- ISO DIS 3938: Hydraulic fluid power contaminant analysis data reporting method.
- ISO STD 4402: Hydraulic fluid power calibration of liquid automatic particle counter instruments—method of using air cleaner fine test dust contaminant.

TESTS FOR ELASTOMER SEALS

- ASTM D 2240-68: Standard method of test for indentation hardness of rubber and plastics by means of durometer.
- ASTM D 395-68: Standard method of test for compression set of vulcanized rubbers.
- ASTM D 412-68: Standard method of tension testing of vulcanized rubber.
- ASTM D 1329-72: Standard method of test for evaluating low-temperature characteristics of rubber and rubber-like materials by a temperature retraction procedure (TR test).

- ASTM D 297-72: Standard methods of chemical analysis of rubber products (section 15: *Determining the specific gravity*).
- ASTM D 471-72: Standard method of test for change in properties of elastomeric vulcanizates resulting from immersion in liquids.
- ASTM D 865-62: Standard method of heat aging of vulcanized rubber by test tube method.
- ASTM D 1414-72: Standard methods of testing rubber O-rings.

TESTS FOR FLAT GASKETS

- ASTM F 36: Standard test method for compressibility and recovery of selected materials.
- ASTM F 37: Standard test method for sealability under pressure.
- ASTM F 38: Standard test method for creep and relaxation of gasket materials.
- ASTM F 39: Effectiveness of asbestos sheet packing.
- ASTM F 145: Evaluation of flat-faced gasketed joint assemblies.

Pump Installation Procedure

1)	Pump and drive motor speed and direction are matched	_____ OK
2)	Pump and drive motor torque are matched	_____ OK
	Pump running torque	_____ lb-ft
	Pump breakaway torque	_____ lb-ft
	Drive motor starting torque	_____ lb-ft
3)	Pump and drive motor mounts are compatible	_____ OK
4)	Pump and drive motor shafts rotate concentrically	_____ OK
5)	Appropriate coupling is selected (flexible__)	_____ OK
6)	Pump and drive motor shafts are aligned (centerline)	_____ OK
7)	Coupling connection does not cause stress on the drive motors or pump shafts	
	Vertical	_____ OK
	Horizontal	_____ OK
	Longitudinal	_____ OK
8)	Suction lift is appropriate (within 2-3 feet)	_____ OK
9)	Suction line is properly sized (velocity within 4-6 ft/sec)	_____ OK
Caution 10)	Seal tape (no dope) used on pipe connections	_____ OK
11)	Discharge line properly sized (velocity within 10-20 ft/sec)	_____ OK
Caution 12)	System lines and fittings are clean inside	_____ OK
Caution 13)	Guards are installed	_____ OK
14)	Fluid level in reservoir checked	_____ OK
Caution 15)	Pump has been primed	_____ OK
16)	Pump turns freely by hand	_____ OK
Caution 17)	Check manufacturer specifications before running variable displacement pumps in the zero flow position	_____ OK

Appendix F

TEE TEST FOR HYDRAULIC CIRCUITS

Test connections (see Figures F-1 and F-2) can be made at A, B, C, or D. When the connection is made at C or D the tester is in series with the control valve and cylinder to check individual components.

First, determine how much fluid should be circulating through the circuit. Then by controlling the pressure with the load valve on the tester, the amount of fluid available through each component can be determined. If the test indicates an insufficient flow when the pressure is raised to specifications for the relief valve, the cause can be pinpointed to such components as:

1)   Slipping pump.

2)   Flowing over a faulty relief valve.

3)   Leaking past control valve spools to the reservoir.

4)   Compensation mechanism malfunction on some variable displacement pumps (pressure or flow compensation).

5)   Leaking past pump or motor parts directly to return without power transfer (low volumetric efficiency).

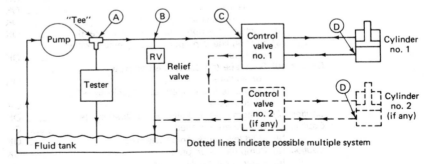

FIGURE F-1: Tee test circuit.

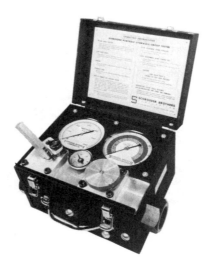

FIGURE F-2: Portable hydraulic tester *(courtesy of Schroeder Brothers Corporation).*

Do not exceed relief valve pressure to prevent damage to the system. Be sure to operate the system long enough to bring the temperature of the fluid within the operating range. If the system is driven by an internal combustion engine, tests should be conducted at near constant rpm.

The tee test checks overall system volumetric efficiency:

TEE TEST

1) Connect the tester at point *A* as shown in Figure F-1. OK _____

2) Run the system until the oil temperature is within the operating range. OK _____

3) Position the control valve to direct fluid to extend the cylinder. OK _____

4) Close the load valve just enough to extend the cylinder. OK _____

5) Release the load valve and record the total flow through the tester flowing at no load and low pressure to the reservoir. _____gpm

6) Close the load valve to bring the pressure to 90% of the relief valve pressure. _____psi

7) Record the flow rate at 90% relief valve pressure. _____gpm

8) Compute the volumetric efficiency (*e*) from

$$e_o = \frac{\text{Flow at 90\% Load}}{\text{Flow at No Load}} \times 100$$

9) If the test is inconclusive, repeat the procedure applying load pressure in 500 lbf/in² increments. Do not exceed manufacturer's specifications or relief valve setting.

Appendix F

FLUID POWER

INDEX

B

Barometer, 392
Bernoulli equation, 36, 48–49
Beta efficiency, 116–17
Beta ratio, 116
Boyle's law, 393
Bulk modulus, of fluids, 102

C

Careers, 8–9
Case drain arrangements, 188
Cavitation, 155, 386
Centrifugal pump:
   components, 129
   diffuser action, 129
Certification, 9
Charles' law, 394–95
Check valve. *See* Valve
Circuits. *See also* Pneumatic circuits:
   accumulator, 291, 346
   basic, 336–44
   bleed-off, 346
   closed loop, 321
   constant flow, 328–29
   constant horsepower, 330–32
   constant pressure, 300–31
   conveyor, 330
   definition, 320
   design and analysis, 321
   fail safe, 342–44
   high-low, 337
   hoist, 323–28
   load demand, 240–41
   load sensing, 332–36
   meter in, 345
   meter out, 345
   one shot, 461–62
   open loop, 321
   open vs closed center, 344–47
   overload protection, 343–44
   regenerative, 232, 339–40

   road grader, 234
   safety, 342–44
   sequence, 337–38
   servo, 321
   speed control, 216
   symbols, 321
   synchronous, 340–42
   timing out, 474–75
   unloading, 336
Closed loop, 321. *See also* Circuits
Components, matching, 349–50
Compressed Gas Association,
   352
Compression set, 266–67
Compressor:
   centrifugal, 416–19
   classification, 413
   controls, 423–24
   horsepower, 420
   positive displacement, 413
   piston, 414
   pulsations, 423–24
   rating, 420
   screw, 415
   sizing, 419–21
   staging, 414–15
   unit, 416
Compressor controls:
   auto start-stop, 423
   constant speed, 424
   dual control, 424
   modulating control, 424
Conditioning, 448
Conductors, 293–97
   sizes of steel tube, 295
   wall thickness, 294–96
Continuity equation, 40, 58
Conversion factors, 13, 486
Critical velocity, 65
Cushioning devices. *See* Cylinder
   cushioning devices
Cylinder:
   acceleration factors, 171–76
   classification, 160–61
   construction, 161–62
   cushioning devices, 170, 178

deceleration factor, 171–76
definition, 161
mounting, 167
piston rod sizes, 166–69
pneumatic, 434–36
rams, 162
rod loading, 166
selection factors, 161
stop tube length, 167
symbols, 449
velocity, 34, 58–59
working pressure, 166
Cycle plot, 350

D

Darcy-Weisbach formula:
    for deriving head losses, 73
    for pressure drop, 74
Dew point, 399–401
Displacement, 56–57
Durometer, 266
Dynamic seals, use of O-rings, 260, 270

E

Elastomer seal. *See* Seals
Electric controls, 312–315
Energy:
    flow, 36
    kinetic, 36, 42, 44–47
    potential, 36, 42, 44, 46
    velocity, 42, 45
Energy conversion, 447
Equivalent length, 86–87

F

Filter:
    full-flow, 111

last chance, 111
location, 114
machine mounted, 111
magnets, 111
performance of, 114
ratings, 115
size of particles, 114
tell-tale, 112
Filter-regulator-lubricator, 450
Filtration:
    alfa ratings, 116
    absolute, 115
    beta efficiency, 116–17
    beta ratio, 116
    multipass test, 116–17
    nominal, 115
    particle counter, 118
    ratings, 116
    ratio, 116
Fire resistant fluids. *See* Fluids
Fittings, 298–99
Fittings, banjo, 435–36
Flow:
    C-coefficients, 78–81
    definition, 56
    f-Factor, 75–78
    K-values, 81–86
    laminar, 65
    losses, 36
    non-compressible, 65
    rate, 23
    steady, 40
    turbulent, 65
    viscous, 65
Flow dividers, 309
Flow energy, 36
Flow in pipes, 56
Flow meters, 305–8
Flow rate, 23
Flow through orifices, 79–81
Fluid:
    acid number, 105
    adding, 109
    anti-wear additives, 106
    anti-wear properties, 94
    API gravity, 94, 100

H

Hagen Poiseuille formula, 73–74
Harris formula, 410–13
Head, 36–37
Heat exchangers, 278–84
Hoist, 447
Horsepower, 21–22
Hose:
  installation, 301–2
  materials, 300
  quick disconnects, 302
Hose compounds:
  Buna-N, 98
  Viton, 98
Hydraulic fluid. *See* Fluid
Hydraulic motors, 181–205
Hydraulic, tester, 308
Hydrostatic transmissions, 310–13

I

Isothermal expansion, 287
Instrumentation, 303–8
Intensifiers, 291
Intercooler, 414, 418

K

Kinetic energy, 36, 42, 45

L

Ladder diagrams, 313–16
Laminar flow, 65
Load vs time plot, 323–24
Logic, 470–75
Logic control, 471
Logic control circuits:
  NOT, 472
  Memory, 458

M

Miniature pneumatics, 8, 476
Modular air components, 476
Moisture content of air, 400
Motor:
  case drain loss, 185
  classification, 182
  continuous rotation, 180
  efficiency, 182, 184–85
  electrohydraulic, 200–202
  gear, 189
  gerotor, 189–90
  high torque-low speed, 193–99
  horsepower, 192
  hydraulic, 181
  limited rotation, 186–87
  piston, 190–91
  power, 182
  reduction, 200
  speed, 182
  stall, 26, 29
  testing, 202–4
  torque, 182
  vane, 189
  wheel, 196
Moving part logic, 8, 476–77
Mufflers, 404–5
Multiplication of force, 19

N

National Fluid Power Association, 8
Neutralization number, 105–6
Noise, 383–86
Noise, permissible levels, 405

O

Occupational titles, 8
Offset bend calculations, 297
Oil. *See* Fluid
Open loop, 321

Orifice. *See* Flow, losses
O-ring 254-50, 269–70
OSHA standards, 352
Overload protection, 343–44

P

Performance testing, 350–51
Petroleum base fluids. *See* Fluid
Pipe:
  American Standard Taper Pipe,
    Thread, 296
  Dryseal American National Stan-
    dard, Taper Pipe Thread, 296
  National Standard Taper Pipe
    Thread, 296
  relative cross-sections, 297
  sizes and pressure ratings, 165
Piston motors, 190–97
Piston pumps, 138–41, 146–51
Pneumatic:
  air consumption, 437
  basics of, 391–406
  cylinders, 434
  distribution system, 409–10
  friction losses, 412
  uses in industry, 391
Pneumatic circuits,
  air-oil bolster, 468–69
  bench press, 468–69
  double acting, 455–58
  interlock, 466–70
  memory, 471–74
  OR, 472
  pressure sequence, 465
  sequence, 459–66
  sequence of operation, 453
  single acting, 452–55
  time based, 474
  timing in, 474–75
Pneumatic cylinders:
  anti-rotation, 434–35
  banjo fittings, 435–36
  construction, 434
Pneumatic logic control, 440–41

Pneumatic symbols, 442–51, 489–509
Pneumatic symbols, valve operators,
  451
Pneumatic valves:
  directional control, 426–31
  fifth port, 455–57
  five way, 427–28, 430–31
  flow regulation, 426
  four way, 427
  operators, 427, 429
  pressure regulator, 425–26
  pressure relief, 425
  sizing, 431–34
Potential energy, 36
Potential energy, 36, 42, 44, 46
Power, 21
Power packages, 309
Power units, 309
Pressure:
  deadhead, 144
  head, 37–39
Pressure gauge, 304
Pressure relief valves, 208–10
Pressure units conversion, 305
Pump:
  cavitation, 155
  centrifugal, 129
  classification, 129
  components, 129
  duty rating, 349
  elements, 123–24
  fixed displacement, 130
  frequency, 154
  gear, 130–32
  gerotor, 133–34
  industrial, 122
  mechanical efficiency, 127
  noise, 154–55
  non-positive, 124, 128–29
  other characteristics, 154–56
  output, 123
  overall efficiency, 124–25, 157
  performance, 150–52
  piggyback, 133
  piston, 138–41, 146–51
  positive displacement, 124, 130–42